Raw-Score Formula for the Regression

$$Y' = r \cdot \frac{S.D._y}{S.D._x} \cdot X + \left(M_y - r \cdot \frac{S.D._y}{S.D._x} \cdot M_x \right)$$

(6.15) page 152

Formula for the Probability of Determining x in n trials.
(The Binomial Distribution)

$$P_x = \frac{n!}{x!(n-x)!} p^x q^{n-x}$$

(7.1) page 167

Formulas for the Mean and Standard Deviation of the Binomial

$$M_b = np$$

(7.8) page 188

$$S.D._b = \sqrt{npq}$$

(7.9) page 188

Formula for the True Standard Deviation of the Sampling
Distribution of Means (Single Group)

$$\sigma_M = \frac{\sigma}{\sqrt{N}}$$

(8.1) page 206

Formulas Relating S.D. and s

$$s = (S.D.) \cdot \sqrt{\frac{N}{N-1}}$$

(8.3) page 210

$$S.D. = s \cdot \sqrt{\frac{N-1}{N}}$$

(8.4) page 210

Formula for the Standard Error of the Mean

$$s_M = \frac{s}{\sqrt{N}}$$

(8.5) page 211

$$s_M = \frac{S.D.}{\sqrt{N-1}}$$

(8.6) page 211

Formula for Standard Score Relative to the t Distribution

$$t = \frac{M - \mu_0}{s_M}$$

(8.7) page 215

Formulas for the 95% and 99% Confidence Intervals
(Single Sample Mean)

95% confidence interval: $M \pm t_{.05} \times s_M$ (8.8a) page 218

99% confidence interval: $M \pm t_{.01} \times s_M$ (8.9a) page 218

ELEMENTARY STATISTICS

A Problem-Solving Approach

ELEMENTARY STATISTICS
A Problem-Solving Approach

SECOND EDITION

Andrew L. Comrey
University of California, Los Angeles

Paul A. Bott
California State University, Long Beach

Howard B. Lee
California State University, Long Beach

wcb

WM. C. BROWN PUBLISHERS
DUBUQUE, IOWA

Book Team

Developmental Editor *Sandra E. Schmidt*
Designer *Julie E. Anderson*
Production Editor *Sherry Padden*
Photo Research Editor *Mary Roussel*
Permissions Editor *Vicki Krug*
Visuals Processor *Joseph P. O'Connell*

wcb group

Chairman of the Board *Wm. C. Brown*
President and Chief Executive Officer *Mark C. Falb*

wcb

Wm. C. Brown Publishers, College Division

President *G. Franklin Lewis*
Vice President, Editor-in-Chief *George Wm. Bergquist*
Vice President, Director of Production *Beverly Kolz*
Vice President, National Sales Manager *Bob McLaughlin*
Director of Marketing *Thomas E. Doran*
Marketing Communications Manager *Edward Bartell*
Marketing Information Systems Manager *Craig S. Marty*
Marketing Manager *Kathy Law Laube*
Manager of Visuals and Design *Faye M. Schilling*
Production Editorial Manager *Colleen A. Yonda*
Production Editorial Manager *Julie A. Kennedy*
Publishing Services Manager *Karen J. Slaght*

Table of Contents

v

Preface

Teaching elementary statistics is particularly challenging because the students are heterogenous in aptitude, preparation, and motivation. At one extreme are the students with little or no background in mathematics who are taking statistics only because it is a required course and who have no postgraduate education plans or no desire to take other statistics courses. At the other extreme are the students who are well prepared in mathematics, who recognize the need for statistics, and who plan to enter a career where statistics are used every day. The challenge, then, is to provide a meaningful learning experience in the same classroom for such a diverse collection of students.

The major objective of *Elementary Statistics: A Problem-Solving Approach* is to provide a connected, orderly presentation of statistics so that all students, particularly those who have little mathematical background and are afraid of statistics or who are convinced that they cannot do them, will be successful. Each of the 54 Problems in the text represents a major concept or commonly used statistical procedure. Students with little mathematical background can easily follow them in a step-by-step fashion. Students can concentrate on learning how to do one problem at a time, achieving success through the attainment of intermediate goals instead of having to confront a monolithic and seemingly insuperable obstacle. For those students with greater motivation and superior preparation, supplementary problems and mathematical proofs are provided to enrich their learning experiences and to present them with challenges commensurate with their capacities.

Although the emphasis in this text is on the application of statistical techniques in the social and behavioral sciences, it is important to note that the same statistical problems occur in engineering, business, biology, and many other disciplines. The basic concepts of statistics and inferential methods apply in all fields.

TO THE INSTRUCTOR

Major Features of
Elementary Statistics

The Problem-Solving Approach   The material in this book has been organized around 54 specific problems in statistics—statistical procedures that are commonly encountered in practice. These 54 problems have been broken down into 50 "number series" problems and 4 "letter series" problems. The 50 number series problems are to be regarded as the primary, most essential problems to be covered. The 4 letter series problems are supplementary problems that are less essential either because they are less commonly used or because they represent merely laborsaving alternatives to other procedures. Letter series problems can be omitted entirely at the option of the instructor, assigned as special problems for extra credit, or, of course, can be required of all students. Letter series problems can be omitted entirely without disturbing the continuity of study with the number series problems. The letter series problems, past any point, can be dropped even if earlier ones have been assigned.

Some of the problems are included more for their conceptual importance than for their wide use as applied methods. Problems in Chapters 1 through 6 are concerned with commonly used "descriptive statistics," statistical procedures that describe what a body of data is like. The remaining problems are concerned with "inferential statistics," procedures that permit the investigator to draw certain conclusions based on the data analyzed. We have attempted to deemphasize seldom-used procedures and obscure theoretical issues to permit greater elaboration with respect to essential concepts and the methods most often used in practice.

Learning Aids

Several pedagogical aids have been built into this text to reinforce the mastery gained by the problem-solving approach. **Marginal definitions** explain key terminology. A special section called **Summary of Steps** appears frequently throughout the text, following the most important problems. This section lists the steps required to perform the statistical operations exemplified by the problem.

Mathematical Notes provide mathematical proofs or demonstrations of the statistical formulas. Because the **Mathematical Notes** contain material that is not essential to mastery of the statistical techniques in this text, they may be omitted at the instructor's discretion. Alternatively, the **Mathematical Notes** can be assigned to enrich the beginning course for students with greater mathematical facility or background. They are also helpful for the instructor who prefers a more rigorous or more theoretical approach to elementary statistics, and can also be assigned for extra credit. However, all math proofs and demonstrations regarded as immediately essential for all students as they progress through the material have been left in the main body of the text.

Another text feature is **Statistics in the World Around You** boxes, which appear in each chapter. These boxes provide a real-world application of the statistical techniques under discussion, giving students an immediate sense of the relevance and the usefulness of statistics.

End of chapter **Exercises** are included at the end of each chapter to give students experience in working statistical problems. Approximately 20 questions are provided for each chapter. Complete solutions and answers for odd-numbered questions, and answers for even-numbered questions, are provided in Appendix C. (Complete solutions and answers for all questions are found in the *Instructor's Manual*.)

Flexible Organization

We have already suggested that some or all of the letter series problems, A, B, C, and D can be omitted from a given course of instruction in elementary statistics. Some instructors might also wish to leave out the problems in Chapter 9 on chi square and Chapters 10 and 11 on analysis of variance and Chapter 12 on nonparametric methods, or all four. The problems you select depends on the length of the course, the number of credit hours, the pace to be maintained, and your own perception of the relative importance of the methods. If you or your course have a heavy experimental orientation, you will probably wish to include some or all of the problems in Chapters 10 and 11 on analysis of variance. If you are interested in sociological or behavioral research, you might consider it imperative to include some or all of the problems from Chapter 9 on chi square and Chapter 12 on nonparametric statistics. A fairly typical one quarter or one semester course usually includes all of the problems in Chapters 1 through 10. A two quarter or two semester course usually includes all of the problems. More specific suggestions are included in the *Instructor's Manual*.

Instructional Aids

A comprehensive *Instructor's Manual/Test Item File* has been prepared by the text authors. The first section of the manual provides sample syllabi for course configurations of varying lengths and emphases. Additional features of the *Instructor's Manual* include performance objectives for each chapter of the text; complete solutions and answers to all of the end-of-chapter exercises; suggested computer laboratory assignments; multiple-choice test items; and additional problems and solutions that represent supplemental examples of the 54 problems covered in the text. These additional examples may be used for classroom demonstrations, assignments, or for exams.

All test items are available on **wcb** TestPak, a free, computerized testing service, which simplifies testing while offering you flexibility. There are two convenient TestPak options available:

—Use your Apple *II*e, IIc, Macintosh, or IBM PC to pick and choose your test questions, edit them, and add your own questions. We can send you program and test item diskettes for this purpose.

—If you don't have a microcomputer, you can still pick and choose your questions via our call-in/mail-in service. Within two working days of your request, we'll put a test master, a student answer sheet, and an answer key in the mail to you. Call-in hours are 8:30–5:00 CST, Monday through Friday.

TO THE STUDENT

Students in the social and behavioral sciences and education are often accustomed to reading material that does not contain numbers or mathematical formulas. This kind of reading material often leads the student to expect to grasp the meaning of a passage in one reading. If this does not occur, students sometimes become dismayed and assume that either the material is incomprehensible or that they are unintelligent. Most students, however, must read mathematical material more than once before it really sinks in. You are urged, therefore, not to be discouraged if the first reading of some passage in this text fails to make everything crystal clear. You should expect, as a matter of course, to read it several times before the pieces all fit together. If you completely understand everything the first time, just consider it an unexpected bonus.

If after several readings the material is still not clear, it is often useful to try a different exposition of the same subject by another author. Where one author's way of explaining a particular point draws a blank, a second author's may hit the mark, particularly when you have already tried to understand another exposition. Your library will probably have several books on elementary statistics that you can consult. A few titles are listed in the references section at the end of the book. Since there are over 100 published texts on elementary statistics that could be of interest to students, the list obviously will exclude many worthwhile books.

You may also have occasion to consult more advanced texts on topics not covered in this or other elementary statistics texts. This could occur as a result of a reference to some technique in the text, a search for a method of treating available data, or just a desire to increase your knowledge of the subject. A very abbreviated list of references on more advanced topics in statistical analysis is given in the references section at the back of the text. Each of those books in turn will contain many more references. Your librarian should be able to help you locate these or other more advanced books on statistical methods if you should need them.

ACKNOWLEDGMENTS   The authors would like to acknowledge the invaluable assistance of our colleagues who reviewed this manuscript at different stages in its development:

Janet Andrew, *Vassar College*

J. P. Behrens, *Pennsylvania State University*

Kathleen Kowal, *University of North Carolina*

Robert Lowder, *Bradley University*

Howard Lyman, *University of Cincinnati*

Paul Moes, *Bordt College*

Thomas Nygren, *Ohio State University*

Robert Pasnak, *George Mason University*

Jacqueline Reihman, *State University of New York–Oswego*

Nicholas R. Santilli, *Augustana College*

David Skotko, *University of Arkansas*

Kay Smith, *Brigham Young University*

James Weyant, *University of California*

ELEMENTARY STATISTICS
A Problem-Solving Approach

Chapter 1

The Importance of Learning Statistics

Many students are enrolled in statistics for only one reason—it is a required course! For many students the study of statistics seems to hold a special fear that borders on the pathological. This fear is often called "math shock" or "math anxiety."

Math anxiety interferes with effective performance. Math-anxious students are convinced that they will have difficulty, and their certainty of this often induces self-defeating behavior, such as avoiding studying, missing classes, and postponing assignments and examinations.

Math anxiety is a product of the typical methods used to teach mathematics in the primary and secondary schools. Mathematics is a subject that is organized vertically rather than horizontally, so, like erecting a building, understanding depends upon building the structure block-by-block with each block firmly in place before the next row of blocks is added. If blocks in the substructure are missing or not firmly in place, the entire structure is apt to come tumbling down.

The teaching procedures presently used in the primary and secondary schools are designed to accommodate only the top 5% or 10% of the students in mathematical aptitude. These are the only students who learn the material in the first course sufficiently to be ready to progress to the second course. Proper management of mathematical instruction would require that all students master the first course with a grade of "A" before they could progress to the second course. The second course, likewise, would have to be mastered to the "A" level of proficiency before going on to the third course, and so on. As it is now, however, "B," "C," and even "D" grade students are passed right on to the next course along with the "A" students. At the second course level, the less talented students in math are competing not only with less talent, but with an improper foundation as well.

A gap in the student's knowledge in verbal fields, such as history, English, or languages is less serious since the student can perform adequately on most tasks, albeit at less than 100% efficiency overall. Background gaps in mathematics, however, quickly accumulate to put students in a position where they cannot understand the next thing to be added to the store of knowledge. When this happens, students typically do not attribute this failure to a lack of background. Instead, they are apt to think they are stupid and tend to develop a deeply entrenched feeling of mathematical inferiority.

The authors are convinced that the vast majority of college students are capable of learning the material in elementary statistics if they do not permit fear of failure to interfere with the learning process. This text was written with the math-anxious student in mind. Any student with enough general intelligence to do well in verbally-oriented courses has the mental capacity to absorb a great deal of mathematics given the correct approach, the right attitude, and sufficient time to master each step in sequence before going on to the next step.

First, the student who brings math shock to the elementary statistics course must change his or her attitude. The student must recognize the value of a knowledge of statistics. Statistical thinking is so pervasive in modern civilization that people cannot really consider themselves educated unless they comprehend at least the elementary statistical ideas and techniques. On almost any job that students eventually hold, circumstances are apt to occur in which they will be able to perform more effectively if they have a knowledge of statistics. The research literature in the social sciences cannot be read without a knowledge of statistics. It is a truly embarrassing predicament to be a college graduate in a social science field without being able to read even an easy article in that field or to understand how researchers make conclusions based on data. Some students delude themselves by saying that they are interested in people, not numbers, and hence do not need to know anything about statistics. To read about what is being done in fields devoted to helping people and especially to participate professionally and to function competently in those efforts does require a knowledge of statistics.

Also, the math-anxious student must change her or his negative attitude about mathematics. The student should try to understand that any difficulty in absorbing mathematics in the past was due to many factors, including improper pacing and an inadequate foundation at the time new material was presented. The student who feels mathematically inferior is urged to try to accept intellectually the idea that he or she has enough ability to learn this kind of material—it will just take more time.

Second, and maybe most important, the student must provide enough study time and must have the patience to go slowly. Each step in building the structure of statistical knowledge is comparatively easy when taken by itself *if* the previous steps have been mastered thoroughly. The material selected for this text and the manner of presenting it were developed with the idea of making this process as easy as possible for all students, no matter how much difficulty they previously have had with mathematics. The last element to insure a student's success is the correct approach.

MAJOR FEATURES OF THIS TEXT

No new text in statistics is completely different from other published texts, but this text has several features, which combined, make it unique among available elementary statistics texts.

Mathematical Knowledge

Although the student is expected to know some arithmetic and a little elementary algebra, every effort was made to minimize the presumption of previous mathematical knowledge. Processes are explained in plain English for the verbally-oriented students so that they do not have to rely exclusively on facility with symbolic nonverbal manipulations. At the same time, the responsibility for teaching students how to use symbolic processes of a mathematical nature was not neglected.

Logic

The authors made a concentrated effort to provide an understanding of the basic logic underlying statistics rather than merely a set of cookbook rules to be applied by rote. Statistics is a subject requiring an extensive knowledge of advanced mathematics for a complete understanding, so some compromises were of necessity required. First of all, mathematical proofs are used in the main body of the text only where the mathematics involved are at a rather elementary level. Second, considerable dependence on intuitive understanding of concepts is substituted for actual rigorous mathematical proof. Language used in demonstrations and proofs was chosen for the purpose of conveying an intuitive grasp of the underlying logic rather than to satisfy the demands of elegant mathematical exposition.

Mathematical Proofs

In deference to students with math shock, most of the proofs and demonstrations are separated from the main text and placed in "Mathematical Notes" sections. Students may omit these sections entirely if they want to learn how to use statistical methods without, for the moment, being concerned about where they come from. It is often helpful for some students to leave theoretical material of this sort until a later stage in their learning. The proofs and demonstrations left in the main body of the text are regarded as immediately essential for all students as they progress through the material. The Mathematical Notes sections also serve as extra material that can be used to enrich the beginning course for students with greater mathematical facility and background.

Useful Techniques

The most commonly encountered and needed statistical ideas and techniques were chosen for this text. The variety and complexity of statistical methods utilized in published research articles in journals today is bewildering. A mastery of all these methods goes well beyond the scope of this text. Fortunately, however, a substantial proportion of the commonly used methods can be accommodated in an elementary textbook. Conceivably, a competent research person could successfully work for a lifetime in research with the statistical methods presented in this introductory text. Most of the more complicated types of statistical treatment encountered in the literature are used only occasionally. In many cases, where these more complicated data treatment methods were used, simpler experiments could have been designed to accomplish the same end result. Thus, although statistics is a very complicated and extensive subject, a surprising amount of what is really essential to the research worker is included here. By confining the material in this text to the essentials, the math-shock student is spared the demoralizing effect of being confronted with an overwhelming mountain of figures, symbols, formulas, and special cases.

Problem-Solving Approach

Perhaps the most unique feature of this text is its manner of presenting the material. Fifty-four "Problems" are identified as commonly used techniques or processes. The student is given these 54 statistical procedures to master. Or, putting it another way, elementary statistics is broken down into 54 distinct, graded units of study to be checked off by the student, one-by-one, as they are mastered. Students gain a discrete increment in their power to understand and apply statistical methods to research data as each successive problem is mastered. Instead of presenting a great mass of abstract theory, this approach presents actual practical applications of statistical methods individually. Theory is introduced in connection with a specific problem only where it is needed to understand the problem. The connection between theory and practice is kept much more immediate in this manner, although perhaps at some cost as far as mathematical elegance is concerned.

This method of material presentation evolved as a way of helping less mathematically inclined students to cope with statistics. Breaking statistics into discrete types of problems proved to be the most effective of the learning methods tried. Some reasons for effectiveness of the problem approach are: (1) Instead of being faced with a monolithic obstacle to overcome, students have only to master a relatively small number of well-defined tasks. Thus, the total job is broken into a small number of manageable units. (2) Many of these units, particularly at the beginning, are very easy, so students gain success, a definite feeling of progress, and pick up confidence that the material can be mastered after all. This success helps to extinguish to some extent the conditioned fear response established by previous failures. (3) With the more positive climate established by these conditions, math-shock students are less likely to engage in self-defeating behavior. Examination anxiety also tends to be reduced, since students feel more confident, permitting a performance that is closer to real capacity.

TWO TYPES OF STATISTICS

Two of the first building blocks of statistics are the distinctions between *descriptive* and *inferential* statistics.

Descriptive Statistics

Descriptive statistics are used when a researcher attempts to summarize and describe the characteristics of the collected data for a study. The Census Bureau, for example, collects data on household income and educational attainment for all residents of the United States. When these data are all combined, the Census Bureau prepares tables that list the numbers of people in the different income and education categories. Some of the tables may contain averages and comparisons with previous census years. Often, percentage changes over a several-year period are shown in an effort to make the data more meaningful than just raw numbers.

descriptive statistics
Statistics concerned only with describing and summarizing sets of numerical data.

Inferential Statistics

inferential statistics
Statistics, which allow an inference to be made about the whole population from information about only a part of the population or a sample.

Inferential statistics are used when a researcher wishes to determine the characteristics of a larger group by collecting data on a smaller group. For example, the Nielson Group, a research organization, regularly monitors the television-watching habits of approximately 2600 American households and uses the results obtained to infer, or generalize, the television-watching preferences of all television watchers. This generalization is made possible by the use of inferential statistics. Both descriptive and inferential statistics are discussed in this text.

MEASUREMENT SCALES AND PARAMETRIC VERSUS NONPARAMETRIC METHODS

Many authors have suggested that certain kinds of statistical methods, called parametric statistics, can only be used if the data are measured on an interval or a ratio scale. The term parametric comes from the word **parameter,** which is a mathematically exact, measurable characteristic within a population. A **population** is any group of observations that share a common characteristic. In statistics, Greek letters are usually used to represent population parameters. In computing an arithmetic mean, for example, the scores are added up and divided by N, or the total number of scores. These operations are valid only when the measurement process is carefully executed to insure that equal differences between scores in different parts of the scale mean the same thing. Such operations are involved in many statistical methods considered in this text. Before dealing with this issue, let us consider the different kinds of measurement scales and their characteristics.

parameter A number derived from knowledge of the entire population, such as the minimum of the population, or the mean of the population.

population Any complete set of objects, usually numbers.

Ordinal Scale

ordinal scale Data measured using this scale represent order or rank.

The **ordinal scale** is the lowest quantitative scale in the hierarchy of measurement scales, although there is an even lower scale, called a "nominal scale," which is not quantitative at all. To be considered "quantitative," a measurement scale must have a numerical basis. By contrast, individuals can be classified on a nominal scale of hair color as "blondes," "brunettes," and "redheads," with the numbers 1, 2, and 3

Glen Dines/Kappan.

assigned to the three categories. Hair color is an example of "qualitative" data, that is, the differences are distinguished by attribute rather than by numerical differences. This does not constitute measurement in any sense, however, but merely classification into categories with numerical labels attached. The ordinal scale, next up in the hierarchy, does involve quantitative information. If a judge ranks a collection of handwriting specimens in order of excellence, assigning the number "1" to the best, the number "2" to the next best, and so on, a "rank-order" or "ordinal" scale is created.

With ordinal or rank-order scales, only certain properties ordinarily associated with numbers also apply to the objects to which those numbers are assigned, namely, the "greater than," "equal to," and "less than" properties. For example, handwriting specimens with lower numbers are better than those with higher numbers. With other ordinal scales, objects with higher numbers assigned are greater than those with lower numbers assigned. Differences in numbers assigned to adjacent specimens are all equal to 1, yet some of these specimens may be very close to each other in quality and others may be rather far apart. In ordinal scales, therefore, it cannot be assumed that numerical difference between scale values are quantitatively meaningful. The reader should be aware that most of the measurements obtained in social science research are on ordinal scales, except perhaps in areas of physiological psychology.

Interval Scale

interval scale   A measurement scale with all the properties of an ordinal scale plus the property that distances between pairs of objects may be calculated and interpreted. Data measured on this scale lacks a meaningful absolute zero point.

An **interval scale** ensures that the different properties ordinarily associated with numbers also apply to the objects to which the numbers are assigned during the measurement process. Perhaps the best-known example of an interval scale is the temperature scale. The centigrade scale, for example, has an arbitrary zero point at the temperature where water freezes. The point at which water boils is assigned the number 100 on the centigrade scale. The points in between can be divided equally using expansion of mercury in a thermometer to represent the scale between 0 and 100 degrees. Equal units along the scale represent equal amounts of expansion of mercury, and hence of temperature. The lack of an absolute zero point prevents statements such as, 50 degrees is twice as hot as 25 degrees. It can be stated, however, that 75 degrees is as much hotter than 50 degrees as 50 degrees is hotter than 25 degrees.

In social science disciplines such as psychology, education, and sociology, it is difficult to establish operations to prove equality of units of measurement as was demonstrated with temperature. The most common procedure used to obtain equal units in these disciplines is to assume equal units under the normal curve are equal.

normal curve   The normal distribution.

The **normal curve** is a theoretical distribution where all the values, when plotted, form a bell-shaped curve. If a measurement procedure is developed that yields a normal distribution of scores in the general population, it is assumed that the yardstick is not a rubber one, that is, one that is stretched too much here and not enough there. Thus, the measuring instrument is presumed to be a good one from a scaling point of view. Then, if scores are converted to a standard-score scale (to be described later in the book), units in different parts of the distribution of the same size will be considered quantitatively equal. While this is a reasonable assumption and equal units in different parts of the scale may be, in fact, approximately equal, there is no actual proof that this is true. Scaling procedures that use the normal curve to obtain interval scales can be treated only as approximations of unknown accuracy. They are, nevertheless, usually superior in this respect to the untreated raw-score measurements. In this sense, then, it can be stated that many measurements obtained in the social science disciplines fall on approximately equal-unit, or interval, scales if the necessary scaling procedures were undertaken.

Ratio Scale

ratio scale   The highest order of measurement scales. It has all the properties of the interval scale plus an absolute zero point.

When numbers have been assigned in measurement in a manner such that the ratio properties ordinarily associated with numbers also apply to the objects to which those numbers have been assigned, we have a **ratio scale.** Few measurement procedures in the social sciences fall in this category. In physical measurement, ratio scaling is ordinarily established through an operation of addition. With weight, for example, an arbitrarily chosen object is assigned the number 1 to represent an object weighing 1 pound or 1 gram. Placing the object on a balance scale pan, another object is found that balances it, and is therefore proclaimed to be equal to it. These two objects are placed together on the same balance pan (addition) and another object found to equal the two objects when it is placed in the other balance pan.

This larger object is labeled by the number 2. By adding units equal to 1, a scale of objects is built with the numbers assigned so that all interval and ratio properties that apply to numbers also apply to the objects with those assigned numbers. The major problem is that in the social sciences, operations of addition cannot be defined. Imagine trying to add together two individuals of I.Q. 100 to equal one individual of I.Q. 200! Beyond this, there are no really satisfactory substitutes for the operation of addition in the social sciences that permits the development of ratio scales by some other method. There have been some partially successful attempts using very complicated scaling methods requiring extensive computation. In practice, however, few variables in the social sciences are measured on anything approximating a ratio scale.

Statistics and Measurement

Since ordinal measurement is used most often in the social sciences, many people conclude that no attempt should be made to apply statistical methods that require the addition and division of numbers. This includes the most commonly used statistical methods such as means, standard deviations, Pearson correlation coefficients, t-tests, and analysis of variance. Practically all of the most powerful statistical methods, the parametric methods, would not be available for use in the social sciences if this injunction was honored. Social science researchers would be confined to using nonparametric methods that make minimal assumptions about distributions and scales of measurement.

What such critics of statistical usage in the social sciences fail to take into account is that parametric statistical methods are derived mathematically on the assumption of normality of population distributions, not on the basis of the type of measurement scale used. Thus, if one is using psychological measurements—however derived—that generate normal distributions in the population, it is entirely appropriate to use such parametric procedures as means, standard deviations, and computing the t-test of the difference between means. The fact that ratio or interval scales are not established is irrelevant as long as the distributions meet the underlying assumptions. Thus, there is no substance to the argument that the social sciences cannot use these more powerful methods because the properties of their measurement scales fail to measure up to the required standards.

To be sure, certain limitations in what can be done with these statistical values from inferior measurement scales still apply. If one mean of a social science variable is 50 and another is 25, it is still improper to make a statement that one is twice the other, or similar statements about differences between values, unless higher measurement scale properties are demonstrated. These limitations are not particularly confining for the research enterprise as long as the most useful statistical methods are applied to aid in reaching sound scientific conclusions. With social science research, it is usually appropriate through refinement of the measurement methods and scaling procedures to obtain variables that have reasonably normal distributions. This makes available the full range of parametric statistical methods. It is only necessary to carefully avoid making statements about the measurements or the statistical constants computed that assume something about the measurement scale itself, which the statistical test, *per se,* does not. It is fully expected, therefore, that

Figure 1.1
Data table for *N* observations
and *k* variables.

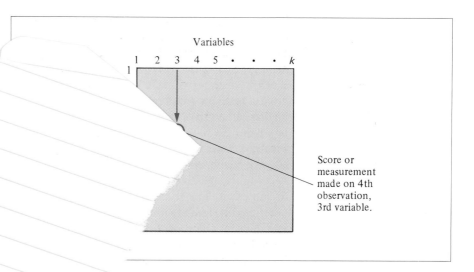

The methods described in this book will be applicable to a wide variety of analyzing social science data, even though ratio and interval scales have been widely established in these disciplines.

In addition to collecting data on the various scales, an important aspect of doing research is to develop a correct picture of how the data should be layed out before processing. The researcher generally collects several pieces of information on each subject. Each piece of information is referred to as a **variable,** and each subject is referred to as an *observation,* or *case.* Throughout this book there are references to *scores.* A score is a measurement made on one subject, case, or observation on one variable. Figure 1.1 is a representation of a typical data table.

Before the advent of high-speed computers, statistical computations were usually done by hand or with a calculator. Computations were long and tedious, often requiring many months to complete. Because of this, much effort was devoted to finding simpler methods of computation and ways of analyzing only part of the data instead of all of it to obtain the desired information. The introduction of the electronic computer caused a drastic change in data-processing methods since the computer performed calculations rapidly and with little error. Calculations that previously required many hours of laborious effort can now be accurately performed in a matter of seconds. The availability of computers enables the research worker to spend more time on creative effort and less time on routine calculations.

At first, the computer was available only to those who knew how to write instructions or programs that the computer could understand. Such programs were written exclusively in "machine language," a difficult medium that few could understand. As computer technology evolved, the writing of computer programs, or "software," was facilitated by the development of more sophisticated languages that permitted the creation of highly complex programs with much less effort.

Computer programming is a technical speciality in its own right. Today, many statistical program packages are marketed that permit individual users to carry out extensive statistical calculations without having to write their own programs. This makes the computer widely available to people who otherwise would never develop the skills necessary to take advantage of these machines. In order to use these programs, users are required only to learn how to provide the program with the information it needs to carry out the calculations, such as how many variables are involved, how many cases, where the data are stored and in what form, and where the results are to be output.

For many years, data were prepared for input to the computer by keypunching. A keypunch machine punched holes in IBM cards to provide the alphabetical and/or numerical data needed by the program. These cards were "read" by the computer, that is, fed into a machine where the holes on the cards would be translated into data for the program to use. Today, data are input directly to the computer from a typewriter-like keyboard and then edited for accuracy before use. This process is sometimes still referred to as "keypunching" even though no actual punching of cards takes place. Lines of "keypunched" data may also be referred to as "cards" of information, or "card images," since a line of data reproduces what was previously punched on a single card.

Today's computers (hardware), like computer software, are much better and more sophisticated than those available even a few years ago. For less than $2000, a computer can be purchased today that sits on a desk top and performs as well as the $1 million mainframe computer of 10 years ago.

As better statistical computing packages, sets of computer programs, are developed, more attention is given to making them "user friendly" or easy for the novice to apply. Some of the early packages were so complicated that the user spent many hours learning how to use the programs. For many modern program packages, "tutorials" are available that teach the user, in easy lessons, how to use the programs. The programs themselves are developed to provide the user with cues on what to do if things go wrong.

Despite the enormous advances in the development of computer hardware and software, knowledge of what statistics to use when and where is a fundamental necessity. Before students can use these powerful computer facilities effectively, they must develop some basic understanding of statistics.

A diskette containing some simple statistical programs is provided with this text. These programs are not designed to compete with sophisticated program packages that permit complicated calculations for a wide variety of statistical methods. Rather, these programs are demonstration programs, designed to give the student an idea of what the computer can do to help apply statistics in an efficient manner without being burdened with an enormous amount of tedious hand calculation. The computer can reduce the routine drudgery involved in statistical computation to a minimum, but the computer cannot do the necessary thinking. The student must know which statistical method to apply to what type of problem before the proper program can be selected to do the necessary computations. Instructions for applying these demonstration programs are also provided with the diskette.

It should be emphasized that it is not always cost effective to use the computer. Sometimes, it is better to do the calculations with a hand-held calculator. If, for example, there is no program readily available so that a new one has to be written, it will usually take less time to do the calculations by hand than to write and "debug" a program (find and correct the errors), unless the amount of data involved is rather extensive. Even if a program does not need to be written, there is often a considerable investment in locating a suitable program and learning how to use it. Where there is a large amount of data, however, particularly if the type of analysis involved is to be run with more than one data set, it would almost always pay to use the computer. Again, the user must do the thinking, not the computer.

Before a general discussion of computer data analysis can proceed, an explanation of some computer terminology and processes is necessary. For the uninitiated, an explanation using the "old" technology as an example might provide an easier-to-understand basis for how personal (and mainframe) computers work and how data are "stored."

IBM Card   Nearly everyone is familiar with the IBM card. The federal government used to print its checks on IBM cards. Many large businesses such as utilities companies printed their bills on IBM cards. Remember the admonition "Do not fold, spindle, staple, or mutilate," that was printed on those cards? There was a reason for this warning. Damaged cards could not be read into the computer without jamming the card reader.

The IBM card (see the example below) is divided into 80 numbered vertical columns running from left to right. Each column is divided into 12 spaces, 10 of which are numbered: 0, 1, 2, 3, 4, 5, 6, 7, 8, 9, starting at the top and going down. Each of those spaces may be used to indicate a bit of data in the form of a number by punching a hole in the card at the appropriate place. The computer, using instructions given to it by a programmer or researcher, would "read" and store each bit of data punched on the card.

"Courtesy International Business Machines Corporation."

The researcher is then able to manipulate the data by rearranging it, performing intricate computations, examining several variables simultaneously, and computing a variety of statistics. It is now possible to input data to the computer using better technology, but the principle remains the same.

Some of the terms that are commonly used in data processing and computer analysis follow. These terms may be found in the context of a research problem in the example below.

Code   A code refers to the numbers assigned to characteristics of the data to be analyzed. For example, a person's gender might be coded for analysis purposes as "1" for "male," or "2" for "female." The gender of every observation, or case, would be coded using the same numbering system.

Case   A case is a unit of analysis. In the social and behavioral sciences, a case is usually a person. Each case may be assigned a number that is used for data sorting and analysis purposes. The number of cases would be the total number of individuals whose characteristics are being analyzed.

Record   A record is the complete set of data for one case. All records for a set of cases would be entered and stored in the computer in the same order.

Byte   A byte is a specific location within a record. A byte is usually the numeric code for one observation of one variable in a case.

Record Length   The length of a record is dictated by the program or software being used to analyze the data. Record length was constrained in the past to multiples of 80, the number of columns on the IBM card. Today, with other methods of computing data, the length of a record is only constrained by practical considerations; for example, the longer the record, the more difficult it is to find errors when checking input.

Example   Assume that you have questioned 100 individuals about a number of characteristics including dietary intake, age, ethnicity, sleeping habits, and exercise. These characteristics are called variables. Each of them is assigned a particular numerical value for each case, or person. For example, males are coded "1," and females are coded "2." Rates of exercise are coded by hours spent or types of exercise engaged in. The coding scheme of a 35-year-old Caucasian male who is a health food consumer and who exercises strenuously every day after getting 8 hours of sleep is:

$$35\ 1\ 1\ 3\ 5\ 1\ 8$$

The "35" is the age, the first "1" is the code assigned to males, the second "1" is the code for Caucasians, and the "3" is the code assigned to the dietary group. The "5" is the code for vigorous exercise, and the last "1" is the code assigned to the frequency of exercise. The final number, the "8," is the number of hours of sleep the individual typically gets.

The coding scheme for another participant, a 26-year-old black female who eats a balanced diet but only exercises mildly and not often, on 7 hours of sleep, is:

26 2 2 1 3 3 7

In this case, the "26" is the age, the first "2" is the code for female, the second "2" is the code for blacks, and the "1" is the code assigned to a balanced diet. The two "3s" are indicators of type of exercise and frequency, and the "7" is the number of hours of sleep.

Each of these records is entered to the computer diskette (or disk) or a computer tape in the same order. A printout of the records would look like the following:

35 1 1 3 5 1 8
26 2 2 1 3 3 7

All 100 cases of data for the one file are entered and stored in the same order using a predetermined coding scheme.

Most computer programs have some limitations on the amount of data per person, the number of persons, and so on, but these are peculiar to each program. Obviously, computer analysis is limited to data that can be numerically codified. Data such as those obtained from open-ended questionnaires that request verbal responses cannot readily be statistically analyzed without translating (coding) the data into numerical form.

Summary

Statistics is a discipline that is organized vertically, much in the same way that buildings are constructed. The upper floors of the statistics house can only be built if there is a strong foundation. Students who feel mathematically inferior must remind themselves to proceed slowly, allowing plenty of study time. To aid the student, the authors selected several unique features for this text, including a minimal presumption of previous mathematical knowledge; a foundation of basic logic; math proofs that are set apart from the main text; the most basic statistical methods; and a problem-solving approach.

Also discussed in this chapter are the types of measurement scales used in statistics and the role of the computer in statistics.

Chapter 2

Data Representation and Organization

T he research worker in any field is commonly confronted with a mass of data, which must be condensed to be useful. A psychologist, for example, might have a dozen or more scores on psychological tests for each of many hundreds of subjects. A stock-market analyst might have months of sales figures for several different key stocks. Poring over page after page of numbers is not a very effective way to obtain a clear picture of the data. One of the most important functions of statistical procedures is to provide effective methods for describing masses of data. In this chapter three elementary statistical procedures of this kind are explained and illustrated. The first of these, the frequency distribution, presents data in tabular form. The data can be in the form of quantitative values, such as scores, cases, units, or other numerical measurements, or in the form of qualitative values, such as attributes. The **frequency distribution** is an orderly arrangement of scores showing the frequency with which each value occurs. The other two procedures, the frequency polygon and the histogram, present data in pictorial form. The frequency distribution and the frequency polygon, which are the more commonly used procedures, are treated in Problems 1 and 2 of the number series of problems. The **histogram** is popular in the behavioral sciences and for making presentations. Since this type of graph is more aesthetic than informative, it is treated in Problem A of the letter series of problems. Letter series problems may be omitted without disturbing the study of the number series problems.

frequency distribution A tabular arrangement or grouping of numerical data into classes and a count of the number of observations that fall into each class.

histogram A diagram that indicates the relative frequency or percentage of different groupings of numbers, called classes, in the population. Often called a bar chart.

PROBLEM 1

Prepare a Frequency Distribution

A frequency distribution is applied when there are so many measurements or scores available in the data that some method is needed to summarize the information. A frequency distribution is also prepared, in many instances, as a preliminary step to the computation of other statistics, such as the mean, median, standard deviation, and the semi-interquartile range.

Preparing a Frequency Distribution

To illustrate the preparation of a frequency distribution consider Data A, a set of test scores, in table 2.1. A count shows that there are 50 scores in Data A, 10 columns of 5 scores each. Although 50 is not a large number of scores, it is still difficult to assimilate the meaning of Data A in the form shown in table 2.1. Preparation of a frequency distribution from Data A provides a better impression of what the data are like.

TABLE 2.1

Data A

90	82	76	135	82	68	77	75	124	89
63	38	65	76	98	68	72	50	80	85
58	111	94	58	63	91	113	12	54	74
87	70	113	75	46	79	60	117	42	90
84	84	48	74	96	59	130	33	69	78

"Tonight, we're going to let the statistics speak for themselves."

Finding the Range   The first step in preparing a frequency distribution from Data A is to locate the highest and lowest scores and to compute the range using these two scores. A more formal discussion of the range is provided in Chapter 4. The highest score, or measurement, in Data A is 135 and the lowest score is 12. The **range** is the difference between these two scores plus one unit, that is, $135 - 12 + 1 = 124$.

One unit is added to the difference between the highest and lowest scores because each score covers a full one unit range, from one-half a unit below the score to one half a unit above the score. The highest score in Data A, therefore, goes from 134.5 to 135.5 and the lowest score goes from 11.5 to 12.5. The range, then, is really from 11.5 to 135.5, rather than just from 12.0 to 135.0. Adding the extra one unit to the difference between the highest and lowest scores takes care of the two half units on either end of the scale needed to reach the exact limits of the highest and lowest scores. Thus, 134.5 and 135.5 are the exact limits of the score 135. A few authors use the convention of treating a score of 135 as going from 135.00 to 135.99. This is not incorrect, but the more common procedure, and the one followed throughout this text, is to treat the score as extending from one-half a unit below the score integer (a whole number) to one-half a unit above the score integer.

range   A measure of variation determined by subtracting the lowest value in a set from the highest.

class interval   The range of actual values that belong to a given class.

Finding the Class Interval, i   After computing the range, the next step in preparing a frequency distribution is to determine the size of the **class interval, i.** The total range is broken into a number of equal steps called class intervals. Each class interval is of width i, where i is usually a number selected from the following possibilities: 1, 2, 3, 5, 10, or a multiple of 10. Only in cases where the scores are decimal fractions will class intervals of other sizes be employed. Using one of these preferred class intervals is only a matter of convention or standard practice. Theoretically, there is no reason why the class interval, i, cannot be 4, 6, 7, 9, 11, or some other number not in the preferred list, but this is not commonly done. The choices of 5 and 10 are preferred because of their convenience in the base 10 number system used in our culture. The values 1, 2, and 3 are also admitted because $i = 5$ might be too large.

The selected value of i depends upon the range. The commonly accepted rule is to select a value of i that produces between 10 and 20 class intervals over the entire range of scores. Thus, each potentially acceptable value of i can be divided into the range to see if it gives a result between 10 and 20. If so, it is an acceptable class interval. If the value of i goes into the range more than 20 times or less than 10 times, it is generally not an acceptable class interval.

With Data A dividing the range, 124, by the various recommended values of i, 1, 2, 3, 5, 10, and so on, only $i = 10$ yields the recommended number of class intervals. With $i = 10$ there are 124/10 or 12 and a fraction class intervals. The fraction, no matter how small, adds another full interval, so there are 13 class intervals for Data A with $i = 10$. The next smaller interval yields 124/5 or 24 and a fraction, or 25 class intervals. This is more than 20, so $i = 5$ is not acceptable. The next larger possible class interval is $i = 20$, which yields 124/20 or 6 and a fraction or 7 class intervals. This is less than 10 class intervals, so $i = 20$ is not acceptable. Often this process yields two values of i that are acceptable. In that event either value is used. Typically, the smaller value is chosen if there are many scores (hundreds), and the larger value is chosen if there are few scores (dozens).

Why should the number of class intervals be between 10 and 20? This rule is somewhat arbitrary but it is based on sound reasoning. When a score of 17 is tallied in the class interval of 10–19, for example, some information is lost because the score is then treated as though it is at the midpoint of the interval, at 14.5 not 17.0. This represents an error of 2.5 points. Other scores tallied in this interval also have their associated errors resulting from being treated as though they fell at the midpoint of the interval. These errors tend to average out since some of the errors are positive and some are negative, but they usually do not average exactly zero. The larger the size of the class interval, the greater the residual error is apt to be. When there are fewer than 10 class intervals, the residual error may be too large to be tolerated since with only a few class intervals i must be large if the entire range is to be covered. On the other hand, if more than 20 class intervals are used, the residual grouping error is reduced to negligible proportions because the intervals are small. The economy and convenience derived from the grouping process, however, is diminished when a large number of class intervals is used. Comparing table 2.2, where $i = 10$ with table 2.3 where $i = 1$, it is clear that table 2.3 is no better than the list of raw data given in table 2.1. Keeping the number of class intervals between 10 and 20 gives an important dividend in economy without giving up too much in the way of accuracy.

Locating the Bottom Class Interval   After determining the size of the class interval, i, the next step is to decide where to start the bottom class interval. The rule followed in this text is to start the bottom class interval on the whole number multiple of i next below the lowest score in the data. In Data A (refer to table 2.1, page 18), $i = 10$ and the lowest score is 12. The bottom class interval does not begin on 12, the lowest score, but begins on 10, which is the first whole number multiple of 10 below the score 12. If the lowest score in the data had been 8 with $i = 10$, the lowest class interval would start at zero. If the lowest score had been 27, the lowest class interval would start at 20. When the value of i is 3, the lowest class interval must start on a value divisible by 3. If $i = 5$, the lowest class interval must start on a value divisible by 5, and so on. The reasoning behind this rule is that frequency distributions look better in the decimal system when class intervals begin with multiples of 10. There is less reason for the rule with values of $i = 2$ or $i = 3$, but to maintain consistency the same rule is followed.

Starting the lowest class interval on a multiple of i effectively increases the range if the lowest score is not divisible exactly by i. With Data A, the range is increased from 124 to 126 because the lowest class interval is started at 10 instead of at 12, the lowest score. The range of 126 divided by 10, the value of i previously selected, still gives the same number of class intervals as 124 divided by 10. Therefore, the number of class intervals is not increased in this instance. In other cases, however, this increment to the effective range might increase the number of class intervals needed by 1, thereby possibly affecting the decision about which value of i to choose for the frequency distribution. This rule might even permit the use of a value of i that would only give 9 class intervals if the lowest class interval is started at the lowest score. Starting the lowest class interval on a lower value that is a multiple of i can increase the number of class intervals to 10, making the value of i acceptable. By the same token, increasing the number of class intervals might push the number from 20 to 21, which would render a previously acceptable class interval unacceptable when the starting point is lowered. In determining the number of class intervals and the value of i to be selected, it is necessary to take into account any increment to the range produced by lowering the starting point for the lowest class interval to a point below the lowest score.

Tallying the Scores   Once the class interval size, i, and the starting point for the bottom class interval are determined, all the class intervals are listed as in column X, table 2.2. In this example, the lowest class interval runs from 10 to 19, the next one runs from 20 to 29, and so on, up to 130 to 139 for the highest class interval. After listing the class intervals in a column headed by X, which is usually how the score is represented, the scores are **tallied** (see tally column). Each score in Data A from table 2.1 results in a tally mark in the tally column of table 2.2.

When the tallying is completed, the number of tallies in each row is entered in the column headed by f, which stands for score frequency. If there are no scores in a class interval, and hence no tallies for that row, a frequency of zero is entered (see class intervals 20–29 and 100–109).

tally   Counting one for each number that falls in a given interval or class.

TABLE 2.2

Frequency distribution with $i = 10$

Class Interval X	Tally	Frequency f
130–139	//	2
120–129	/	1
110–119	////	4
100–109		0
90–99	///// /	6
80–89	///// ////	8
70–79	///// ///// /	11
60–69	///// //	7
50–59	/////	5
40–49	///	3
30–39	//	2
20–29		0
10–19	/	1
N		50

TABLE 2.3

Frequency distribution with $i = 1$

Score X	Tally	Frequency f	Score X	Tally	Frequency f
135	/	1	76	//	2
130	/	1	75	//	2
124	/	1	74	//	2
117	/	1	72	/	1
113	//	2	70	/	1
111	/	1	69	/	1
98	/	1	68	//	2
96	/	1	65	/	1
94	/	1	63	//	2
91	/	1	60	/	1
90	//	2	59	/	1
89	/	1	58	//	2
87	/	1	54	/	1
85	/	1	50	/	1
84	//	2	48	/	1
82	//	2	46	/	1
80	/	1	42	/	1
79	/	1	38	/	1
78	/	1	33	/	1
77	/	1	12	/	1
			N =		50

The sum of the frequencies is entered at the bottom of the frequency column (f) in table 2.2. This is the total number of scores, N, in the distribution. The N is shown on the same row as the numerical total of the frequencies. In table 2.2, the N appears at the bottom of the X column, opposite 50, the number of scores in the distribution. The sum of the frequencies, N, is checked by counting the number of scores in the original data. To be sure there is no error, however, it is necessary to do the entire frequency distribution over again, checking it line-by-line with the original.

Note that the interval widths appear to be only 9 units, 10–19, 130–139, and so on. The exact limits of these intervals, however, are not as shown in table 2.2. The bottom class interval has exact limits that run from 9.5 to 19.5, an interval of 10 units as called for by the choice $i = 10$ for the class interval size. The extra unit comes from the fact that the score 10 goes from 9.5 to 10.5 and the score 19 goes from 18.5 to 19.5, giving a half unit below 10.0 and a half unit above 19.0. The written **class** interval **limits** in the frequency distribution are whole numbers rather than the exact interval limits. This not only gives a neater frequency distribution as far as appearances are concerned, but also helps to avoid errors in tallying.

class limits   The highest and lowest recorded values that can go into a specific class.

If the bottom two intervals were written as 9.5–19.5 and 19.5–29.5, respectively, in tallying a score of 19 it would be easy to mistakenly tally a score of 19 in the interval 19.5–29.5 instead of in the interval 9.5–19.5 where it belongs. The appearance of 19 as part of the exact lower limit of the interval 19.5–29.5 could trigger this type of error. The only problem with using integers or whole numbers as the written class interval limits in the frequency distribution is that the reader must remember to add one unit to the difference between the upper and lower class interval written limits to obtain the proper class interval size, i.

SUMMARY OF STEPS

Preparing a Frequency Distribution

The procedures just described for preparing a frequency distribution from a set of raw scores or measurements are summarized as follows:

Step 1   Find the range of scores by adding one unit to the difference between the highest and the lowest score in the body of data.

Step 2   Select a class interval, i, from the recommended values of i (1, 2, 3, 5, 10, or a multiple of 10) so that the range divided by i yields at least 10 class intervals but not more than 20 class intervals.

Step 3   Start the lowest class interval on the first whole number multiple of i immediately below the lowest score in the data set. Check to see if this increase in the effective range affects the value of i selected and readjust if necessary.

Step 4   List the class intervals in a column, the lowest at the bottom and the highest at the top, and tally the scores in a tally column.

Step 5   Enter the sum of the row tallies in the frequency column, f, and sum up the frequencies to find N.

Step 6   Check to see that N agrees with the number of scores in the data.

PROBLEM 2

Prepare a Frequency Polygon

It is often convenient to display the information from a frequency distribution in pictorial form rather than in tabular form, as shown in the frequency distribution in table 2.2. Two common methods for presenting results pictorially are the **frequency polygon** and the histogram, which will be treated in Problem A. The frequency polygon contains no information beyond that found in the frequency distribution. The information merely is presented in a different form. Thus, given a frequency distribution for a body of data, the frequency polygon can be constructed without additional information.

frequency polygon   A graphical representation of a frequency distribution in which frequencies are plotted by placing a dot above the class mark and connecting these dots with straight lines.

Preparing a Frequency Polygon

The frequency polygon is a graph with the score values laid out along the horizontal axis (the abscissa) and the frequency dimension laid out along the vertical axis (the ordinate). One point is located on the graph for each class interval in the frequency distribution. The point for a given interval is located above the midpoint of the interval at a height corresponding to the frequency for that interval.

The frequency polygon for Data A is shown in figure 2.1. It was prepared from the information contained in the frequency distribution for Data A shown in table 2.2. Note that the point for the class interval 10–19 (*) in table 2.2 is plotted above the point 14.5 on the score axis (abscissa) and opposite 1 on the frequency axis (ordinate). This is because 14.5 is the midpoint for this class interval, that is, halfway between 9.5 and 19.5, and 1 is the frequency for this class interval in the frequency distribution for Data A (table 2.2). Each of the other class intervals is represented by a point determined by the class interval midpoint, which fixes the horizontal axis coordinate, and the class interval frequency, which fixes the vertical axis coordinate.

Note that the points for the intervals 20–29 and 100–110 fall on the horizontal score axis (at zero on the vertical axis). This is because there are no scores in these intervals.

When preparing a frequency polygon, it is customary to add an interval below the bottom interval containing scores and give it a zero frequency. Another interval with zero frequency is added above the top interval containing scores. Since these added intervals have zero frequencies, their points in the frequency polygon are plotted on the base line, the horizontal axis, at the midpoints of these intervals. This has the effect of bringing the graph back down to the base line on either side of the figure to produce a polygon (a many-sided figure). Without these two end points, the figure is left hanging in the air. In figure 2.1, the two end points are at 4.5 and 144.5.

After all the points are plotted on the graph, they are connected with straight lines (fig. 2.1).

Proportions of the Figure   The horizontal and vertical axes in figure 2.1 are on different scales so that the height of the frequency polygon bears a reasonable proportional relationship to the width. If the same scale is used for both the score and the frequency axes the figure, in this instance, will be extremely flat. In other instances it might be extremely tall and narrow. By choosing the correct scales for the

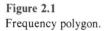

Figure 2.1
Frequency polygon.

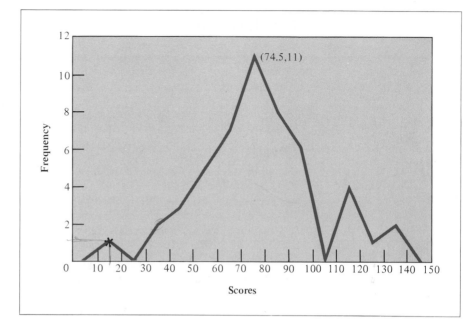

two axes, the resulting figure is pleasing to the eye and effectively displays important features of the data. This is important because the main reason for using a frequency polygon is to make the information easier for the reader to examine and assimilate than it is in the frequency distribution itself. For individuals accustomed to reading frequency distributions, there is little point in preparing a frequency polygon. Many people, however, are confused by tabular presentations and prefer information presented in pictorial form. A frequency polygon presents a clear picture of large amounts of data.

Labeling the Numerical Axes   Even if the class interval, *i*, is a number like 2 or 3, which produces class intervals beginning with numbers other than multiples of 10, the numbers listed below the horizontal axis and beside the vertical axis to mark off the scales are often multiples of 10, depending on the range to be covered. If the horizontal score axis is to cover 100 points, for example, only the numbers at multiples of 10 are written below the axis. The points in between are marked by dots or small vertical lines touching the horizontal axis, but the actual numbers are omitted to give a more pleasing figure. Only in the case of a very narrow range of scores are all the integers in the entire range written below the score axis. The same rules are applied to the vertical frequency axis. In figure 2.1, since the maximum frequency is small, every other integer is listed. If the maximum frequency is 45, for example, only the numbers 5, 10, 15, and so on, up to 45 are used to label the frequency axis. With a maximum frequency of 100, only the values 10, 20, 30, and so on, up to 100 are listed with small marks at the points halfway between these values. In some cases, additional dots at the single integer division are included if a high degree of precision is needed.

STATISTICS IN THE WORLD AROUND YOU

In the world of business the frequency distribution, frequency polygon, and histogram are used to convey information quickly. By examining a histogram of sales, an executive gains a perspective of the progress of a business with one glance. Often executives make informal inferences by viewing a frequency polygon. Recently, an announcement was made in the news media about the difficulties of the stock brokerage firm of E. F. Hutton. The frequency polygon gives a picture of the financial status of the company to the general public. On a frequency polygon, sharp rises and drops are readily recognizable as drastic occurrences. With some analysis, the cause of the fluctuations can be determined. The graph below shows the performance of Hutton's stock for a 24-month period. The sharp decline in October of 1987 was attributed to the semi-crash of the stock market. The decline, compounded by other financial errors, led to the sale of the company.

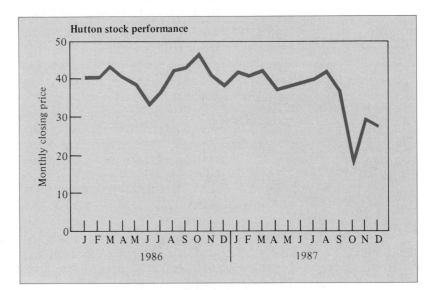

Hutton stock performance

Business executives have limited time and cannot read and assimilate long reports or complex numerical tables. Good research might be overlooked, or their suggestions not implemented, just because the studies are poorly presented. Brief reports, graphs of the results, and a memo summarizing the information are recommended tools for those behavior scientists who venture into the business world.

SUMMARY OF STEPS

Preparing a Frequency Polygon

Step 1   Lay out a chart or graph with the score axis along the horizontal dimension and the frequency axis along the vertical dimension. Choose scales that accommodate the full range of scores plus at least one extra class interval and the full range of frequencies.

Step 2   Label the axes numerically, using numbers that are multiples of 10 if the range of values is large enough, numbers that are multiples of 5 as the next best choice, or with individual integers if the range of values is too small for anything else.

Step 3   Plot one point on the graph for each class interval, placing the point above the exact midpoint of the interval and at a height equal to the frequency of scores for that interval.

Step 4   Place a point on the score axis, that is, with zero frequency, at the midpoint of the interval next below the lowest class interval and at the midpoint of the interval next above the highest class interval.

Step 5   Connect all points by drawing straight lines between the points for adjacent class intervals.

PROBLEM A

Prepare a Histogram

Problem A is the first of the supplementary letter series problems. These problems may be omitted without interfering with an understanding of the numbered problems. Problems in the number series are concerned with more commonly used statistical procedures than those in the letter series.

A histogram is merely an alternate type of figure to the frequency polygon for pictorially presenting the information contained in the frequency distribution. A histogram can be prepared from the frequency polygon or from the frequency distribution just as the frequency distribution and the frequency polygon can be reproduced from the histogram.

Preparing a Histogram   The histogram for Data A from table 2.1 is shown in figure 2.2. The histogram basically consists of a series of bars touching each other (usually with the lines where they touch removed) to give a "downtown skyline" effect. If drawn using a computer software program, the lines are not removed. Each class interval has a bar above it that has a width equal to i, the class interval size. The base of the bar runs from the exact lower limit of the interval to the exact upper limit of the interval. The height of the bar corresponds to the frequency for the class interval. Since the bars extend to the exact interval limits, note in figure 2.2 that the bars begin at 9.5, 19.5, 29.5, and so on, not at 10, 20, 30, the written limits. The bars end at 19.5, 29.5, 39.5, and so on, not at 19, 29, and 39, the written limits.

a. Find the range (highest score − lowest score + 1).
b. Select the appropriate class interval size i.
c. Find the lowest class interval.
d. List all class intervals in column X, tally scores, and sum the tallies in column f.

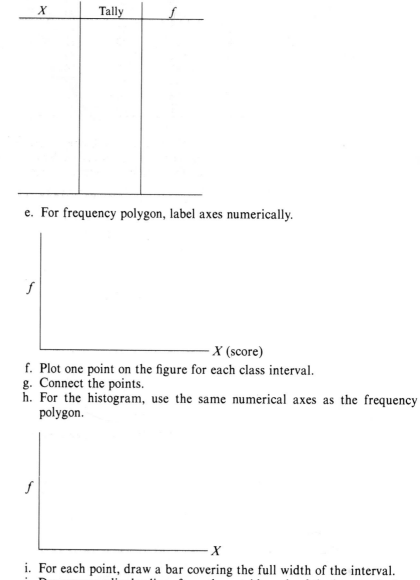

X	Tally	f

e. For frequency polygon, label axes numerically.

f

X (score)

f. Plot one point on the figure for each class interval.
g. Connect the points.
h. For the histogram, use the same numerical axes as the frequency polygon.

f

X

i. For each point, draw a bar covering the full width of the interval.
j. Draw perpendicular lines from the outside ends of the bars to the horizontal axis.

2-2.   Prepare a frequency distribution, a frequency polygon, and a histogram (if Problem A is assigned) for the class examination scores shown below.

Class examination scores

75	84	69	77
85	69	80	71
79	95	50	49
76	61	64	90
83	58	70	83
91	72	41	74
82	63	37	66
76	79	76	88
87	58	70	83
70	58	66	75

a. Find the range (highest score − lowest score + 1).
b. Select the appropriate class interval size *i*.
c. Find the lowest class interval.
d. List all class intervals in column *X*, tally scores, and sum the tallies in column *f*.

X	Tally	*f*

e. For frequency polygon, label axes numerically.

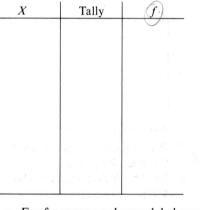

f

X (score)

f. Plot one point on the figure for each class interval.
g. Connect the points.
h. For the histogram, use the same numerical axes as the frequency polygon.

i. For each point, draw a bar covering the full width of the interval.
j. Draw perpendicular lines from the outside ends of the bars to the horizontal axis.

2–3.   Prepare a frequency distribution, a frequency polygon, and a histogram (if Problem A is assigned) for the liquor store data shown below.

Number of liquor stores per selected census tract

4	8	3	18	33	11
20	7	5	11	10	17
2	10	9	8	6	12
9	30	7	6	14	9
14	28	14	9	16	8

a. Find the range (highest score − lowest score + 1).
b. Select the appropriate class interval size i.
c. Find the lowest class interval.
d. List all class intervals in column X, tally scores, and sum the tallies in column f.

X	Tally	f

e. For the frequency polygon, label the axes numerically.

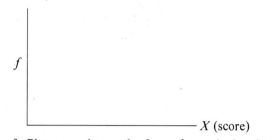

f. Plot one point on the figure for each class interval.
g. Connect the points.
h. For the histogram, use the same numerical axes as the frequency polygon.

i. For each point, draw a bar covering the full width of the interval.
j. Draw perpendicular lines from the outside ends of the bars to the horizontal axis.

2–4.   Prepare a frequency distribution, a frequency polygon, and a histogram (if Problem A is assigned) for the sleep data shown below.

Reported hours of sleep for previous night

8	7	8	6	8
5	7	7	8	10
9	6	9	9	7
7	2	4	5	9
8	8	7	6	8
8	9	6	7	8
9	12	8	8	7
7	10	8	9	6

a. Find the range (highest score − lowest score + 1).
b. Select the appropriate class interval size i.
c. Find the lowest class interval.

d. List all class intervals in column *X*, tally scores, and sum the tallies in column *f*.

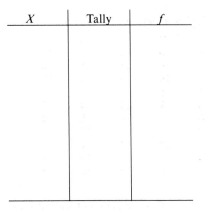

X	Tally	f

e. For the frequency polygon, label the axes numerically.

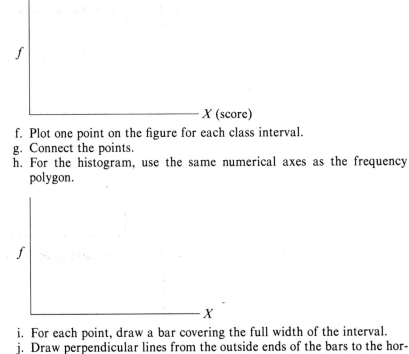

f. Plot one point on the figure for each class interval.
g. Connect the points.
h. For the histogram, use the same numerical axes as the frequency polygon.

i. For each point, draw a bar covering the full width of the interval.
j. Draw perpendicular lines from the outside ends of the bars to the horizontal axis.

2–5.   Prepare a frequency distribution, a frequency polygon, and a histogram (if Problem A is assigned) for the typing error data shown below.

Typing errors per subject in a speed trial

4	8	7	15	10
7	10	13	5	6
20	16	3	19	17
11	5	12	6	8
6	9	23	10	7
10	8	10	15	9
9	10	7	25	5
7	22	13	4	11
6	8	7	12	8
4	6	9	10	3

a. Find the range (highest score − lowest score + 1).
b. Select the appropriate class interval size i.
c. Find the lowest class interval.
d. List all class intervals in column X, tally scores, and sum the tallies in column f.

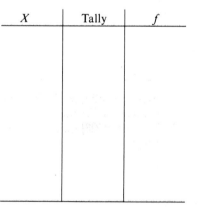

X	Tally	f

e. For the frequency polygon, label the axes numerically.

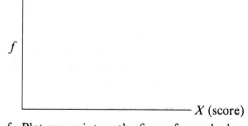

f

X (score)

f. Plot one point on the figure for each class interval.
g. Connect the points.

h. For the histogram, use the same numerical axes as the frequency polygon.

i. For each point, draw a bar covering the full width of the interval.
j. Draw perpendicular lines from the outside ends of the bars to the horizontal axis.

2–6.   Prepare a frequency distribution, a frequency polygon, and a histogram (if Problem A is assigned) for the data shown below.

100 yard dash times of college sprinters

9.8	9.8	9.9	9.4
9.9	9.5	9.7	9.6
9.7	9.6	9.5	9.8
9.8	9.3	9.9	9.7
9.6	10.1	9.4	9.8
10.0	9.7	9.9	9.7
9.7	9.6	9.7	9.7
9.5	9.5	10.0	9.6
9.6	10.2	9.2	9.8
9.4	9.6	9.9	9.7

a. Find the range (highest score − lowest score + .1).
b. Select the appropriate class interval size i.
c. Find the lowest class interval.
d. List all class intervals in column X, tally scores, and sum the tallies in column f.

X	Tally	f

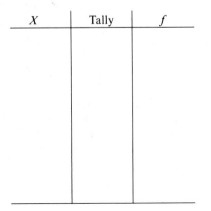

e. For frequency polygon, label the axes numerically.

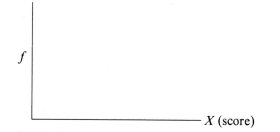

f. Plot one point on the figure for each class interval.
g. Connect the points.
h. For the histogram, use the same numerical axes as the frequency polygon.

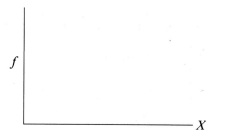

i. For each point, draw a bar covering the full width of the interval.
j. Draw perpendicular lines from the outside ends of the bars to the horizontal axis.

2–7.   Prepare a frequency distribution, a frequency polygon, and a histogram (if Problem A is assigned) for the sample of grade point averages below.

A sample of grade point averages			
2.0	2.5	2.4	2.9
2.2	2.8	2.7	3.1
2.8	2.7	2.5	2.1
3.1	3.5	3.0	2.4
3.2	1.9	2.9	2.8
2.5	2.4	3.7	
2.4	2.3	2.2	
2.2	2.8	2.1	
2.7	2.7	2.5	
3.0	3.3	2.5	

a. Find the range (highest score − lowest score + .1).
b. Select the appropriate class interval size i.
c. Find the lowest class interval.

d. List all class intervals in column X, tally scores, and sum the tallies in column f.

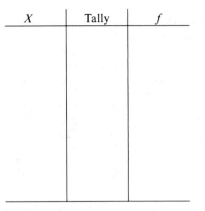

X	Tally	f

e. For the frequency polygon, label the axes numerically.

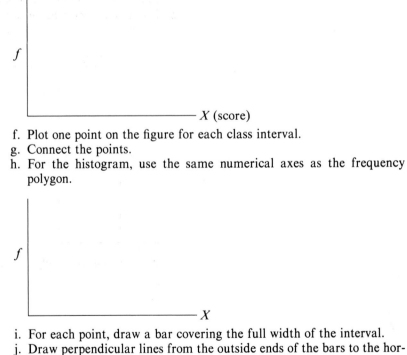

f. Plot one point on the figure for each class interval.
g. Connect the points.
h. For the histogram, use the same numerical axes as the frequency polygon.

i. For each point, draw a bar covering the full width of the interval.
j. Draw perpendicular lines from the outside ends of the bars to the horizontal axis.

2-8.   Prepare a frequency distribution, a frequency polygon, and a histogram (if Problem A is assigned) for the sales data shown below.

A sample of gross sales in thousands of dollars

300	420	260	420	425
320	480	250	465	275
305	490	285	430	335
225	325	325	340	350
285	250	300	360	360
355	265	330	290	375
455	230	345	285	330
500	240	360	290	315
520	310	375	300	290
475	305	380	305	285

a. Find the range (highest score $-$ lowest score $+$ 1).
b. Select the appropriate class interval size i.
c. Find the lowest class interval.
d. List all class intervals in column X, tally scores, and sum the tallies in column f.

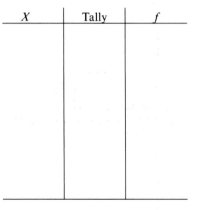

X	Tally	f

e. For the frequency polygon, label the axes numerically.

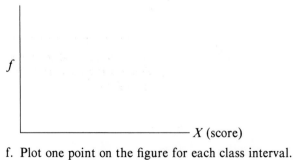

f. Plot one point on the figure for each class interval.
g. Connect the points.

h. For the histogram, use the same numerical axes as the frequency polygon.

i. For each point, draw a bar covering the full width of the interval.
j. Draw perpendicular lines from the outside ends of the bars to the horizontal axis.

2–9. Find the range for the weights in the data below.
a. Find the smallest weight.
b. Find the largest weight.
c. Compute the range.

Player	Weight	Minutes	Player	Weight	Minutes	Player	Weight	Minutes
1	180	15	16	215	0	31	205	19
2	220	20	17	170	0	32	200	17
3	200	29	18	185	50	33	195	15
4	175	8	19	155	2	34	220	49
5	195	0	20	230	46	35	185	11
6	200	38	21	225	38	36	190	5
7	240	6	22	180	25	37	200	18
8	190	21	23	175	10	38	175	12
9	205	33	24	200	25	39	190	20
10	230	40	25	195	31	40	200	17
11	160	8	26	210	35			
12	180	0	27	190	21			
13	185	13	28	200	26			
14	200	19	29	195	16			
15	210	28	30	185	12			

2–10. Find the lower and upper bounds for the following numbers.

a. 5
  1. Subtract 1/2 unit.
  2. Add 1/2 unit.
b. −2.3
  1. Subtract 1/2 unit.
  2. Add 1/2 unit.
c. 101
  1. Subtract 1/2 unit.
  2. Add 1/2 unit.
d. 76.32
  1. Subtract 1/2 unit.
  2. Add 1/2 unit.
e. −61.0
  1. Subtract 1/2 unit.
  2. Add 1/2 unit.

2–11.   Find the midpoints for each of the following intervals.
a. 10–20
  1. Add the two limits.
  2. Divide the number obtained by 2.
b. 16.2–16.45
  1. Add the two limits.
  2. Divide the number obtained by 2.
c. 100–107
  1. Add the two limits.
  2. Divide the number obtained by 2.
d. −5 to −1
  1. Add the two limits.
  2. Divide the number obtained by 2.
e. 0.25–0.50
  1. Add the two limits.
  2. Divide the number obtained by 2.

2–12.   Compute the range for the following data set: 7, 7, 7, 7, 7, 7, 7.
a. Find the lowest measurement.
b. Find the highest measurement.
c. Subtract lowest from the highest.
d. Add 1 point to the number obtained in (c).

2–13.   Compute the range for the following data set: 100, 3, −1, 2, 6, 0.
a. Find the lowest measurement.
b. Find the highest measurement.
c. Subtract lowest from the highest.
d. Add 1 point to the number obtained in (c).

2–14.   Compute the range for the following data set: −2, −5, −6, −9, −15, −36.
a. Find the lowest measurement.
b. Find the highest measurement.
c. Subtract lowest from the highest.
d. Add 1 point to the number obtained in (c).

2–15.   Prepare a histogram for the frequency distribution given below.

X	f
60–51	10
50–41	11
40–31	17
30–21	5
20–11	20
10–1	2
0–(−10)	4

a. Set up the axes as above.
b. Place a point on the figure for each class interval.
c. For each point, draw a bar covering the full width of the interval.
d. Draw perpendicular lines from the outside ends of the bars to the horizontal axis.

2–16. Develop a frequency distribution for the minutes data given in question 2–9.

a. Find the range.
b. Select the appropriate class interval i.
c. Find the lowest class interval.
d. List all class intervals in column X, tally scores, and sum the tallies in column f.

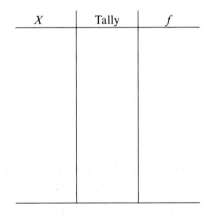

X	Tally	f

2–17. Develop a frequency distribution for the weight data given in question 2–9.

a. Find the range.
b. Select the appropriate class interval i.
c. Find the lowest class interval.

d. List all class intervals in column *X,* tally scores, and sum the tallies in column *f.*

X	Tally	f

Use the following data for questions 2–18 through 2–20.

**Aggression scores for students
at a large university**

17	42	10	49
12	8	25	14
29	36	26	40
36	20	33	31
16	44	39	26

2–18. Find the range.
    a. Find the lowest score.
    b. Find the highest score.
    c. Subtract lowest from the highest and add 1 point.

2–19. Find the appropriate class interval size *i.*
    a. Use the range found in question 2–18.
    b. Divide the range by class interval sizes 1, 2, 3, 5, and 10.
    c. Choose the class interval size that gives a number between 10 and 20.

2–20. Find the lowest class interval.
    a. Use the class interval size found in question 2–19.
    b. Find the multiple of *i* just below the lowest score.

2–21. Using correct English, explain why in finding the range, 1 point is added to the difference between the highest measurement and the lowest measurement. (Hint: read pages 19–20.)

2–22. Why would we want to have only 10 to 20 class intervals? Explain using well-written English statements. (Hint: read pages 20–21.)

Chapter 3

Measures of Central Tendency

T he frequency distribution and its graphical representations, the frequency polygon and histogram, provide the research worker with a much better impression of the data than can be obtained from inspection of mere columns of randomly arranged numbers. These methods of data representation do not provide, however, a single precise numerical figure to represent the entire distribution of scores as far as overall size is concerned. Measures of central tendency are designed to provide a single numerical index that will tell, in general, how high or how low the scores are on the average for the entire distribution.

The measure of central tendency packs a great deal of information about a distribution of scores into a single number. However, it also omits a great deal of information. One of the things not described by a measure of central tendency is how the scores are spread in the distribution. Measures of dispersion (see Chapter 4), are used to compare distributions where the scores are very close together, widely spread apart, or somewhere in between. Two single numbers, one a measure of central tendency, and another a measure of dispersion, go a long way toward describing what really needs to be known about an entire distribution involving perhaps hundreds of individual scores.

The three measures of central tendency described in this chapter are the mode, the median, and the mean. Methods for computing these statistics are described and illustrated. For some of the measures of central tendency, different methods are presented for handling grouped and ungrouped data. In general, researchers apply the ungrouped methods to their raw data directly. The grouped methods are presented for the sake of completeness and to enable the researcher to compute measures of central tendency in those situations where a frequency distribution is given.

PROBLEM 3

Determine the Mode of a Distribution

nominal scale   Objects are assigned numbers for the purpose of identification only. So-called measurements under this scale are used for naming or labeling only. No meaningful arithmetic operations can be performed on these measurements.

mode   The most frequently occurring number in a population or sample. The mode of a population is a parameter.

The mode is the easiest of all measures of central tendency to obtain and also the only one appropriate for data measured on a **nominal scale.** The **mode** is the most frequently occurring value. If data are not grouped in class intervals, merely find the value that appears most often (fig. 3.2). Or, if the data are grouped in class intervals of size 1, the mode is the value with the greatest frequency. In case the data are grouped in class intervals and the class interval, i, is greater than 1, the mode is the midpoint of the interval containing the greatest number of values.

Table 3.1 represents a frequency distribution where the data are grouped into class intervals with a class interval width, i, of 1 point. The mode of the distribution in table 3.1 is 87, since the frequency for the value of 87 is 25, and 25 is higher than any other frequency in the distribution.

Table 3.2 represents a frequency distribution where the data are grouped into class intervals with $i = 5$. The mode of the distribution in table 3.2 is 47.0, the midpoint of the interval with the largest frequency. This is the interval 45–49 with exact interval limits of 44.5 to 49.5 and a frequency of 25. The mode, 47.0, lies halfway between the exact limits, $44.5 + 1/2(5)$, or halfway between 44.5 and 49.5. It is also halfway between the class interval written limits, 45 and 49.

TABLE 3.1

Frequency distribution

X	f
95	3
94	5
93	7
92	10
91	13
90	15
89	16
88	20
87	25
86	22
85	19
84	16
83	14
82	11
81	8
80	3
79	2
N	209

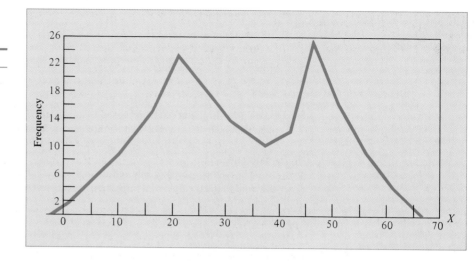

Figure 3.1
A bimodal distribution
frequency polygon.

TABLE 3.2

Frequency distribution

X	f
60–64	3
55–59	8
50–54	15
45–49	25
40–44	12
35–39	10
30–34	13
25–29	18
20–24	23
15–19	16
10–14	10
5– 9	5
0– 4	2
N	160

Table 3.2 is an unusual frequency distribution because there are two distinct modes rather than just one. According to the definition of the mode as the midpoint of the interval with the largest number of cases, the mode is 47.0 and technically there is only one mode. On the other hand, there is another large frequency for the interval 20–24, which is only two values less than the frequency for the interval containing the mode. Further, the frequencies fall off in magnitude on either side of both these peak frequency intervals. Such a distribution is referred to as **bimodal** (with two modes), even though the two peak frequencies are not identical. The frequency polygon for the frequency distribution in table 3.2 is shown in figure 3.1. The two separate peaks in the curve clearly reveal the bimodal character of the distribution. In case there are two definite and distinct modes, both should be reported.

In searching for the mode of a given distribution, there might be two frequencies for adjacent intervals that are equal and larger than any of the other frequencies. If that occurs the mode is at the point halfway between the midpoints of these adjacent intervals, at the exact interval limit separating the two intervals.

When to Use the Mode

Research workers can, at their discretion, choose any of the measures of central tendency to describe their particular data set. This choice is ordinarily dictated by the nature of the situation that is being dealt with. The mode, for example, is chosen in those cases where the most common value, or score, is desired. A manufacturer who is deciding how many pairs of shoes to produce of a certain size is interested in what sizes occur most frequently in the shoe-wearing population. If size nine occurs more often than any other size, the company wants to make more size nine shoes than any other size. Any measure of central tendency that failed to yield a value closer to size nine than to any other shoe size is misleading information.

Descriptive statistics have been used to advance the cause of the civil rights movement. Several years ago, a major union joined forces with a number of female job applicants to bring suit against a large employer and a state agency for sex discrimination.

The women were not allowed to apply for a job in the large company because of a rule that all employees must be five feet, seven inches tall. The union cited a survey by a federal government agency that indicated that the five foot, seven inch requirement eliminated more than 94% of all females aged 18 to 79.

The suit was finally settled when the company agreed to test the applicants on their ability to perform a given job task, which was previously thought could only be performed by those people over five feet, seven inches tall. The lone government statistician strikes again!

Another situation where the mode is useful is to find the class size a student might expect to encounter at a state university. At a university, many classes are small but there are also some very large classes. The most likely class to be encountered randomly is a small class. Using the modal class size is a good measure of central tendency to provide an indication of what the student is most likely to find. Choosing a different measure of central tendency could result in a class size figure that would never actually be encountered.

The mode is also used to obtain a very quick estimate of the central tendency of the distribution because it is determined merely by inspection instead of by performing lengthy computations. The mode is not, however, a very stable measure since it can change considerably with the addition or deletion of a few scores, and for interval type data, the changing of class interval sizes can affect the mode.

PROBLEM 4

Determine the Median from Ungrouped Data

median   The number such that half the population (or sample) is larger and half is smaller.

The **median** of a distribution of measurements is the point on the measurement scale below which 50% of the measurements fall. It follows logically that it is also the point on the scale above which 50% of the measurements fall. The abbreviation for the median is *Mdn*. Roughly speaking, it is the middle value in a collection of measurements. This definition must be interpreted carefully because the middle value is not necessarily in the center of the numerical measurement range. If there are some extreme measurements, either high or low, this lengthens the measurement range in the direction of the extreme values so that the median will probably not fall halfway between the bottom and top measurements. In fact, the extreme values

Figure 3.2
An illustration of the median using data from a small fictitious company.

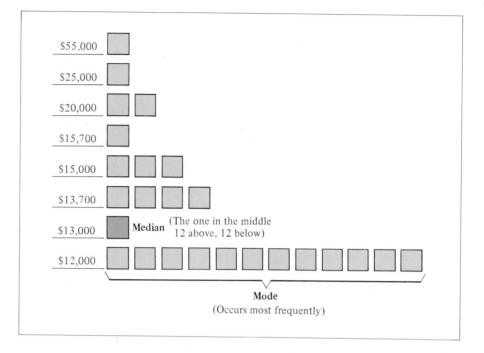

In figure 3.2 the median salary is $13,000 per year. Twelve people earn less than $13,000, and twelve people earn more. If the person making the most ($55,000) earned $200,000 per year, the median remains the same. Likewise, if one or more of the people making $12,000 per year was dropped to $10,000 per year, the median remains the same.

at one end of a distribution can be made even more extreme without affecting the median at all. The measurement in the middle remains in the same place even if the extreme measurements are altered to be drastically more extreme (fig. 3.2).

Computing the Median from Ungrouped Data

When measurements are not grouped into class intervals and a median is to be computed, the first step is to arrange the values in order of size from lowest to highest. Thus, the scores 10, 9, 7, 7, 11, 14, 14, 12, and 6 are arranged as follows: 6, 7, 7, 9, 10, 11, 12, 14, and 14. Counting the scores shows that $N = 9$. This is an odd number of scores, so it is possible to have a middle score. The middle score is 10 so the median is 10.0. This is true in this distribution of scores because there are no other scores of size 10 except the 1 score. If there had been another score of 10, that is, no 9 but two 10s, (6, 7, 7, 10, 10, 11, 12, 14, 14) then 10.0 would not be the median. An example of this more complicated situation will be given later.

How does a median of 10.0 for the 9 scores fit the original definition of the median given at the start of Problem 4? Half the scores, 9/2 or 4 1/2 scores, should be below 10.0 and 4 1/2 scores should be above 10.0. Below the score of 10 are the scores of 6, 7, 7, and 9, or 4 scores instead of 4 1/2 scores. Also, above the score of 10 are the scores of 11, 12, 14, and 14; again 4 scores instead of 4 1/2 scores (fig. 3.3).

Figure 3.3
Determination of the median
when N is odd.

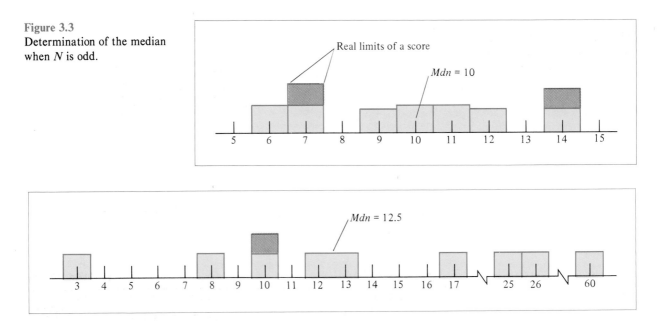

Figure 3.4
Determination of the median
when N is even.

Note that the score of 10 falls in an interval that runs from 9.5 to 10.5. It does not fall precisely at 10.0. This single score in the interval 9.5 to 10.5 is treated as though it was evenly spread throughout the entire interval from 9.5 to 10.5. This means that half of the score lies between 9.5 and 10.0 and half lies between 10.0 and 10.5. If the median is 10.0, then not only do the 4 scores of 6, 7, 7, and 9 fall below the median, but so does the half score that is the portion of the score 10 that falls between 9.5 and 10.0. This gives the total of 4 1/2 scores that fall below the median as demanded by the definition. At the same time, above 10.0 are the scores of 11, 12, 14, and 14 plus the half score that is the portion of the score 10 that falls between 10.0 and 10.5. This gives the total of 4 1/2 scores that lie above the median.

Example   Consider the scores 3, 8, 10, 10, 12, 13, 17, 25, 26, and 60, already arranged in order from lowest to highest. Since there is an even number of scores, $N = 10$, there can be no middle score. The 2 single scores 12 and 13 fall in the middle. In this case the median falls at 12.5, the exact interval limit between the scores of 12 and 13. Below 12.5 there are 10/2 or 5 scores, 3, 8, 10, 10, and 12, since all of the scores fall below the upper limit of the interval for 12, namely, 12.5. Above 12.5 are the 5 scores of 13, 17, 25, 26, and 60. Thus 12.5 fits the definition of the median because 50% of the scores are above it and 50% of the scores are below it (fig. 3.4).

When There Are Multiple Scores in the Region of the Median   In the two previous examples the median was either equal to a single middle score with an odd number of scores, or between two single middle scores with an even number of scores. This is unusual in real life, especially when statistics must be calculated by hand! It is more common to find that there is no single middle score with an odd number of

Figure 3.5
Finding the median when the
frequency of the middle value
is greater than 1.

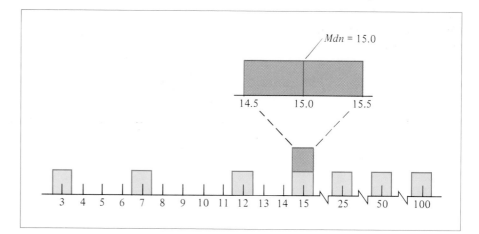

FUNKY WINKERBEAN   **BY TOM BATIUK**

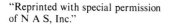

scores or two single middle scores with an even number of scores. Consider the following scores already arranged in order of size from lowest to highest: 3, 7, 12, 15, 15, 25, 50, and 100. In this case $N = 8$, so 4 scores must fall below the median and 4 scores must fall above the median. The value 15.0 is the only point in the distribution for which this is true, hence 15.0 is the median for this group of scores. Note that there are 2 scores in the interval 14.5 to 15.5, which contains the median. One score falls between 14.5 and 15.0 and the other score falls between 15.0 and 15.5. Below 15.0 is the 1 score between 14.5 and 15.0, plus the 3 scores 3, 7, and 12 (fig. 3.5).

The determination of the median becomes more complicated if there are several identical scores for the 1 point interval in which the median must fall. Consider the following example of scores already arranged in order of increasing size: 3, 7, 12, 15, 15, 15, 15, 25, 50, 75, 90, and 100. In this example $N = 12$, so 6 scores fall below the median and 6 scores fall above the median. Beginning at the bottom, 3 scores are counted below 14.5 and below 15.5 a total of 7 scores are counted, the 3 scores below the interval 14.5 to 15.5 plus the 4 scores of 15 contained in the interval 14.5 to 15.5. Since 3 scores are less than the needed 6 scores, the median must be above 14.5. Also, because 7 scores are more than the needed 6 scores, the median must be below 15.5. This places the median somewhere between 14.5 and 15.5.

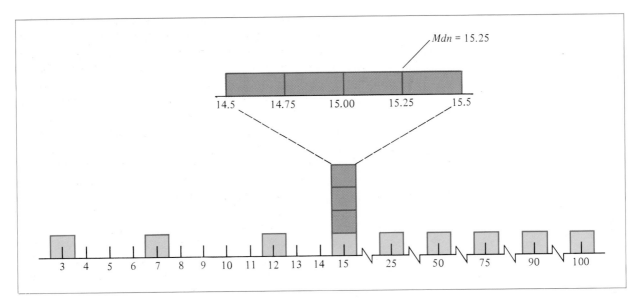

Figure 3.6
Finding the median when the frequency of the middle value is greater than 1.

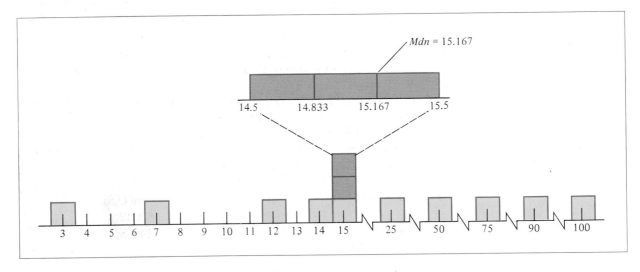

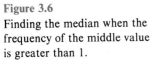

Figure 3.7
Finding the median when the frequency of the middle value is greater than 1.

There are 3 scores below 14.5 and 6 are needed before reaching the median point since the median has 6 scores below it in this example. With 4 scores in the interval 14.5 to 15.5, three-quarters of the scores in that interval must fall below the median and only one-quarter above it. Since it is assumed that the scores are spread evenly throughout the interval, it is necessary to rise above 14.5 a distance equal to three-quarters of the interval before coming to the point below which 3 of the scores in the interval will fall, plus the 3 below this interval, making a total of 6. The interval from 14.5 to 15.5 is 1 score point, hence the median is three-quarters of 1 point, or 0.75, above 14.5. In other words, the median in this case equals 14.5 + 0.75 = 15.25. Exactly 6 scores then fall below 15.25 and 6 scores fall above 15.25, thereby meeting the required definition of the median (fig. 3.6). A formula approach to finding the median is described in Problem 5.

If 1 of the scores of 15 is changed to 14, the result is 3 scores of 15. There are now 4 scores below 15 and 5 above. Then the interval 14.5 to 15.5 is divided into thirds to find the median, where one-third is above the median and two-thirds below (fig. 3.7). The median is 15.167 because 6 scores are below it (3, 7, 12, 14, 14.5–14.833, 14.833–15.167) and 6 scores are above it (15.167–15.5, 25, 50, 75, 90, 100).

MATHEMATICAL NOTES

The median of 15.25 is rounded off to one decimal place, or 15.2. There is usually little reason to report measures of central tendency to a greater level of precision than one decimal place where the scores are integers (whole numbers). In rounding, the rule is to round up if the second and third digits to the right of the decimal place give a number larger than 50; round down if they give a number less than 50. Thus 6.651 becomes 6.7 because the 51 is greater than 50; 6.649 becomes 6.6 because 49 is less than 50; 100.104 becomes 100.1; 20.380 becomes 20.4; 8.92 becomes 8.9; 4.29 becomes 4.3; 9.97 becomes 10.0; 0.02 becomes 0.0. In the case that the second and third digits give a number exactly equal to 50 and any additional digits to the right are also zero, the rule does not apply. To determine the direction of rounding in this case, round always to an even number rather than to an odd number. Thus 13.450 becomes 13.4 because the second digit to the right is 5 and all digits to the right of that are zero plus the fact that 4 is an even number; 107.75 becomes 107.8 because the second digit to the right is 5 and 7 is an odd number so the rounding is up to the even digit, 8; 9.95 becomes 10.0; 6.55 becomes 6.6; 8.850006 still rounds up to 8.9, however, since 50006 is greater than 50000. This last case does not fit the requirements for the immediately preceding examples because all the digits to the right of the 5 are not zeroes. If the number had been 8.850000 the rounded number would have been 8.8.

When Several Medians Exist   When the median falls at the exact limit between two intervals so that one or more score intervals on either side is empty, there might be many possible score points to satisfy the requirements for the median. Consider the following scores: 3, 4, 8, 9, 9, 11, 15, 25, 50, and 100. With $N = 10$ there must be 5 scores below the median. Counting up from the bottom shows that exactly 5 scores are covered by the time the point 9.5 is reached. It might be concluded, therefore, that 9.5 is the median, especially since there are also 5 scores above 9.5. Note, however, that *any* point between 9.5 and 10.5 also has 5 scores above it and 5 scores below it. This is because there are no scores of 10 in this distribution and hence no additional scores are added to the 5 already passed in moving upward from 9.5 to 10.5. Where a range of such values exists, all of which will satisfy the definition of the median, choose the median to be the midpoint of this range. In this example the median is at 10.0, halfway between 9.5 and 10.5. Sometimes this range of points, all of which satisfy the median, extends over an interval of several score points, or over more than one interval.

SUMMARY OF STEPS

Computing the Median from Ungrouped Data

Step 1   Arrange the scores in order of increasing size from the lowest to highest, repeating equal scores.

Step 2   Count the number of scores in the distribution, N, and find N divided by 2.

Step 3   Count from the bottom until a score interval is reached that has $N/2$ scores or less below its exact lower limit and $N/2$ scores or more below its exact upper limit.

Step 4   Take the difference between $N/2$ and the number of scores below the exact lower limit of the score interval containing the median, the interval located in Step 3. Divide the difference by the number of scores in the score interval. The result is the proportion of cases from the interval that is below the median or the number of scores that are passed in the interval before reaching the median moving up from the bottom of the distribution.

Step 5   Multiply the proportion of scores in the interval below the median by the value of i, the class interval size, which is always "1" for ungrouped data. Add the product to the exact lower limit of the interval containing the median to obtain the median itself.

Step 6   If there is a range of points that satisfy the definition of the median, place the median at the midpoint of this range.

When to Use the Median   Since the median is the point in the score distribution below which (and above which) 50% of the scores fall, it is determined by the location of the scores in the middle of the distribution, not by the scores at the extremes of the distribution. Consequently the extreme scores in a distribution can be altered drastically without affecting the median. The highest score in a distribution can be increased by 1000 points, or a million, without changing the median of the distribution.

The median is a good measure to use where there are some extreme scores at either end of the distribution, which should not be allowed to influence the central tendency value obtained. One example is determining a measure of central tendency for reaction times. Some reaction times tend to be abnormally long because the subject was inattentive. These extreme measures are not valid indications of reaction time but rather are due to extraneous factors. By selecting the median as the measure of central tendency, the extreme scores are not allowed to falsely raise the summary reaction time figure obtained.

Another situation when the median is the preferred measure of central tendency is where the distribution is truncated (when the scores at one end of the distribution are not actually precisely known). For example, if subjects are given arithmetic problems under timed conditions and the time allotted is too generous, some subjects may finish the problems before the time runs out. These subjects could have earned higher scores if more problems had been included. Since these individuals will almost always have scores well above the median, it is possible to calculate the median even though some of the highest scores in the distribution are not exactly determined.

PROBLEM 5

Determine the Median from Grouped Data

median   The number such that half the population (or sample) is larger and half is smaller.

In most real-life data collection situations, the number of scores available are too large to permit calculation of the median by the procedures described in Problem 4. In such cases, scores are first arranged in a frequency distribution using the methods described in Problem 1. Whether the median is calculated from ungrouped data, as in Problem 4, or from grouped data, as in this problem, the basic principles used are the same. In both cases the **median** is the point below which 50% of the scores fall and in both cases it is assumed that scores in a given interval are spread evenly throughout that interval for purposes of determining a precise location for the median. Perhaps the major difference between the procedures in Problem 4 and those in Problem 5 is that in Problem 4 scores are spread over intervals of only 1 point while in Problem 5 scores are spread evenly over class intervals with widths equal to i points, where i is the class interval size. In those instances where the researcher has a large number of data points available and does not want to group them but wants to find the median, a modification of the formula used for grouped data is given in the later portion of Problem 5. The grouped procedure is given for the sake of completeness in dealing with frequency distribution.

Some accuracy is lost when computing the median from grouped data because grouping the data into intervals loses some information. For example, scores in the interval 5 to 9 could be 5s, 6s, 7s, 8s, or 9s, or any combination of these scores. Since the particular distribution of scores in this interval is not known in grouped data, the scores are treated as though they are spread evenly throughout the interval, from 4.5 to 9.5, the exact limits of the interval.

Computing the Median from Grouped Data

To illustrate the procedure for calculating a median from grouped data, some examples are worked out in detail. Consider the frequency distribution in table 3.3. This situation is very similar to calculating the median from ungrouped data, Problem 4, since the class interval size is $i = 1$. The only difference is that instead of stringing the scores out horizontally from smallest to largest, writing the identical scores as many times as they are repeated, the scores are listed vertically, and they are only listed once with a frequency column included to indicate how many times that particular score is repeated (table 3.3). With so many scores, it is much easier to compute the median from a frequency distribution than to compute it by writing out the scores in order of size, as in Problem 4.

The median is the point below which 50% of the scores fall, so 50% of $N = 209$ is determined first to be $.50 \times 209 = 209/2 = 104.5$. So the median is the point in the distribution below which there are exactly 104.5 cases. This demonstrates that scores are divided into fractions and spread evenly over intervals when calculating the median. Scores are not treated as individual indivisible units.

The last column in table 3.3, headed by cf, gives the cumulative frequencies for the frequency distribution. These are obtained by starting with the lowest interval and placing its frequency, shown in column f, as the lowest cumulative frequency in column cf. The value represents the number of scores in the distribution below the exact upper limit of that interval. In table 3.3 the cumulative frequency (cf) for the bottom interval is 2, indicating that there are 2 scores below 79.5, the exact upper limit of the interval 78.5 to 79.5. Since there are no intervals below this one containing nonzero frequencies, the frequency from column f and the cumulative frequency from column cf are identical.

To obtain the cumulative frequency for the next interval in table 3.3 (the interval corresponding to the score 80), add the frequency, 3, to the cumulative frequency, 2, and place the results, $3 + 2 = 5$, in the cf column to correspond with the score 80. For the next higher cumulative frequency, add the frequency for the next higher interval, 8, to 5, the cumulative frequency for 80, to get the cumulative frequency, 13, for the score 81. The cumulative frequency for score 82 is $11 + 13 = 24$, for 83 it is $14 + 24 = 38$, and so on, up to the last interval where the cumulative frequency for the score 95 is $3 + 206 = 209$. Note that the top cumulative frequency equals N, the number of cases. This is expected since all the scores in the distribution must fall below the upper limit of the highest interval. To avoid errors in calculating the median, it is essential to remember that the cumulative frequency represents the number of cases below the exact *upper* limit of the interval, not the midpoint or the lower limit.

After determining that there are $N/2 = 104.5$ scores below the median for the frequency distribution in table 3.3, the cumulative frequencies in the cf column are inspected to locate the interval that must contain the median. This must be the interval 86.5 to 87.5, corresponding to the score 87, since below 86.5 there are only 95 cases, which is less than 104.5, and below 87.5 there are 120 cases, which is more than 104.5. Consequently the point below which there are 104.5 cases must lie somewhere between 86.5 and 87.5.

TABLE 3.3		
Frequency distribution		
X	f	cf
95	3	209
94	5	206
93	7	201
92	10	194
91	13	184
90	15	171
89	16	156
88	20	140
87	25	120
86	22	95
85	19	73
84	16	54
83	14	38
82	11	24
81	8	13
80	3	5
79	2	2
N	209	

There are 25 scores in the interval 86.5 to 87.5, the frequency from column f for the score 87. The 25 scores are considered to be spread evenly throughout the interval. A portion of the 25 scores must fall below the median since there are 104.5 scores altogether below the median and there are only 95 scores below 86.5. The difference needed is $104.5 - 95.0 = 9.5$ scores. Of the 25 scores in the interval 86.5 to 87.5, 9.5 scores must fall below the median. When moving up from 86.5 toward the median a total of 9.5 of the 25 scores in the interval for the score 87 must be passed before the median is reached.

Since the 25 scores in the interval 86.5 to 87.5 are considered to be spread evenly over the interval, and 9.5 scores must be passed before the median is reached, the distance between 86.5, the lower limit of the interval, and the median must be $9.5/25$ of the interval width, which is 1 point in this case. So, the distance between 86.5 and the median is equal to $(9.5/25)(1) = 9.5/25$. The calculations to obtain the median for the frequency distribution in table 3.3 are summarized as follows:

$$Mdn = X_{LL} + \left(\frac{N/2 - \text{Cum } f_{LL}}{f} \right) i \qquad (3.1)$$

$$Mdn = 86.5 + \frac{104.5 - 95}{25} \times (1)$$

This interpolative process is expressed in words as follows:

$$\text{Median} = \begin{bmatrix} \text{Exact lower} \\ \text{limit of interval} \\ \text{containing the median} \end{bmatrix} + \frac{\dfrac{N}{2} - \begin{bmatrix} \text{Number of scores} \\ \text{below lower limit} \\ \text{of interval holding } Mdn \end{bmatrix}}{\begin{bmatrix} \text{Number of cases in} \\ \text{the interval that} \\ \text{contains the median} \end{bmatrix}} \times (i)$$

where $N =$ the number of scores in the distribution, and
$i =$ class interval size.

Completing the computations for the median of the frequency distribution in table 3.3 gives:

$$Mdn = 86.5 + \frac{9.5}{25} \times (1) = 86.5 + 0.38 = 86.88 = 86.9$$

Note that the median is rounded off to one decimal place. Figure 3.8 shows how to obtain the values for the components in Formula 3.1 using the data in table 3.3.

Figure 3.8
Identifying the components of
a frequency distribution for
Formula 3.1 using data from
table 3.3.

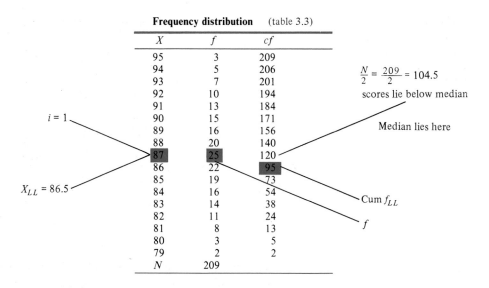

Frequency distribution   (table 3.3)

X	f	cf
95	3	209
94	5	206
93	7	201
92	10	194
91	13	184
90	15	171
89	16	156
88	20	140
87	25	120
86	22	95
85	19	73
84	16	54
83	14	38
82	11	24
81	8	13
80	3	5
79	2	2
N	209	

$i = 1$

$X_{LL} = 86.5$

$\frac{N}{2} = \frac{209}{2} = 104.5$

scores lie below median

Median lies here

Cum f_{LL}

f

TABLE 3.4

**Frequency
distribution
(see table 3.2)**

X	f	cf
60–64	3	160
55–59	8	157
50–54	15	149
45–49	25	134
40–44	12	109
35–39	10	97
30–34	13	87
25–29	18	74
20–24	23	56
15–19	16	33
10–14	10	17
5– 9	5	7
0– 4	2	2
N	160	

To illustrate the computation of a median with grouped data where $i = 5$, calculate the median for the frequency distribution in table 3.4. There are 160 cases in the frequency distribution in table 3.4, so $N/2 = 160/2 = 80$ scores must fall below the median. The desired point on the score scale has 80 scores below it and 80 scores above it. Examine the cumulative frequencies in column *cf* of table 3.4. Note that there are 74 scores below 29.5 and 87 scores below 34.5, so the median must lie in this interval. Using the same principles and Formula 3.1 that were applied in the previous example for table 3.3, the median for the data in table 3.4 is calculated as follows:

$$Mdn = 29.5 + \frac{80 - 74}{13} \times (5)$$

$$= 29.5 + \frac{6}{13} \times (5) = 29.5 + \frac{30}{13}$$

$$= 29.5 + 2.31 = 31.81 = 31.8$$

These values were obtained from the data in table 3.4, and are further illustrated in figure 3.9.

In the example above, 29.5 is the exact lower limit of the interval found to contain the median; 80 is the number of cases below the median; 74 is the number of cases found below the interval containing the median; and 5 is *i*, the class interval

Figure 3.9
Identifying the components of
a frequency distribution for
Formula 3.1 using data from
table 3.4.

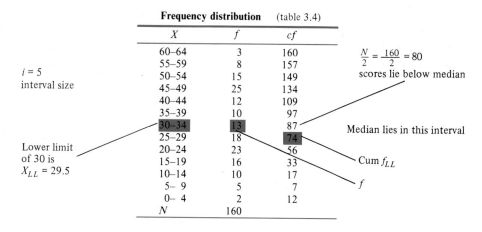

Frequency distribution (table 3.4)

X	f	cf
60–64	3	160
55–59	8	157
50–54	15	149
45–49	25	134
40–44	12	109
35–39	10	97
30–34	13	87
25–29	18	74
20–24	23	56
15–19	16	33
10–14	10	17
5– 9	5	7
0– 4	2	12
N	160	

$i = 5$
interval size

Lower limit
of 30 is
$X_{LL} = 29.5$

$\frac{N}{2} = \frac{160}{2} = 80$
scores lie below median

Median lies in this interval

Cum f_{LL}

f

size. The numerator of the fraction, $80 - 74 = 6$, represents the number of cases needed from this interval to reach the median. Dividing this figure by the frequency for the interval gives the proportion of the class interval width that is added to the lower limit of the interval containing the median to reach the median itself.

Formula 3.1 can be used for ungrouped data with some slight modifications. The class interval size is set equal to 1. The values used to compute the median in Formula 3.1 no longer pertain to intervals. Consider the set of scores:

3, 7, 12, 15, 15, 15, 15, 25, 50, 75, 90, 100

Figure 3.6 illustrates where the median lies. However, Formula 3.1 can be applied in the following way. The number of scores, N, is 12, $N/2$ equals 6. The median is a value between the 6th and 7th scores. The 6th and 7th scores are both equal to 15. The lower limit of 15 is 14.5. Hence, $X_{LL} = 14.5$. The number of scores counted before reaching the value is 3 (3, 7, 12). Therefore, Cum $f_{LL} = 3$. There are 4 scores with a value of 15, so f is set equal to 4. Placing these values in Formula 3.1 gives:

$$Mdn = 14.5 + \frac{6 - 3}{4} \times (1)$$

$$= 14.5 + 3/4 = 15.25$$

Consider another set of data used earlier in this chapter: 6, 7, 7, 9, 10, 11, 12, 14, 14. By inspection, it can be determined that the median is 10. Using Formula 3.1, $N/2$ is 4.5, $X_{LL} = 9.5$, $f = 1$, $i = 1$, Cum $f_{LL} = 4$.

$$Mdn = 9.5 + \frac{4.5 - 4}{1} \times (1) = 9.5 + .5 = 10$$

SUMMARY OF STEPS

Computing the Median from Grouped Data

Step 1   Find $N/2$, or 50% of N, the number of scores or cases falling below the median.

Step 2   Write down the cumulative frequencies for the frequency distribution and using these, locate the interval that contains the median, the point with $N/2$ scores below it.

Step 3   Determine the fraction of the class interval width that must be covered to reach the median by taking the ratio of the number of cases needed from the interval to the number of cases in the interval containing the median.

Step 4   Multiply this fraction by the class interval width, i, and add the result to the exact lower limit of the interval containing the median to obtain the value of the median itself.

Step 5   Round off to one decimal place.

PROBLEM 6

Compute a Mean from Ungrouped Data

mean   The average of a collection of numbers, obtained by adding all the numbers and then dividing by the total number of numbers.

average   One of the measures of central tendency—mean, median, or mode.

The **mean,** M, for a set of scores is merely the arithmetic **average** of those scores. That is, the scores are added up and the total is divided by the number of scores added. If the differences between each score and the mean are added up, the sum would be equal to zero. For the scores 3, 8, 8, 10, 12, 5, and 1, for example, the sum is 47. Dividing 47 by 7, the number of scores, N, gives $47/7 = 6.71 = 6.7$. The mean is rounded off to one decimal point in most situations, as with the other measures of central tendency already discussed. The differences of each score from the mean are -3.7, 1.3, 1.3, 3.3, 5.3, -1.7, and -5.7. The sum of these differences equals zero.

There is no need to arrange the ungrouped raw scores in order of increasing size, since the scores are merely summed as they are. The order of addition of a set of scores does not affect the total. With ungrouped data the mean is easier to calculate than the median. This advantage is not very important unless there are many scores in the frequency distribution. This particular method of calculating the mean is commonly used when N is small. It is also used when N is large and electronic calculating equipment is available. With N large, it is preferable to break the scores into groups of smaller size so the sums of scores can be obtained and checked within each group separately. Because it is difficult to add a long column of numbers without an error, scores are best broken into groups. After the sums of the separate groups are computed, they are added to obtain an overall sum. This overall sum is divided by the total number of scores, N, to obtain the final mean value.

Raw-Score Formula for the Mean

The process used above for calculating the mean from ungrouped data is represented symbolically by the following raw-score formula for the mean:

$$M = \frac{\Sigma X}{N} \qquad (3.2)$$

where M = the mean,

ΣX = the sum of the raw scores, and

N = the number of scores in the distribution.

The symbol M was selected to represent the sample mean because it is clear and easy to remember. Other texts and some articles may use $\overline{X}$ or $\overline{Y}$ to represent the sample mean.

The symbol X, already used in earlier tables, represents raw scores and N represents the number of cases. The new symbol on the right of Formula 3.2 is the capital Greek letter sigma, Σ, which is used as a summation sign. This symbol represents the summation of the elements that appear to the right of the sign. In this formula, only X appears to the right and X stands for a raw score from the distribution, so ΣX is the sum of all the X scores in the distribution.

The Summation Sign, Σ   Given that the ungrouped raw scores in a distribution are 4, 7, 10, 20, 16, 7, 5, and 2, what does ΣX for this distribution stand for? The results of this operation are shown as follows:

$$\Sigma X = 4 + 7 + 10 + 20 + 16 + 7 + 5 + 2 = 71$$

The symbols ΣX are actually an abbreviation for the more complete symbolic representation of:

$$\Sigma X = \sum_{j=1}^{N} X_j = X_1 + X_2 + X_3 + \cdots + X_N$$

In this more complete notation, X_j represents an indefinite raw score, that is, score j is any one of the scores $X_1, X_2, X_3, \cdots X_N$. In the example given in this section $X_1 = 4, X_2 = 7, X_3 = 10, X_4 = 20, X_5 = 16, X_6 = 7, X_7 = 5$, and $X_8 = 2$ with $N = 8$. X_j can stand for any of these 8 scores. The symbols $\sum_{j=1}^{N} X_j$ are interpreted verbally as follows: sum the scores for which the general term is X_j, starting with $j = 1$ and going up to $j = N$. In this particular example, it is $\sum_{j=1}^{8} X_j$, with the last term in the summation being the eighth term because $N = 8$. Abbreviations for formulas involving summation signs, that is, without putting the limits on the summation signs and subscripts on the variable terms, are used in this text wherever possible to make the formulas more readable. The subscripts and summation limits are implied even if they are not actually included in the formulas. If the subscript was not implied, X would be merely one score, albeit an unknown one, and the symbol ΣX would mean:

$$X + X + X + \cdots + X = NX$$

where N values, all equal to X, are added up. The implied subscript on the X makes it X_j, which represents a *variable*, not a constant. The sum of a variable term gives:

$$\sum_{j=1}^{N} X_j = \Sigma X = X_1 + X_2 + X_3 + \cdots + X_N$$

where at least some of the values of X_j are not equal to each other. If they are all equal to each other and to the number symbolized by X (without a subscript), then the result of the summation is to merely add up the same number, a constant, N times to get $N \times X$, or NX. The ΣX, therefore, denotes a sum of different terms only where X is a variable, not a constant. This implies that X really stands for X_j but the subscript j is deleted for convenience. In this text, X is used to represent a variable term rather than a constant term.

constant   A quantity that retains the same value throughout a series of calculations.

The summation of a **constant** term can be indicated by the symbol Σa, where a is merely a number such as 5. If the summation is over N terms, then Σa really stands for $\displaystyle\sum_{j=1}^{N} a$. There is no subscript on the constant term, a, because it is just a number, 5 in this case. The result is:

$$\Sigma a = \sum_{j=1}^{N} a = a + a + a + a + \cdots + a = N \times 5 = 5N$$

If N is 10, then the final result of the summation is 50. The difference between summations of constant terms, variable terms, and mixtures of the two is important in later problems so it is essential to understand the meaning of the summation sign and its use before proceeding further.

Computing the Mean from Grouped Data

If the data are already grouped in class intervals, Formula 3.2 is still used to calculate the mean. The class interval midpoint is first found and then the midpoint of each interval is multiplied by the frequency for that interval. These products are summed to obtain ΣX. The sum is divided by N, the sum of the frequencies to obtain the mean. In some texts, a mean calculated this way is considered as a weighted mean. All scores in the interval are assumed to have the value of the midpoint.

When to Use the Mean

The mean is the most stable of the measures of central tendency because it is influenced by every score in the distribution. It is a good measure of central tendency to use if accuracy and stability are important. If the distribution has many extreme scores at one end, however, the mean is pulled in the direction of these scores, giving

Figure 3.10
Relationship between the
measures of central tendency.

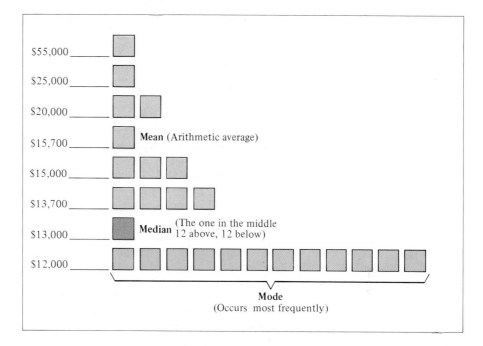

a misleading indication of the typical score. When computing a mean for reaction time scores, for example, very long reaction times due to inattention can add substantially to the central tendency value obtained even though they are not representative reaction times.

COMPARISON OF THE MEAN, MEDIAN, AND MODE

The mean is by far the most accurate, stable, and the most commonly used measure of central tendency. It is ordinarily the one chosen, especially if certain additional statistical computations such as standard deviation, correlation coefficient, or tests of significance are to follow. If the mean is not chosen, it is likely to be for one of the following reasons:

1. If only a very crude and quickly determined measure of central tendency is desired, the mode or possibly the median is preferred.
2. If some scores are not known, but only identified as either very large or very small, the mean cannot be computed. In this case the median is usually chosen, but occasionally the mode is preferred.
3. If the distribution of the scores is badly skewed (fig. 3.10), that is, with a frequency polygon showing a long tail either to the left or to the right, the median is preferred to give a more realistic figure for the central tendency of the distribution of scores. This is particularly true if the accuracy of the extreme scores in the tail of the distribution is suspect. The phenomenon of **skewness** is discussed further in Chapter 4.
4. If only the most common score is desired, the mode is chosen as the preferred measure of central tendency.

skewness   Departure from symmetry. The distribution with extreme scores at one end.

STATISTICS IN THE WORLD AROUND YOU

After being badly jolted during a takeoff when the pilot took evasive action to miss a small plane, a newspaper reporter decided to find out how frequently such near mishaps occur. The Federal Aviation Administration (FAA) keeps such information for every airport in the country. The reporter randomly selected ten airports and compiled the following data:

Airport	Last Year	This Year
Miami	22	24
New York (Kennedy)	19	20
Los Angeles	14	14
Chicago	13	15
Sioux Falls	9	10
Kansas City	8	9
Atlanta	7	9
Seattle	6	9
Washington (Dulles)	5	6
Denver	5	5

To write an article on the number of near misses, the reporter must calculate the mean, median, and modal number of the compiled data.

Why might the reporter choose the mean as the measure of central tendency for this article? Is there a way to tell if the differences in frequency of near misses from one year to the next mean anything?

TABLE 3.5

Frequency distribution

X	f	cf
110–119	3	161
100–109	8	158
90– 99	12	150
80– 89	20	138
70– 79	36	118
60– 69	33	82
50– 59	24	49
40– 49	16	25
30– 39	7	9
20– 29	2	2
N	161	

TABLE 3.6

Frequency distribution

X	f	cf
18–19	1	82
16–17	6	81
14–15	10	75
12–13	15	65
10–11	18	50
8– 9	12	32
6– 7	9	20
4– 5	6	11
2– 3	3	5
0– 1	2	2
N	82	

Exercises

3–1. What is the mode of the frequency distribution in table 3.5?

3–2. What is the mode of the frequency distribution in table 3.6?

3–3. Find the median of the frequency distribution in table 3.5.

3–4. Find the median of the frequency distribution in table 3.6.

$$\frac{n+1}{2} \qquad \frac{10+1}{2} = \frac{11}{2} = 5.5 \quad \leftarrow$$

3–5. Find the median of the scores: 3, 15, 15, 25, 25, 25, 60, 75, 100, and 200.

3–6. Find the median of the scores: 20, 21, 23, 23, 23, 26, 26, 29, 30, and 31.

3–7. Find the median of the scores: 400, 800, 35, 30, 35, 900, 200, 20, 20, 35, 14, and 35.

3–8. Find the median of the scores: −11, −10, −8, −4, −4, 0, 1, 2, 2, 10, and 30.

3–9. Compute the median for the frequency distribution given as the answer to question 2–1 in Appendix C. Compute the median working up from the bottom and also working down from the top of the distribution.

3–10. Calculate the median both from the bottom up and the top down for the frequency distribution given as the answer to question 2–3 in the back of the book.

15.

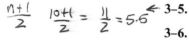

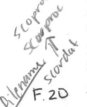

F. 20

F. 30

3–11. Find the mean of the scores: 4, 8, 10, 8, 7, 6, 12, 2, 9, 7, 8, and 8.
  a. Find the sum of all the scores/measurements.
  b. Divide the sum by the number of measurements.

3–12. Find the mean of the scores: 90, 110, 95, 98, 102, 108, 125, 68, 89, 112, 100, 92, and 94.
  a. Find the sum of all the scores/measurements.
  b. Divide the sum by the number of measurements.

3–13. Find the mean of the frequency distribution in table 2.2.
  a. Multiply the midpoint of each interval by the frequency.
  b. Add all the numbers created in (a).
  c. Divide the number from (b) by the number of scores.

3–14. Find the mean of the frequency distribution in table 3.2.
  a. Multiply the midpoint of each interval by the frequency.
  b. Add all the numbers created in (a).
  c. Divide the number from (b) by the number of scores.

3–15. Calculate the mean of the frequency distribution in table 3.5.
  a. Multiply the midpoint of each interval by the frequency.
  b. Add all the numbers created in (a).
  c. Divide the number from (b) by the number of scores.

3–16. Calculate the mean of the frequency distribution in table 3.6.
  a. Multiply the midpoint of each interval by the frequency.
  b. Add all the numbers created in (a).
  c. Divide the number from (b) by the number of scores.

Use the following data for questions 3–17, 3–18, and 3–19. Frequency distribution for memory scores in an anxiety situation.

X	f
70–74	1
65–69	3
60–64	6
55–59	6
50–54	5
45–49	8
40–44	9
35–39	8
30–34	12
25–29	10
20–24	7
15–19	5

3–17.   What is the mode of the memory scores above?

3–18.   What is the median of the memory scores above?

3–19.   What is the mean for the memory scores data?
    a. Multiply the midpoint of each interval by the frequency.
    b. Add the numbers created in (a).
    c. Divide the number from (b) by the number of scores.

3–20.   Using the data given in question 2–9, find the mean, median, and mode for the weights and minutes played. For each set of scores:
    a. Find the sum of all the scores.
    b. Divide the sum by the number of scores.
    c. Order all scores from smallest to largest.
    d. Choose the score equidistant from each end.
    e. Find the score that occurs most often.

Chapter 4

Measures of Variability

he measures of central tendency described in the previous chapter provide an indication of the general level of scores in a body of data but they do not give any indication of how "spread out" the scores are. Two distributions can both have a mean of 100, for example, but one can have a range of scores from 90 to 110 and the other can have a range of scores from 10 to 1,000.

Figure 4.1 shows an example of how two distributions with the same central tendency have different degrees of dispersion. To compare only the measure of central tendency between two dispersions can be misleading about how different or similar the two distributions are.

It is just as important to have some method for determining the amount of spread in a group of scores as it is to have a method for determining the central tendency. The most commonly used methods for determining the amount of spread in a body of scores, called "measures of variability or dispersion," are the semi-interquartile range (Q), the standard deviation (S.D.), and the Index of Dispersion (D). This chapter is concerned with the calculation of these statistics for both ungrouped and grouped data.

Generally, if raw data are given, the value of variability is computed directly using the raw-score formula. The grouped method for finding variability is used in those situations where a frequency distribution is already created and submitted for further analysis or where the researcher found it necessary to create a frequency distribution from raw data. The range, which is another measure of variability, was presented in Chapter 2 in the section on the creation of frequency distributions. Although it is a simple statistic to calculate, the range is an unstable measure. With different end points, a different range value is obtained, even though the values between the two points remain the same. Essentially the range ignores these "in-between" values, which may contain valuable information. In some rare instances

Figure 4.1
Frequency polygons of the I.Q.'s for two groups of students with the same central tendency but different dispersions.

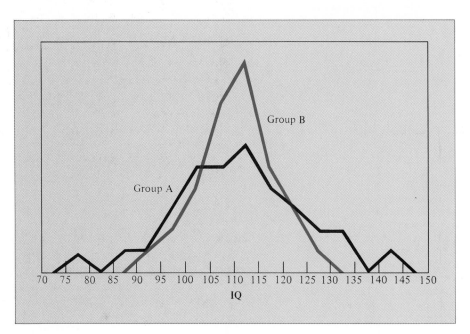

STATISTICS IN THE WORLD AROUND YOU

The three measures of central tendency, the mean, median, and mode, denote some kind of average or center point. The use of an average can be deceptive. For example, a newspaper article in the business section or sports section sometimes refers to statistics and averages. A report out of Washington, D.C. mentions that the median cost of homes has declined 10.1% from the previous month to $122,400. Those unfamiliar with statistics might be inclined to believe that they can now purchase a home for that amount. This average amount does not tell the reader about the variability. The average amount is only a center point about which the real costs vary. Such misbeliefs can be eliminated if the writer includes some idea of variability and, perhaps, a range of costs. The potential buyers can estimate whether they can afford the most costly of homes.

In sports articles, athletes' salaries are often a point of interest. Many sports writers mention "average" salary per professional sport. This average might show that basketball players make more money than baseball players and football players. The use of average without defining which specific measure is being used, is very misleading. Salaries generally follow a skewed distribution where only a relative few have very high salaries and a majority have lower amounts. With a skewed distribution, the median is the most appropriate measure of average. However, with just the average figure, the same problem exists as with the cost of homes. People tend to believe that every ball player makes "X" amount of dollars. The introduction of a measure of variability provides more useful and accurate information.

---

where the distribution of values has a large number of entries and the distribution closely resembles a bell-shaped curve, the range divided by 4 is used as a rough estimate of the standard deviation.

PROBLEM 7

Determine the Semi-Interquartile Range (Q) from Ungrouped Data

quartile One of the division points between four equal-sized pieces of the population when the population is arranged in numerical order. The second quartile is the number such that 2/4 of the population is smaller and 2/4 is larger.

Before presenting a formula for the semi-interquartile range (Q) it will be necessary to define the three **quartile** points, Q_1, Q_2, and Q_3. Q_2 is easiest to understand at this point because it is the median, the point below which 50% of the cases fall. Q_1 is the point below which 25% of the cases fall and Q_3 is the point below which 75% of the cases fall. Q_1 and Q_3 are needed to calculate Q, the semi-interquartile range. Note that Q, the semi-interquartile range has no subscripts. It is a width or distance measure. Q_1, Q_2, and Q_3 represent specific points on the score scale, not widths or distances. Q_1 and Q_3 are just like Q_2 because they are defined as points on the scale below which a specified percentage of cases fall. Q_1 and Q_3 are calculated in a manner very similar to that used for calculating Q_2 or the median (Mdn). If the methods for calculating the median described in Problems 4 and 5 are well understood, calculating Q will be easy; if not, these methods should be reviewed before proceeding. The formula for the semi-interquartile range (Q) is:

$$Q = \frac{Q_3 - Q_1}{2} \qquad (4.1)$$

where Q = the semi-interquartile range,
Q_3 = the third quartile point, the point below which 75% of the scores fall, and
Q_1 = the first quartile point, the point below which 25% of the scores fall.

semi-interquartile range   A measure of variability that is insensitive to extreme values. Generally calculated as ($Q_3 - Q_1$)/2, where Q_3 and Q_1 are 3d and 1st quartiles. Used along with the median for describing a distribution of scores.

Inspection of Formula 4.1 reveals that Q is just one-half of the distance between Q_3 and Q_1, the third and first quartile points. It is for this reason that it is called the **semi-interquartile range.** $Q_3 - Q_1$ is the interquartile range, but this distance is rarely ever used as a measure of variability.

Computing Q from Ungrouped Data

To find Q, first find Q_3 and Q_1 and substitute into Formula 4.1. Since Q_3 and Q_1 are like Q_2, the median, in their definition, the same general method used for calculating the median of ungrouped data (Problem 4) is used for finding Q_3 and Q_1 except that 75% and 25% play the same role in calculating Q_3 and Q_1 that 50% plays in calculating the median, Q_2. Recall that when calculating the Mdn, the first step is to find $N/2$, or 50% of the total number of cases, N. The remaining steps are devoted to finding the point with this many cases below it. When calculating Q_3 and Q_1, the first step is to find 75% and 25% of the total number of cases. The remaining steps are the same as in calculating the median, that is, finding the point with that many cases below it in the distribution of scores.

Example   Consider the following ungrouped scores (reaction times in seconds) already arranged in order of increasing size:

1, 10, 25, 25, 30, 35, 35, 60, 60, 60, 75, 100.

The total number of cases is $N = 12$. If the value of the median, which is equal to Q_2, is desired, it is obtained as follows using Formula 3.1.

$$Mdn = X_{LL} + \left(\frac{N/2 - Cum\, f_{LL}}{f} \right) i \qquad (3.1)$$

$12 \times 0.50 = 6$ (50% N)

$$Q_2 = Mdn = 34.5 + \frac{6 - 5}{2} \times 1$$

$$= 34.5 + 0.50 = 35.00 = 35.0$$

The lower limit of the interval containing Q_2, the point below which 6 cases fall, is 34.5. Below 34.5 there are 5 cases, so $6 - 5 = 1$, which represents the number of cases still needed from this interval. There are 2 cases in the interval so this value of 1 is divided by 2 to get the proportion of the score interval that must be counted to get up to Q_2. Since the score interval is only 1 point, this fraction of 1/2 is multiplied by 1 and added to 34.5, the lower limit of the interval to get the value of Q_2. The only difference between calculating Q_2 and the median is that two decimal places are retained for Q_2, whereas for the median the result is rounded off to one decimal place. Carrying two decimal places for the quartile points insures sufficient accuracy to obtain a value of Q correct to one decimal place when Q is rounded off. Rounding off is applicable only in cases where computations are performed by hand.

The Formula 3.1 for finding the median, Q_2, can be modified to find Q_3 and Q_1. Since Q_3 is the value that separates the bottom 75%, or 3/4, from the top 25%, or 1/4, the appropriate formula is:

$$Q_3 = X_{LL} + \left(\frac{3N/4 - \text{Cum } f_{LL}}{f}\right)i \qquad (4.2a)$$

For Q_1, the value sought divides the distribution of scores so that 25%, or 1/4, is below Q_1 and 75%, or 3/4, is above Q_1. The formula for finding Q_1 is:

$$Q_1 = X_{LL} + \left(\frac{N/4 - \text{Cum } f_{LL}}{f}\right)i \qquad (4.2b)$$

where

$X_{LL} =$ the lower limit of the interval containing Q_3 in Formula 4.2a and Q_1 in Formula 4.2b,

$\text{Cum } f_{LL} =$ the number of scores below the lower limit,

$f =$ the number of scores in the class interval containing Q_3 or Q_1, and

$i =$ the class interval size.

For ungrouped data, i is set equal to 1, and f is the number of occurrences of the value at position $3N/4$ for Q_3 and $N/4$ for Q_1. This is graphically illustrated below for the example on page 70.

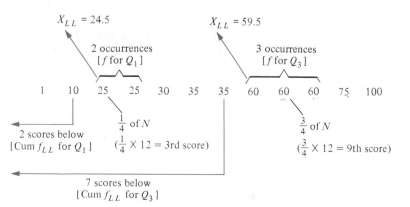

Calculating Q_3, the third quartile point for the sample data, gives:

$$12 \times 0.75 = 9 \ (75\% \text{ of } N)$$

$$Q_3 = 59.5 + \frac{9-7}{3} \times (1)$$

$$Q_3 = 59.5 + .67 = 60.17$$

For Q_3, the interval 59.5 to 60.5 holds the point below which 9 cases (75% of N) fall. Seven cases are below 59.5 and 3 cases lie in the interval so $(9-7)/3$ is the fraction of the interval width (1 point) that must be added to 59.5 to obtain Q_3. For Q_1, the first quartile point, the calculations are:

$$Q_1 = 24.5 + \frac{3-2}{2} \times (1)$$

$$Q_1 = 24.5 + 0.50 = 25.00$$

Three cases must fall below Q_1 and the interval 24.5 to 25.5 must hold this point since below 24.5 there are 2 cases, and below 25.5 there are 4 cases. Since 2 cases lie below 24.5, only $3 - 2$, or 1 case, is needed from this interval. Since there are 2 cases in this interval, $1/2$ of the interval width, or $(1/2) \times 1$, must be added to 24.5, the lower limit of the interval to obtain Q_1. Substituting Q_3 and Q_1 in Formula 4.1 gives:

$$Q = \frac{60.17 - 25.00}{2} = \frac{35.17}{2}$$

$$Q = 17.585 = 17.6$$

In the calculations for Q, at least two decimal places are carried until Q is obtained, but in reporting the final answer, Q is usually rounded off to one decimal place following the rounding rules used for measures of central tendency.

Interpreting Q

Many students have difficulty grasping the meaning of Q because of its abstract nature. When Q is determined to be 17.6, the student asks, "Seventeen point six what?" The answer, "Points!", is often not very satisfying. To provide meaning to this figure, it is necessary to consider several distributions, all with different values of Q. Then this distribution with $Q = 17.6$, can be compared with others and adjudged to be more or less variable, depending on the values of Q for the other distributions. Distributions can be rank ordered for the amount of variability of their scores by putting them in order of size with respect to Q. Determining Q permits a comparison of one distribution with another, but it does not permit any intuitively obvious statements about one distribution by itself.

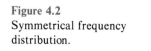

Figure 4.2
Symmetrical frequency
distribution.

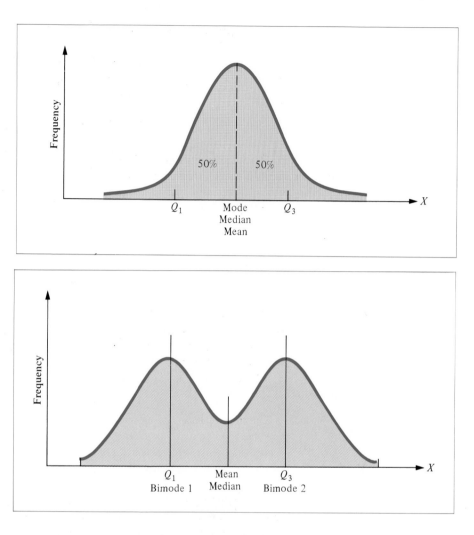

Figure 4.3
Symmetrical frequency
distribution with two modes.

There are a few statements, although cumbersome, that might help. For example, the interval $2Q$ covers the middle 50% of the scores if the interval is properly located on the score scale. Of course, the proper location calls for placing the lower end of the interval at Q_1 and the upper end at Q_3. The statement then must be true because Q_3 has 75% of the cases below it and Q_1 has 25% below it, so the middle 50% falls between Q_3 and Q_1, which is an interval of $2Q$ points.

If the distribution is symmetrical, not skewed, then it can be stated that the interval $Mdn \pm Q$ (median $\pm Q$) encompasses the middle 50% of the cases. A symmetrical distribution has no long tail on either side. If the distribution frequency polygon is cut in half by a vertical line, the left-hand side is a mirror image of the right-hand side.

Figures 4.2 and 4.3 show two different types of symmetrical distributions. Figure 4.2 is a bell-shaped curve, which has the property that all three measures of central tendency lie at the same point. In figure 4.3 the mean and median are at the same point, but there are two modes.

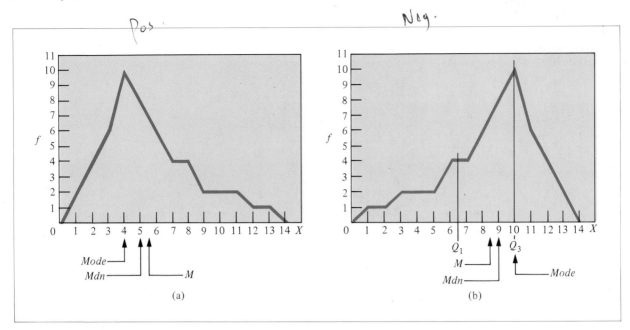

Pos

Neg.

(a)

(b)

Figure 4.4
Location of the mean, median, and mode for skewed distributions. (a) Positively skewed, (b) negatively skewed.

The rule for determining symmetry is not particularly useful because most distributions are not exactly symmetrical so the *Mdn* is not exactly between Q_3 and Q_1.

Continuing from the example on pages 70–72, $Q_1 = 25.00$, $Q_2 = 35.00$, and $Q_3 = 60.17$. The distance from Q_1 to Q_2 is 10 points while the distance from Q_2 to Q_3 is 25.17 points, over twice as great an interval. This shows that there is a longer tail on the right-hand side of the frequency polygon, or of the distribution, and hence the distribution is *positively* skewed. Figure 4.4a is an example of a positively skewed distribution. Thus, if $Q_3 - Q_2$ is greater than $Q_2 - Q_1$, the distribution is positively skewed, with a longer tail to the right. If $Q_2 - Q_1$ is greater than $Q_3 - Q_2$, the distribution is negatively skewed, with a longer tail to the left. Figure 4.4b is an example of a *negatively* skewed distribution. The greater the difference between $Q_3 - Q_2$ and $Q_2 - Q_1$, the greater the skew in the distribution. If the distribution is symmetrical, there is no skew and $Q_3 - Q_2 = Q_2 - Q_1$, although finding $Q_3 - Q_2 = Q_2 - Q_1$ does not guarantee that the distribution will be symmetrical. Thus $Q_3 - Q_2 = Q_2 - Q_1$ is a necessary but not sufficient condition to assure symmetry in the distribution. Additional examples of skewed distributions are shown in figures 4.5 and 4.6.

Figure 4.5
A positively skewed
distribution.

$Q_3 - Q_2 >$
$Q_2 - Q_1$

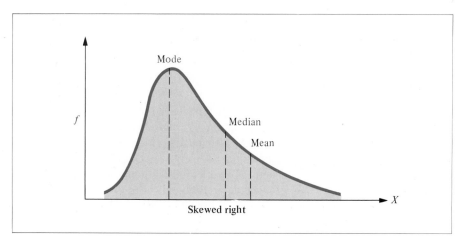

Figure 4.6
A negatively skewed
distribution.

$Q_2 - Q_1 >$
$Q_3 - Q_2$

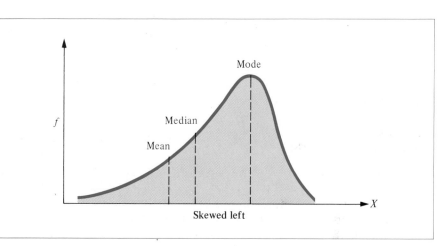

When to Use Q

Q has the same advantages as a measure of variability that the median has as a measure of central tendency. That is, the semi-interquartile range is free of influence by the extreme scores. The scores below Q_1 and above Q_3 do not affect the size of Q at all, so if such extremes are exaggerated, inaccurate, or not precisely known, Q is still computed and reported as a statistic relatively unaffected by these extraneous influences. A generally accepted rule is that when the median is selected as the most appropriate measure of central tendency for describing a given body of data, Q is usually selected as the most appropriate measure of variability.

PROBLEM 8

Determine the Semi-Interquartile Range (Q) from Grouped Data

The principle difference between computing Q by the method described in Problem 7 and computing Q from grouped data is that now the data are arranged in a grouped frequency distribution. In both cases Q_1 and Q_3 must be computed and substituted in Formula 4.1 to obtain Q, $Q = (Q_3 - Q_1)/2$. In Problem 7, Q_3 and Q_1 were computed by methods similar to those used to compute the median (Q_2) from ungrouped data (Problem 4). In Problem 8, Q_3 and Q_1 are computed by methods similar to those used to compute the median (Q_2) from grouped data (Problem 5). Regardless of the quartile point being located and the form in which the data are arranged, the task is to find a point on the score scale with a specified percentage of cases below it. In each case, this is accomplished using the same basic method.

Computing Q from Grouped Data

The method of finding the semi-interquartile range from grouped data is illustrated by showing the computations to obtain Q for the data in table 4.1. First, for computing Q_3, it is necessary to find the point below which 75% of the cases fall, 209 $\times 0.75 = 156.75$. Thus 156.75 cases must fall below Q_3 in the distribution of scores. Reference to the cumulative frequency column in table 4.1 (which is identical to table 3.3) shows that 156 cases fall below 89.5, the upper limit of the interval 88.5

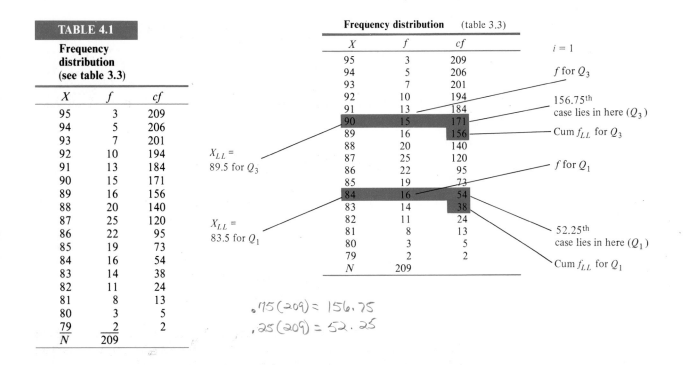

TABLE 4.1		
Frequency distribution (see table 3.3)		
X	f	cf
95	3	209
94	5	206
93	7	201
92	10	194
91	13	184
90	15	171
89	16	156
88	20	140
87	25	120
86	22	95
85	19	73
84	16	54
83	14	38
82	11	24
81	8	13
80	3	5
79	2	2
N	209	

Frequency distribution   (table 3.3)

X	f	cf
95	3	209
94	5	206
93	7	201
92	10	194
91	13	184
90	15	171
89	16	156
88	20	140
87	25	120
86	22	95
85	19	73
84	16	54
83	14	38
82	11	24
81	8	13
80	3	5
79	2	2
N	209	

$i = 1$

f for Q_3

156.75th case lies in here (Q_3)

Cum f_{LL} for Q_3

f for Q_1

52.25th case lies in here (Q_1)

Cum f_{LL} for Q_1

$X_{LL} =$ 89.5 for Q_3

$X_{LL} =$ 83.5 for Q_1

$.75(209) = 156.75$
$.25(209) = 52.25$

to 89.5. This is less than the 156.75 needed. Below 90.5, the upper limit of the next interval, there are 171 cases, which is more than 156.75. Therefore Q_3 must lie in the interval 89.5 to 90.5. Using Formula 4.2a:

$$Q_3 = 89.5 + \frac{156.75 - 156}{15} \times (1)$$

Q_3 is the lower limit of the interval containing Q_3 plus the class interval multiplied by the ratio of the number of cases needed from this interval frequency.

$$Q_3 = 89.5 + \frac{0.75}{15} = 89.5 + .05 = 89.55$$

Q_1, on the other hand, has only 25% of the scores below it, $209 \times 0.25 = 52.25$ cases. The cumulative frequency column in table 4.1 shows 38 cases below 83.5 and 54 cases below 84.5 so using Formula 4.2b:

$$Q_1 = 83.5 + \frac{52.25 - 38.00}{16} \times (1)$$

$$= 83.5 + \frac{14.25}{16} = 83.5 + .89 = 84.39$$

Substituting Q_3 and Q_1 into Formula 4.1 gives:

$$Q = \frac{Q_3 - Q_1}{2} = \frac{89.55 - 84.39}{2} = \frac{5.16}{2}$$

$$Q = 2.58 = 2.6$$

To obtain some indication concerning the degree of skew in the distribution, compute Q_2 as follows:

$$Q_2 = 86.5 + \frac{104.5 - 95}{25} \times (1)$$

$$= 86.5 + \frac{9.5}{25} = 86.5 + 0.38 = 86.88$$

Then

$$Q_3 - Q_2 = 89.55 - 86.88 = 2.67$$

$$Q_2 - Q_1 = 86.88 - 84.39 = 2.49$$

Thus $Q_2 - Q_1$ is less than $Q_3 - Q_2$ showing a shorter tail toward the left, making the frequency distribution in table 4.1 positively skewed. The difference is less than 1 point, so the amount of skew relative to the range is very slight.

TABLE 4.2

Frequency distribution

X	f	cf
60–64	3	160
55–59	8	157
50–54	15	149
45–49	25	134
40–44	12	109
35–39	10	97
30–34	13	87
25–29	18	74
20–24	23	56
15–19	16	33
10–14	10	17
5– 9	5	7
0– 4	2	2
N	160	

Example   The example just shown involves a frequency distribution with a class interval of $i = 1$. Now compute Q for the frequency distribution in table 4.2 (which is identical to table 3.4), where $i = 5$. Computing Q_3 and Q_1:

$$\text{Number of cases below } Q_3 = 160 \times 0.75 = 120$$

$$Q_3 = 44.5 + \frac{120 - 109}{25} \times (5)$$

$$= 44.5 + \frac{55}{25} = 44.5 + 2.20 = 46.70$$

$$\text{Number of cases below } Q_1 = 160 \times 0.25 = 40$$

$$Q_1 = 19.5 + \frac{40 - 33}{23} \times (5) = 19.5 + \frac{35}{23} = 19.5 + 1.52 = 21.02$$

Then

$$Q = \frac{Q_3 - Q_1}{2} = \frac{46.70 - 21.02}{2} = \frac{25.68}{2} = 12.84$$

$$Q = 12.8$$

The distribution in table 4.2 is much more variable where $Q = 12.8$ than the distribution in table 4.1 where $Q = 2.6$. This is also apparent in comparing the ranges for the two distributions. For table 4.1 the range is 17.0 and for table 4.2 the range is 65.0.

PROBLEM **9**

Compute the Standard Deviation (S.D.) by the Deviation-Score Method

standard deviation   A measure of spread, applied either to a population or to a sample. Also, the square root of variance.

The **standard deviation** is a measure of variability that is affected by every score in the distribution. A deviation score, $x = X - M$, is computed for every raw score, X, in the distribution. In other words, each individual score is examined to determine how much it deviates from the mean score. These deviation scores are then squared so that the deviations do not sum to zero, since all scores below the mean have negative deviation scores. Then the squared deviation scores are averaged. The square root of the average squared deviation score is the standard deviation (S.D.) of the distribution. A score far away from the mean has a large squared deviation score, elevating the S.D. A score closer to the mean has a small squared deviation score, tending to restrict the size of the S.D. This definition of the standard deviation is expressed symbolically by Formula 4.3. The deviation-score formula for the standard deviation (S.D.) is:

$$\text{S.D.} = \sqrt{\frac{\Sigma x^2}{N}} \qquad (4.3)$$

where   S.D. = the standard deviation,
   x = a deviation score, $X - M$, and
   N = the number of scores in the distribution.

For simplicity, in Formula 4.3, the limits are omitted from the summation sign and the subscript from the variable term, x. With these restored, Σx^2 becomes $\sum_{i=1}^{N} x_i^2$, or with a different subscript, $\sum_{j=1}^{N} x_j^2$. In Formula 4.3 it is understood that the summation is from 1 to N and that x is variable term. The symbol S.D. is used in this text to represent the standard deviation. Other texts and articles in the literature may use s'.

Computing the S.D. by the Deviation-Score Formula

Formula 4.3 represents the simplest statement of the definition for the standard deviation and is therefore typically employed in algebraic formulations. Formula 4.2 is not an easy formula to use for computing the standard deviation because it requires the actual computation and squaring of deviation scores. Deviation scores are typically decimal fractions, since M is usually reported to one decimal place, and hence are difficult to work with. Problem 10 presents a more practical method for computing the S.D., but the method of computing the size of the S.D. by using Formula 4.3 is shown here with a simple example to insure that its meaning is clear.

Consider the following 10 raw scores (X values):

11, 12, 10, 9, 9, 5, 14, 13, 7, and 10.

Adding these scores gives $\Sigma X = 100$ and hence $M = 10$. Computing the deviation scores (x values) corresponding to these 10 raw scores gives:

1, 2, 0, -1, -1, -5, 4, 3, -3, and 0.

If these deviation scores are added up, the sum is zero. This is always true if the computations are done correctly. In fact, the mean (M) is further defined as the point about which the sum of the deviation scores equals zero.

Formula 4.3 calls for the sum of the squared deviations, however, rather than for the sum of the deviations themselves, so the squared deviation scores are:

1, 4, 0, 1, 1, 25, 16, 9, 9, and 0.

Summing these values gives a total of 66. Substituting this total and $N = 10$, in Formula 4.2 gives S.D. $= \sqrt{66/10} = \sqrt{6.6} = 2.57$, which becomes 2.6 when rounded off to one decimal place. Rounding is only a recommended practice for hand calculations. Researchers can establish their own rule on rounding, but they must be consistent in applying the rule.

Variance

variance The specific measure of variation that computes mean-squared deviations from the mean. The square of the standard deviation.

The term **variance** appears frequently in statistical material, unfortunately with varying connotations. The simplest and most direct usage of the term is in the following sentence: "The variance of this sample of scores is 4.0." In this usage, the *variance* of the distribution is merely the square of the standard deviation. Thus, when S.D. equals 2, the variance, S.D.^{2}, equals 4.

PROBLEM 10

Compute the Standard Deviation (S.D.) by the Raw-Score Method

Formula 4.3 is a conceptually simple, but not computationally simple, formula for calculating the standard deviation. Formula 4.3 involves deviation scores (x), which are obtained by subtracting the mean (M) from the raw scores (X). It is more convenient to compute the standard deviation using a formula that involves only raw scores, because it is laborious to obtain and then square deviation scores. The raw-score formula for the standard deviation (S.D.) is:

$$\text{S.D.} = \frac{1}{N}\sqrt{N\Sigma X^2 - (\Sigma X)^2} \qquad (4.4)$$

where $X =$ a raw score,
$N =$ the number of scores, and
S.D. $=$ the standard deviation.

Formula 4.4 is mathematically equivalent to Formula 4.3, but all the X values are uppercase, representing raw scores, not deviation scores. Thus it is merely necessary to obtain the sum of the raw scores, ΣX, and the sum of the squares of the raw scores, ΣX^2. Note that ΣX^2 means the raw scores are squared and *then* summed. $(\Sigma X)^2$, the sum of X the quantity squared, indicates that the X-scores are summed first and then the entire quantity is squared. As in Formula 4.3, the limits on the summation signs in Formula 4.4 are implied rather than explicitly stated. The limits on the summation are from 1 to N and X is a variable term rather than a constant.

Computing the S.D. by the Raw-Score Formula

To show that the same value for the S.D. is obtained whether it is computed by the deviation-score formula or the raw-score formula, the same data used for demonstration purposes in Problem 9 will also be used here. The raw scores, to repeat, are:

11, 12, 10, 9, 9, 5, 14, 13, 7, and 10.

The sum of these scores is $\Sigma X = 100$. The squared scores, X^2, are:

121, 144, 100, 81, 81, 25, 196, 169, 49, and 100,

respectively. Summing these squared raw scores gives $\Sigma X^2 = 1066$. The number of cases, N, is 10. Substituting these values in Formula 4.4 gives:

$$\text{S.D.} = \frac{1}{10}\sqrt{10(1066) - (100)^2}$$

$$= \frac{1}{10}\sqrt{10660 - 10000}$$

$$= \frac{1}{10}\sqrt{660}$$

$$= \frac{1}{10}(25.7) = 2.57 = 2.6$$

The same results are obtained from Formula 4.3 and Formula 4.4.

Because of the small amount of data and an integral value of the mean, computation of the previous example deviation scores is simple and involves about the same amount of work for either method. For most examples, however, the methods of Problem 10 involve substantially less work. This is particularly true if a calculator or computer is available to square the raw scores and accumulate the sum of X and the sum of X^2. Then, computing the S.D. is merely a matter of plugging these totals and N into Formula 4.3. There are some hand calculators now available that will compute the S.D. using Formula 4.3 once the raw scores have been entered by merely pushing a key. In some cases however, what is computed by the calculator is $\sqrt{\dfrac{\Sigma x^2}{N-1}}$, which is an estimate of the population standard deviation instead of $\sqrt{\dfrac{\Sigma x^2}{N}}$ which is S.D., the standard deviation of the sample.

The use of $N-1$ versus N is addressed in more detail in Chapter 8. N, however, should be used when the standard deviation is used to describe a set of measurements. Many simple computer programs use this formula.

SUMMARY OF STEPS

Computing the Standard Deviation (S.D.) by the Raw-Score Method

Step 1   Sum the raw scores to get ΣX.

Step 2   Sum the squared raw scores to get ΣX^2.

Step 3   Square ΣX, the sum of the raw scores, to get $(\Sigma X)^2$.

Step 4   Multiply the sum of the squared raw scores by N, the number of cases, to get $N\Sigma X^2$.

Step 5   Subtract Step 3 from Step 4 to get $N\Sigma X^2 - (\Sigma X)^2$.

Step 6   Take the square root of Step 5 to get $\sqrt{N\Sigma X^2 - (\Sigma X)^2}$.

Step 7   Divide the answer in Step 6 by N to get the S.D.

PROBLEM B

Computing the Index of Dispersion

The semi-interquartile range and the standard deviation are used to measure the variability for continuous numerical data only. For such data as eye color, where numbers are used only for identification purposes (nominal scale), the computation of Q and S.D. are either not appropriate or are misinterpreted. When data are categorical or at best represent ranks, the **Index of Dispersion** may be a more proper representation of variability. The Index of Dispersion is the ratio of the variability that exists within the set of data to the maximum variability that could exist. The formula for the Index of Dispersion is:

index of dispersion   A measure of variability used for qualitative data (categorical).

$$D = \frac{k\,(N^2 - \Sigma f_i^2)}{N^2\,(k-1)} \tag{4.5}$$

where $k =$ the number of categories,
$N =$ the total number of cases or observations, and
$f_i =$ the frequency in each category.

The value of D has a range from zero (minimum variability) to 1.00 (maximum variability). When there are an equal number of cases in each category, the value of D will be 1.00. If all the measurements are located in only one category, the value of D will be zero.

Example   A random sample of 200 people was surveyed to determine whether their trust in the mayor had changed over the past year, and if so, in what direction. The tabulated data were:

    Increased trust   = 20
    Decreased trust   = 60
    No change   = 120

Here the *f* values are 20, 60, and 120 respectively. *N* is 200 and k is 3. Using Formula 4.5 gives:

$$\Sigma f_i^2 = (20)^2 + (60)^2 + (120)^2$$

$$= 400 + 3600 + 14400 = 18400$$

$$D = \frac{k\,(N^2 - \Sigma f_i^2)}{N^2\,(k-1)} = \frac{3\,(200^2 - 18400)}{200^2\,(3-1)}$$

$$= 64800/80000 = .810$$

The value of this index shows the researcher that the response diversity is 81% of maximum.

To derive any meaning from the Index of Dispersion or other measures of variability, compare the value computed from one sample to a value from another sample. In the previous example the mode is "No Change." In another sample of 100 people the tabulated data were

    Increased trust   = 7
    Decreased trust   = 10
    No change   = 83

In this second sample, the mode is also No Change, however the Index of Dispersion is:

$$\Sigma f_i^2 = (7)^2 + (10)^2 + (83)^2 = 7038$$

$$D = \frac{3\,(100^2 - 7038)}{100^2(3-1)} = \frac{8886}{20000} = .4443$$

The index shows that the response diversity is only 44.43% of maximum. Hence, the variability in the first sample is higher than the second.

SUMMARY OF STEPS

Computing the Index of Dispersion

Step 1   Count the number of categories. Let this number be equal to k.

Step 2   Count the number of observations, the total number of frequencies. This number is N.

Step 3   Count the number of observations in each category. Let the number of observations in the first category be set equal to f_1. Let the number of observations in the second category be set equal to f_2, and so on.

Step 4   Square N, the number of cases, to get N^2.

Step 5   Square each f_i and then add them together.

Step 6   Subtract the result in Step 5 from Step 4.

Step 7   Multiply the result in Step 6 by the value of k.

Step 8   Multiply the result in Step 4 by (k − 1).

Step 9   The value of D is the value in Step 7 divided by the value obtained in Step 8.

MATHEMATICAL NOTES

This section may be skipped by the reader who is not interested in the origin of Formula 4.4. The following demonstrates the algebraic derivation of Formula 4.4 from Formula 4.3.

$$\text{S.D.} = \sqrt{\frac{\Sigma x^2}{N}} \tag{4.3}$$

but $x = X - M$, hence

$$\text{S.D.} = \sqrt{\frac{\Sigma (X - M)^2}{N}}$$

$$\text{S.D.} = \sqrt{\frac{\Sigma (X^2 - 2MX + M^2)}{N}}$$

$$\text{S.D.} = \sqrt{\frac{(\Sigma X^2 - \Sigma(2MX) + \Sigma M^2)}{N}}$$

Constants can be placed in front of the summation sign in the sum of a product of constant and variable terms; also, the summation of a constant term is N times the constant. The expression above, therefore can be rewritten as:

$$\text{S.D.} = \sqrt{\frac{(\Sigma X^2 - 2M\Sigma X + NM^2)}{N}}$$

Multiplying numerator and denominator under the radical by N and removing $1/N$ from under the radical gives:

$$\text{S.D.} = \frac{1}{N}\sqrt{N\Sigma X^2 - 2(NM)\Sigma X + (NM)^2}$$

but $NM = N(\Sigma X/N) = \Sigma X$, so

$$\text{S.D.} = \frac{1}{N}\sqrt{N\Sigma X^2 - 2(\Sigma X)(\Sigma X) + (\Sigma X)^2}$$

$$= \frac{1}{N}\sqrt{N\Sigma X^2 - 2(\Sigma X)^2 + (\Sigma X)^2}$$

$$\text{S.D.} = \frac{1}{N}\sqrt{N\Sigma X^2 - (\Sigma X)^2}$$

Exercises

4-1.   Find the semi-interquartile range, Q, for the following divergent production scores on Guilford's *Structure of Intellect:* 3, 10, 10, 20, 25, 25, 25, 31, 31, 31, 40, 50.

  a. Find Q_1 (bottom 25%).
  b. Find Q_3 (top 25%).
  c. Use Formula 4.1 to compute Q.

4-2.   Find the semi-interquartile range, Q, for the following math test scores: 100, 110, 110, 119, 130, 135, 135, 140, 140, 145, 145, 150, 160.
  a. Find Q_1 (bottom 25%).
  b. Find Q_3 (top 25%).
  c. Use Formula 4.1 to compute Q.

4-3.   Find Q for the following scores on pigeon pecks per minute in a Skinner box: 2, 4, 4, 4, 4, 8, 8, 8, 8, 12, 12, 12, 13, 14.
  a. Find Q_1 (bottom 25%).
  b. Find Q_3 (top 25%).
  c. Use Formula 4.1 to compute Q.

4-4.   Find Q for the following scores on Comrey's *Personality Scales, Empathy vs. Egocentrism:* 30, 35, 40, 42, 42, 42, 42, 42, 50, 55, 60, 65, 65, 70, 71.
  a. Find Q_1 (bottom 25%).
  b. Find Q_3 (top 25%).
  c. Use Formula 4.1 to compute Q.

Handwritten annotations:

$Q = \dfrac{Q_3 - Q_1}{2} \sim \dfrac{50\%}{2}$

$Q = \dfrac{31.67 - 3.17}{2} = .25$

$Q = \dfrac{31.17 - 15}{2}$

$Q = 8.08$

3
10
10
20
25
25
25 — 25 median
25
31
31
31 ↑ 31.17
40
50

$\text{median} = \dfrac{n+1}{2} = \dfrac{12+1}{5} = 6.5$

$Q_3 = X_{LL} + \left(\dfrac{3N/4 - \text{cum } f_{LL}}{f}\right)i$

lowest #

$Q_3 = 30 + \left(\dfrac{3 \cdot 12/4 - 1}{12}\right) = 3 + \dfrac{8}{12} = \boxed{30.67}$

$Q_1 = X_{LL} + \left(\dfrac{N/4 - \text{cum } f_{LL}}{f}\right)i = Q_1 = 3 + \left(\dfrac{3-1}{12}\right)1 = 3 + (.17) = \boxed{3.17}$

4–5.   Compute Q for the frequency distribution given as the answer to question 2–1 in the back of the book. Is the distribution skewed? How? Why?
    a. Find Q_1 (bottom 25%).
    b. Find Q_3 (top 25%).
    c. Use Formula 4.1 to compute Q.

4–6.   Compute Q for the frequency distribution given as the answer to question 2–3 in the back of the book. Is the distribution skewed? How? Why?
    a. Find Q_1 (bottom 25%).
    b. Find Q_3 (top 25%).
    c. Use Formula 4.1 to compute Q.

4–7.   Compute Q for the frequency distribution in table 4.1.
    a. Find Q_1 (bottom 25%).
    b. Find Q_3 (top 25%).
    c. Use Formula 4.1 to compute Q.

4–8.   Compute Q for the frequency distribution in table 4.2.
    a. Find Q_1 (bottom 25%).
    b. Find Q_3 (top 25%).
    c. Use Formula 4.1 to compute Q.

4–9.   What is the semi-interquartile range of the memory scores shown in questions 3–17 to 3–19?
    a. Find Q_1 (bottom 25%).
    b. Find Q_3 (top 25%).
    c. Use Formula 4.1 to compute Q.

4–10.   Find the S.D. of the following scores using the deviation-score method: 2, 5, 6, 6, 8, 8, 10, 10, 11, 14.
    a. Find the mean of the scores.
    b. Subtract the mean from each score.
    c. Square each result in (b).
    d. Total all the scores in (c).
    e. Divide total in (d) by N.
    f. Find the square root of the result of (e).

4–11.   Find the S.D. of the following scores using the raw-score method: 2, 5, 6, 6, 8, 8, 10, 10, 11, 14.
    a. Add all the scores.
    b. Square each score and then add them.
    c. Note the number of scores.
    d. Use Formula 4.4.

4–12.   Add 2 points to each score in question 4–10 and recompute the M and the S.D. using the deviation-score method.
    a. Find the mean of the scores.
    b. Subtract the mean from each score.
    c. Square each result in (b).
    d. Total all the scores in (c).
    e. Divide the total in (d) by N.
    f. Find the square root of (e).

4–13.   Add 2 points to each score in question 4–11 and recompute the *M* and the S.D. using the raw-score method.
      a. Add all of the scores.
      b. Square each score and then add them.
      c. Note the number of scores.
      d. Use Formula 4.4.

4–14.   Multiply each score in question 4–10 by 3 points and recompute the *M* and the S.D. using the deviation-score method.
      a. Find the mean of the scores.
      b. Subtract the mean from each score.
      c. Square each result in (b).
      d. Add all the scores in (c).
      e. Divide the total in (d) by *N*.
      f. Find the square root of (e).

4–15.   Multiply each score in question 4–11 by 3 points and recompute the *M* and the S.D. using the raw-score method.
      a. Add all the scores.
      b. Square each score and then add them.
      c. Note the number of scores.
      d. Use Formula 4.4.

4–16.   What is the effect of adding or multiplying by a constant on the *M* and the S.D.?

4–17.   Find the S.D. of the following scores using the deviation-score method: 1, 1, 3, 3, 3, 4, 5, 5, 6, 6, 6, 7. Divide each score by 2 and recompute the S.D. What is the effect on the S.D. of dividing by a constant?
      a. Find the mean of the scores.
      b. Subtract the mean from each score.
      c. Square each result in (b).
      d. Add all the scores in (c).
      e. Divide the total in (d) by *N*.
      f. Find the square root of (e).

4–18.   Find the S.D. of the following scores using the raw-score method: 1, 1, 3, 3, 3, 4, 5, 5, 6, 6, 6, 7. Divide each score by 2 and recompute the S.D. What is the effect on the S.D. of dividing by a constant?
      a. Add all the scores.
      b. Square each score and then add them.
      c. Note the number of scores.
      d. Use Formula 4.4.

4–19.   In a survey of 100 engineers, the following job classification distribution was found:

Title	Count
Senior Engineer	5
Engineer	10
MTS 2	25
MTS 1	60

    a. Find the mode.

    b. Find the Index of Dispersion.

        1. Find k, the number of categories.

        2. Find N, the number of observations.

        3. Find f_1 through f_4, the number of observations in categories 1–4.

        4. Square N.

        5. Square $f_1 - f_4$ and add together.

        6. Subtract Step 5 results from Step 4 results.

        7. Multiply results of Step 6 by k.

        8. $D = N^2 (k-1)$ divided into Step 7 results.

4–20.   In a random sample of 500 people, the following characteristics were tabulated:

Eye color	Count	Sex	Count	Soft drink brand	Preference count
Brown	250	Male	225	A	200
Blue	100	Female	275	B	156
Green	50			C	144
Gray	40				
Hazel	60				

Compute the index of dispersion for each category.

    a. Eye Color

        1. Find k, the number of categories.

        2. Find N, the number of observations.

        3. Find f_1 through f_5, the number of observations in categories 1–5.

        4. Square N.

        5. Square $f_1 - f_5$ and add together.

        6. Subtract Step 5 results from Step 4 results.

        7. Multiply results of Step 6 by k.

        8. $D = N^2 (k-1)$ divided into Step 7 results.

    b. Sex

        1. Find k, the number of categories.

        2. Find N, the number of observations.

        3. Find f_1 through f_2, the number of observations in categories 1–2.

        4. Square N.

        5. Square $f_1 - f_2$ and add together.

        6. Subtract Step 5 results from Step 4 results.

        7. Multiply results of Step 6 by k.

        8. $D = N^2 (k-1)$ divided into Step 7 results.

  c. Soft Drink Preference
   1. Find k, the number of categories.
   2. Find N, the number of observations.
   3. Find f_1 through f_3, the number of observations in categories 1–3.
   4. Square N.
   5. Square $f_1 - f_3$ and add together.
   6. Subtract Step 5 results from Step 4 results.
   7. Multiply results of Step 6 by k.
   8. $D = N^2 (k-1)$ divided into Step 7 results.

4–21. Find the mean and standard deviation using the raw-score method for the following sample: 17, 24, 9, 33, 16.
  a. Add all the scores.
  b. Square each score and then add them.
  c. Note the number of scores.
  d. Use Formula 4.4.

4–22. Find the mean and standard deviation using the raw-score method for the following sample: -5, -6, -11, 0, 3, 14, -2.
  a. Add all the scores, remembering to observe the sign.
  b. Square each score and then add them.
  c. Note the number of scores.
  d. Use Formula 4.4.

4–23. Using the data in question 4–21, find the mean and variance of the data set after 10 has been added to each score using the raw-score method.
  a. Add all the scores.
  b. Square each score and then add them.
  c. Note the number of scores.
  d. Use Formula 4.4.

4–24. Using the data in question 4–22, find the mean and variance of the data set after 10 has been added to each score using the raw-score method.
  a. Add all the scores.
  b. Square each score and then add them.
  c. Note the number of scores.
  d. Use Formula 4.4.

4–25. When would it be more appropriate to use the Index of Dispersion?

4–26. In the raw-score formula for computing S.D., when (in what situation) will $\Sigma X^2 = (\Sigma X)^2$?

4–27. Is it feasible to compute the Index of Dispersion for a frequency distribution? If the answer is yes, find the Index of Dispersion for the frequency distribution given in table 4.2.

4–28. In a very, very skewed distribution how many (what percentage) of the measurements fall below Q_2?

4–29. In a very, very skewed distribution what percentage of the measurements lie above and below Q_1?

Chapter 5

Scaled Scores and the Normal Curve

R esearch measurements typically yield numbers called raw scores or **raw data** on a rank-order scale. For instance a raw score on a test is calculated as the number of test items the subject correctly answers. This number is on a rank-order scale of measurement because the scores merely rank the subjects in order of their ability rather than specifying a precise quantity of ability. Subjects who score 20 do not necessarily have twice the ability as those who score 10, although they presumably have more ability. The test scores only rank people within the limits of the error characteristics of the test as applied to a group of individuals at a given administration of the test.

raw data   Statistical data in its original form, before any statistical techniques are used to refine, process, or summarize.

The drawbacks of raw scores have prompted researchers to find ways of converting them to other types of scores with more desirable properties. One of the difficulties with using raw scores is that they cannot be interpreted without additional information. For example, if a subject has a raw score of 70 on a test, the information is useless without other subjects' scores to provide a standard of comparison. Only then is it possible to know if a raw score of 70 is high, medium, or low. Raw scores can be converted to various scaled scores that have a built-in standard of comparison so that they can be directly interpreted without additional information. This chapter is devoted to describing several kinds of scaled scores and to the normal curve, which is used to obtain some of them.

PROBLEM 11

Given a Raw Score, Find Its Centile Rank

Conversion to centile ranks is one of the most common types of raw-score transformations. There are 99 centile ranks (the numbers from 1 to 99). In some research papers and test manuals the centile ranks are referred to as **percentile** ranks. They represent percentages of subjects falling below the subject whose score is being examined. For example if a raw score of 70 corresponds to a centile rank of 44, it indicates that 44% of the subjects in the reference group have raw scores of 70 or below. The centile ranks corresponding to raw scores are always with reference to one particular group of subjects. A raw score of 70 might be at the 44th centile rank for a group of high school seniors, but only at the 22nd centile rank for a group of college students. The advantage of knowing a subject's centile rank instead of just the raw score is that the centile-rank score immediately specifies the percentage of subjects whose raw scores are lower. The usefulness of the centile rank is diminished, however, if the reference group is not known. An I.Q. test score that falls at the 75th centile rank for elementary school children clearly is less impressive than a 75th centile rank for Phi Beta Kappa college graduates.

percentile   One of the division points between 100 equal-sized pieces of the population when the population is arranged in numerical order. The 78th percentile is the number such that 78% of the population is smaller and 22% of the population is larger.

To convert a raw score to a centile rank first find the number of scores in the reference group that fall below the given raw score. Then divide by the number of scores in the reference group to obtain the proportion of scores in the reference group below this one. This proportion is multiplied by 100 to convert it to a percentage, which is rounded to the nearest whole number value from 1 to 99. Note that centile ranks of zero and 100 are not used. Any percentage less than 1 is rounded up to 1 and any percentage above 99 is rounded down to 99.

Figure 5.1
Example of score limits.

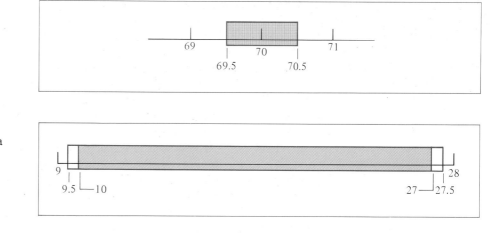

Figure 5.2
Example of the limits for a
range of scores.

When making the calculations remember that a given score covers a range of a full point. A score of 70 goes from 69.5 to 70.5, with a midpoint of 70.0. When computing the number of scores below 70 half of the subject's own score is below 70 (69.5 to 70.0) and half is above 70 (70.0 to 70.5). So the number of scores below 70 will include all subjects below 70 plus half of the subject's own score which is at 70. If there are other subjects whose scores are also at 70, half of these will be below 70.0 and half will be above 70.0. The rule then is to take the number of scores below the given score plus half of those at or equal to the given score to determine the number below that score (fig. 5.1).

Example   Consider the following raw scores representing the number of items correct on a history examination for a class of 15 students: 10, 15, 20, 20, 25, 25, 26, 27, 28, 28, 29, 32, 33, 35, and 42. What is the centile rank for a score of 28?

First count the number of scores below 28. There are 8 scores between 9.5 and 27.5, or below the score of 28, which runs from 27.5 to 28.5. Now add half the scores at 28 (one-half of 2), or 1. The results are 8 + 1 = 9 scores below 28.0, the midpoint of the range for the score 28. Dividing 9 by 15 (which is N) gives 0.60 as the proportion of scores below 28.0. This proportion is multiplied by 100 to convert it to a percent, 0.60 × 100 = 60.0, which is rounded off to 60. The centile rank corresponding to a raw score of 28, then, is 60 in this reference group (fig. 5.2).

An Example Using a Frequency Distribution   Suppose the same history examination is given to a larger class in a different school, yielding the frequency distribution shown in table 5.1. What is the centile rank for a raw score of 28 for the reference group?

TABLE 5.1		
History examination scores		
X	f	cf
50–54	1	45
45–49	2	44
40–44	5	42
35–39	9	37
30–34	10	28
25–29	8	18
20–24	6	10
15–19	2	4
10–14	1	2
5–9	1	1
N	45	

Since the score of 28 is contained within a class interval in table 5.1, it is necessary to use an interpolative process to calculate the number of scores below 28.0. Below the interval 24.5 to 29.5, which contains the score of 28, there are 10 cases, as shown in the cumulative frequency column. To these 10 scores add the portion of the 8 scores in the interval 25–29 (exact limits 24.5 to 29.5), which falls below 28.0. This is computed as follows:

$$10 + \frac{(28.0 - 24.5)}{5} \cdot (8) =$$

$$10 + (0.7) \cdot (8) =$$

$$10 + 5.6 = 15.6$$

The interval from the lower limit up to the score point 28.0 is divided by the class interval width ($i = 5$) to give the proportion of the interval below 28.0, or 0.7. Since the cases are assumed to be evenly spread throughout the interval, this is also the proportion of cases below 28.0 in the interval. Multiplying 0.7 by 8, the total number of cases in the interval, gives 5.6 as the number of cases in the interval below 28.0. Adding 5.6 to 10 gives a total of 15.6 cases below 28 in the entire distribution. Dividing 15.6 by 45 gives 0.3467 as the proportion of cases below 28. Multiplying by 100 and rounding off to the nearest whole number gives 35 as the centile rank corresponding to a raw score of 28 on the history examination. This is the general formula for computing the percentile rank:

$$\text{PR} = \frac{\text{Cum } f_{LL} + \left(\frac{X - X_{LL}}{i}\right) f}{N} (100) \qquad (5.1)$$

FUNKY WINKERBEAN

BY TOM BATIUK

where   Cum f_{LL} = the cumulative frequency at the lower true limit of the class interval containing X,

X = the given score,

X_{LL} = the score at the lower true limit of the interval containing X,

i = the width of the interval, and

f = the number of cases within the interval containing X.

The centile rank for a score of 28 in one group was 60. The same score in another group gave a centile rank of 35. Obviously a centile rank requires knowledge of the reference group for which it was computed. Publishers of tests typically provide tables of centile ranks corresponding to raw scores on their tests for different normative groups, such as high school students, college students, general population males, and females in the work force. When a particular person's test results are evaluated, their centile-rank scores are determined according to the available comparison "norm" group that is most appropriate. The centile-rank scores reveal directly how the single individual compares with this group.

There are other names for the centile-rank score, such as centile, or the centile score. At times an individual's raw score, or even the individuals themselves, are described as being at the 75th centile, for example. This term is abbreviated as P75 or P_{75}. The P is derived from the term percentile, which ultimately may be replaced by the term centile.

SUMMARY OF STEPS

Given a Raw Score, Find Its Centile Rank

Step 1   Find the class interval containing the score.

Step 2   Find the number of scores, ns, below this class interval.

Step 3   Find the number of scores within the class interval that contains the score, *n*.

Step 4   Find the lower limit, *LL*, of the class interval containing the score.

Step 5   Compute the class interval, *i*.

Step 6   Compute using Formula 5.1.

Step 7   Divide the result of Step 6 by the total number of scores, round to decimal places, and multiply by 100.

PROBLEM 12

Given a Centile Rank, Find the Corresponding Raw Score

Instead of finding the centile rank corresponding to a particular raw score, it is often more pertinent to find the raw score that corresponds to a particular centile rank. Previous examples are with the median equal to the 50th centile, P_{50}; and the two quartile points Q_1 and Q_3, equal to the 25th centile, P_{25}, and the 75th centile, P_{75}. The methods used to solve Problems 4, 5, 7, and 8 can be used to solve Problem 12. To obtain the raw score for a centile rank of P_{68}, for example, find the raw-score point below which 68% of the cases fall. Thus, N is multiplied by 0.68 to find the number of scores below the desired point on the raw-score scale.

Formula 5.1 can be modified so that a percentile rank (PR) can be converted to a raw score. With the same description of notation as given with Formula 5.1, the conversion of percentile rank to raw-score formula is:

$$\text{raw score } (X) = X_{LL} + \frac{\left[\dfrac{PR}{100}(N) - \text{Cum } f_{LL}\right] i}{f} \tag{5.1a}$$

Example   To demonstrate the application of Formula 5.1a, consider the distribution of history test scores in table 5.1. Compute the raw score corresponding to a centile rank of 68, the 68th centile.

First find 68% of N, $45 \times 0.68 = 30.6$. Below the point corresponding to P_{68} there are 30.6 cases. Using the methods of Problem 6, this point is found as follows (see table 5.1, page 94):

$$P_{68} = 34.5 + \frac{(30.6 - 28)}{9} \cdot (5)$$

$$P_{68} = 34.5 + \frac{2.6}{9} \cdot (5)$$

$$P_{68} = 34.5 + \frac{13.0}{9}$$

$$P_{68} = 34.5 + 1.44$$

$$P_{68} = 35.94 = 35.9$$

The answer is usually rounded off to one decimal place if the raw scores are integers or whole numbers.

Norms

norm A numerical standard used for the interpretation of test scores, usually based on scores derived from a nationwide sample.

Norms for published measuring instruments typically include tables, which list the possible raw scores and the corresponding centile ranks for various norm groups. These tables can be constructed, with appropriate rounding off procedures, using either the methods of Problem 11 or Problem 12. In some instances the table is constructed by reading from a graph, which was prepared using only a certain proportion of the centile points or raw scores. For example, the raw-score equivalents for every 5th centile rank, 5, 10, 15, etc., can be computed and plotted on a chart. Then a smooth curve is drawn through these points and extrapolated at either end to the 1st and 99th centile ranks. The centile ranks corresponding to all possible raw scores are read off this graph to construct the desired table. Raw scores above and below the end points are merely listed as being either at the 99th or the 1st centile.

CHARACTERISTICS OF THE CENTILE-SCORE SCALE

Conversion of raw scores to centile scores is a **nonlinear transformation** that does not preserve the spacing characteristics of the raw-score scale. Consider Formula 5.2:

$$P = b \cdot R + a \tag{5.2}$$

nonlinear transformation Transformation of scores that changes the scale of the scores (mean and S.D.) and also the shape of the score distribution.

where P = the centile-rank score,

b = a constant (some number),

R = the raw score, and

a = a constant (some number).

If it were possible to find values for *a* and *b* in Formula 5.2 that would convert raw scores to the corresponding centile scores, then conversion to centile scores would be a linear transformation and a graph of the centile scores plotted against the raw scores would give a straight line. This is not the case, however, since the values for *a* and *b* in Formula 5.2 cannot be found to effect this transformation.

The centile-score transformation distorts the spacing of individuals in comparison with the original raw-score scale. It tends to spread the scores in the middle of the score scale and bunch them together at the ends of the scale. This is because an equal step on the centile scale represents 1% of the scores. A score at the 51st centile has 51% of the scores at or below it and a score at the 50th centile has 50% of the scores at or below it. Typically, there are more scores in the middle of the distribution because 1% covers a smaller interval on the raw-score scale in the middle of the distribution than at the ends. To improve from the 50th to the 53rd centile might require getting only one point more on a test, but to go from the 96th to the 99th centile might require an increment of 10 points. The magnitude of this distortion is revealed by the following centile scores that represent, very roughly, equidistant points on the raw-score scale: 1, 3, 16, 50, 84, 97, and 99. The difference between the 1st and 3rd centiles is roughly equivalent to the difference between the 50th and 84th centile. When using the centile scale remember that the differences in the middle of the scale do not mean as much as at the end of the scale. The real amount of performance difference between the score at the 40th and 60th centile, for example, is not very great even though there is a 20 point separation. A separation half this large at the extremes of the distribution is of greater significance. Figure 5.3 shows that transforming scores between raw scores and percentile scores alters the shape of the distribution.

Usefulness of Centile Scores

Raw scores are converted to centile scores because raw scores do not have a standard interpretable meaning. For instance, if a subject has 80 items correct on one test and 60 items correct on another test, it is difficult, using the raw scores alone, to determine which score is better.

However, if the subject has a centile score of 60 on the first test and a score of 10 on the second test, it is immediately obvious that the subject did a bit above average on test one and rather poorly on test two (relative to the reference group for which the centile equivalents were obtained). Centile scores indicate the percentages of cases in the normative group that failed to do as well on the test as the subject under consideration.

These features of centile scores are popular for normative purposes. When a new test instrument is developed, centile norms are prepared for the test for various reference groups, such as high school males, limited-English speaking persons, high school females, college males, college females, and male industrial workers. Each set of norms consists of the possible raw scores on the test with their corresponding centile equivalents. Individuals' raw scores are used to enter the norm table most relevant for them and their centile scores are taken from the table. The centile score shows where individuals stand in relation to their particular reference group on the test. The information is used for such practical goals as vocational and educational guidance or personnel selection.

Figure 5.3
(a) Graph of a distribution of
I.Q. scores, (b) graph of the
distribution of percentile ranks.

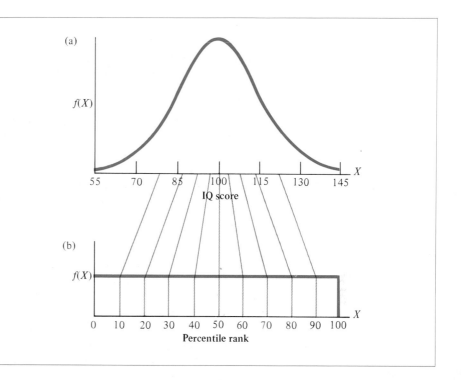

SUMMARY OF STEPS

Given a Centile Rank, Find the Corresponding Raw Score

Step 1   Multiply the centile rank by the total number of scores and divide by 100.

Step 2   Find the class interval where the total cumulative scores lies.

Step 3   Determine the lower limit, *LL*, of the class interval found in Step 2.

Step 4   Determine the number of scores, ns, below the class interval found in Step 2.

Step 5   Determine the number of scores, *n*, in the class interval found in Step 2.

Step 6   Determine the class interval size, *i*.

Step 7   Compute using Formula 5.1a.

Step 8   Round the number found in Step 7 to two decimal places.

PROBLEM **13**

Convert Raw Scores to Standard Scores
Using a Linear Transformation

A common transformation of raw scores that preserves the spacing proportions among the scores is the conversion of raw scores to standard scores using the following linear transformation:

$$Z = \frac{X - M}{\text{S.D.}} \qquad (5.3)$$

where Z = the standard score,

           X = the raw score,

           M = the mean, and

      S.D. = the standard deviation.

In terms of Formula 5.2, Formula 5.3 is rewritten as:

$$Z = \frac{1}{\text{S.D.}}(X) + \frac{-M}{\text{S.D.}} \qquad (5.4)$$

$$b = \frac{1}{\text{S.D.}} \text{ and } a = \frac{-M}{\text{S.D.}}$$

linear transformation   A transformation of data using a linear equation that changes only the scale of the scores (mean and/or S.D.) and not the shape of the distribution. The relative positions of the scores are retained.

Thus constants are found that convert raw scores to standard scores by means of a linear transformation. The transformation in Formula 5.4 is "linear" because the variable X appears in the equation only to the first degree, not X^2, X^3, and so on. If linearly transformed scores are plotted against the original scores, the points all fall on a straight line; hence the term **linear transformation.** If the standard scores obtained by Formula 5.3 are plotted against the corresponding raw scores, the resulting graph is a straight line. This indicates that if two raw scores are close together, their standard scores are likewise relatively close together. If two raw scores are far apart, their standard scores are also far apart on the standard-score scale. The distortions in the scale produced by the raw score to centile-rank transformation are absent in the linear transformation of raw scores to standard scores.

Since the relative spacing of raw scores is preserved in the linear transformation of raw scores to standard scores, it follows that whatever shape was characteristic of the raw-score distribution is also found in the transformed scores. Thus, if the raw scores are badly skewed in the positive direction, the linearly transformed standard scores are also badly skewed in the positive direction. Transformation of raw scores to standard scores by Formula 5.3 do not produce a different shaped distribution of scores than the original scores. Only the scale of measurement is changed.

For example, a person's height can be measured in inches or centimeters. Regardless of which scale of measurement is used, the actual physical height of the person does not change. The only change is the scale used to measure the height of the person. The Z-score, or transformation, only changes the scale about how something is measured. These linearly transformed Z-scores are normally distributed only if the original X-scores are normally distributed.

Z-score The distance of the given value from the mean of the distribution or population, when measured in standard deviations.

A distribution of scores transformed to standard scores using Formula 5.3 always has a mean of zero and a standard deviation of 1. Interpret the **Z-score** as the number of standard deviation units a raw score is away from the mean. If an individual's standard score is -1, it is immediately evident that the person's raw score falls at 1 S.D. below the mean of the reference group from which the score transformation is obtained. If the standard score is near zero, it is known immediately that the raw score falls near the mean of the distribution for the reference group. Linearly transformed standard scores preserve the spacing among scores and reveal something about the individual's standing relative to the reference group, but they do not give the percentage of cases falling below the score that is so conveniently available with the centile score.

Example   Application of Formula 5.3 merely requires knowing the mean and standard deviation of the reference group as well as the raw score that is to be transformed. In a distribution with $M = 25$ and S.D. $= 5$, a raw score of 21 yields a standard score by Formula 5.3 as follows:

$$Z = \frac{21 - 25}{5} = \frac{-4}{5} = -0.8 \tag{5.3}$$

Standard scores for the vast majority of the cases lie between -3.0 and $+3.0$. They are typically reported to two decimal places.

Usefulness of Standard Scores

It has been pointed out already that centile scores are distorted in the sense that subjects in the middle of the distribution have centile scores that vary over a wide range while subjects at the extremes have scores that are bunched together. This type of distortion is eliminated in all standard-score scales or a bell-shaped distribution. Subjects with centile-scale scores from 50 to 84 are spread over the standard scores of 0.0 to 1.0 if the distribution of scores is bell-shaped (normal); subjects with centile scores from 84 to 97 are spread over the standard scores of 1.0 to 2.0; and subjects with centile scores from 97 to 99 are spread over the standard scores of 2.0 to $+\infty$, although standard scores beyond 3.0 are rare. Centile scores from 1 to 50 give negative standard scores in a mirror image fashion to the positive standard scores for centile scores from 50 to 99. When test norms are put into standard-score form instead of centile-score form, it is to eliminate the distortions found in the centile-score form while still retaining the advantage of being able to determine, from the score itself, where in the distribution the person's score falls. This cannot be done with raw scores.

MATHEMATICAL NOTES

The reader may have wondered why the mean and standard deviation of standard scores are zero and 1. To find the mean of Z-scores, substitute Formula 5.3 for Z into the formula for the mean as follows. Given:

$$M = \frac{\Sigma X}{N}, Z = \frac{X - M}{S.D.} \qquad (5.6), (5.3)$$

Substituting Z for X gives:

$$M_Z = \frac{\Sigma \left(\dfrac{X - M}{S.D.} \right)}{N}$$

$$M_Z = \frac{\Sigma \left(\dfrac{X - M}{S.D.} \right)}{N} = \frac{1}{S.D.} \Sigma \left(\frac{X - M}{N} \right) = \frac{1}{S.D.} \left[\frac{\Sigma X}{N} - \frac{\Sigma M}{N} \right]$$

$$M_Z = \frac{1}{S.D.} \left(\frac{\Sigma X}{N} - \frac{N \cdot M}{N} \right)$$

$$M_Z = \frac{1}{S.D.} (M - M) = 0$$

Since the mean of the Z-scores is zero, they are deviation scores. To find the standard deviation of Z-scores the deviation-score formula for the standard deviation can be used.

$$S.D. = \sqrt{\frac{\Sigma x^2}{N}} \qquad (5.7)$$

Substituting $Z = \dfrac{X - M}{S.D.}$ for x gives:

$$S.D._Z = \sqrt{\frac{\Sigma \left(\dfrac{X - M}{S.D.} \right)^2}{N}}$$

$$S.D._Z = \sqrt{\frac{\Sigma \left(\dfrac{x^2}{S.D.^2} \right)}{N}}$$

$$S.D._Z = \sqrt{\frac{\Sigma \left(\dfrac{x^2}{N} \right)}{(S.D.)^2}}$$

$$S.D._Z = \sqrt{\frac{(S.D.)^2}{(S.D.)^2}} = 1$$

PROBLEM **14**

Convert Standard Scores to Scaled Scores with a Different *M* and S.D.

Standard scores are rather inconvenient for use in clinics, schools, and industry because about half of the scores are negative and all scores are decimal fractions rather than whole numbers. To eliminate negative numbers and fractions, standard scores are commonly converted to a more convenient scale by means of a linear transformation.

Converting Standard Scores to Scaled Scores with a Different *M* and S.D.

The linear transformation that converts standard scores to a more convenient scale is given by Formula 5.8:

$$S = (S.D.) \cdot (Z) + M \qquad (5.8)$$

where S = the scaled score,

S.D. = the desired standard deviation in the scaled scores,

Z = the standard score to be transformed to a more convenient scale, and

M = the desired mean of the scaled scores.

Using Formula 5.8, it is merely necessary to select a new mean and standard deviation to avoid the problems caused by the zero mean and unit standard deviation of the standard-score scale. The standard score is multiplied by the new standard deviation and the result added to the new mean score to get the scaled score.

t-score McCall scores rescaled to have a mean of 50 and a standard deviation of 10.

Example   A commonly used example of this type of scaled score is the McCall *T*-Score Scale. This score scale uses a mean of 50 and a standard deviation of 10. To convert standard scores to this scale, each standard score is multiplied by 10 to get rid of the decimal fraction and then added to 50 to eliminate the negative signs. The *T*-score scale is shown in figure 5.4 just below the standard-score scale. A *T*-score of 20 corresponds to a standard score of -3, a *T*-score of 30 corresponds to a standard score of -2, a *T*-score of 60 corresponds to a standard score of $+1$, and so on. The McCall *T*-Score is a popular scale used on personality tests.

CEEB scores Linearly transformed scores with a mean of 500 and a standard deviation of 100. CEEB stands for College Entrance Examination Board.

Just below the McCall *T*-Score Scale in figure 5.4 is the **CEEB** Scale used for the Graduate Record Examination scores. In that score scale the mean is 500 and the standard deviation is 100. For this score scale, standard scores must be determined to two decimal places since they are multiplied by 100 before adding to the new mean, 500. The CEEB Scale gives more points on the scale, differentiating more finely between individuals, reducing the number of tied scores. Whether such precision is justified by the accuracy of the measuring instrument is another question. In most cases, *T*-scores provide as much discrimination between subjects as the measuring instrument can justify.

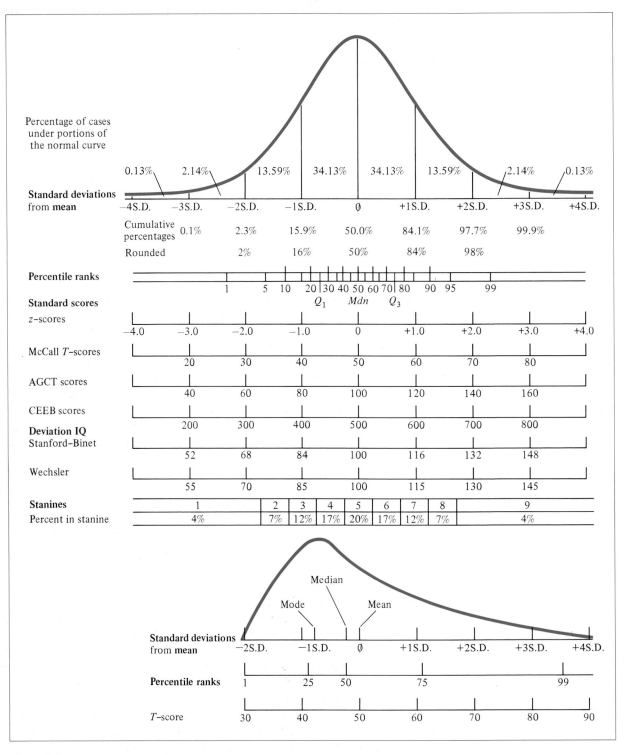

Figure 5.4

The normal curve and scaled scores.

Test Service Bulletin, Volume 48. The Psychological Corporation, San Antonio, Texas. Reproduced by permission of the Publisher.

TABLE 5.2	
Stanine-Standard Score equivalents	
Standard-score equivalents	Stanine scores
Above +1.75	9
+1.25 to +1.75	8
+0.75 to +1.25	7
+0.25 to +0.75	6
−0.25 to +0.25	5
−0.75 to −0.25	4
−1.25 to −0.75	3
−1.75 to −1.25	2
Below −1.75	1

Stanine One digit normalized scores with a mean of 5 and a standard deviation of approximately 2. The distribution of stanines are approximately normal.

Another popular score scale used in elementary and secondary schools is the Stanine-Standard Score Scale. This scale has a mean of 5 and a standard deviation of 2.0. The conversion to this score scale takes place somewhat differently, however, than for the *T*-score scale. A table of Stanine-Standard Score Scale equivalents is used to make the conversion rather than a formula transformation like Formula 5.8. The table of equivalents is shown in table 5.2. The Stanine-Standard Score Scale only has 9 points on it, which requires lumping scores over a half standard deviation interval into the same **Stanine** score. At the extremes, for Stanine scores of 1 and 9, a much larger range of standard scores is lumped into one Stanine. The 9 point Stanine-Standard Score Scale offers adequate differentiation of subjects for most practical personnel decisions and represents as much precision as most tests actually warrant.

PROBLEM **15**

Use the Normal Curve Table to Find Numbers of Cases Expected in Different Parts of the Distribution

normal distribution The pattern that was discovered from repeated measurements and that has the shape of the bell-curve.

Many variables studied by researchers have population frequency distributions that follow the bell-shaped normal curve when plotted for visual display (fig. 5.4), although sample distributions may vary substantially from this form. Consider the distribution of height for all adult females in the United States. The frequency polygon based on the frequency distribution for this variable has a shape similar to that for the distribution in figure 5.4. Such a distribution is called a **normal distribution** or normal curve. It has a precise mathematical equation (see Mathematical Notes, page 117) that describes the shape exactly. Of course the curve of heights for all women in the United States might vary slightly from the theoretical normal curve shape, but generally the actual distribution for this variable is normal in form.

If a random sample is drawn from this very large population, expect the frequency distribution for the sample to differ from that of the population as a result of random sampling fluctuations. When drawing a random sample, each person in the population must have an equal chance of selection for inclusion in the sample.

random sample   A subset of the population where each element had an equal probability of being chosen.

If all adult women in the United States are listed alphabetically by name and then the names are rearranged by some method and every 100,000th person selected from the list, a **random sample** results.

Even in a random sample where each person in the population has an equal chance of selection, there still may be a disproportionate number of individuals in a specific category selected by chance. Too many tall women might be accidentally chosen, resulting in a departure of the sample curve from the population curve. Most research workers rarely work with population distributions because it is costly and impossible to measure everybody or everything in a large population on any variable. This was very evident in the 1980 U.S. Census, which supposedly tried to measure every household in the United States. Usually the researcher works with data only from a sample. Often the sampling procedures are not exactly random in character. Researchers might have information leading them to think the population is normally distributed for the variable under investigation, but they are not sure the sample was drawn randomly from that population. Even if the sample is drawn randomly from a normally distributed population, the sample distribution departs from normality to some extent. The smaller the sample, the greater the likelihood of departure from normality. If the sample was not drawn randomly, there is an even greater chance that certain segments of the population are overrepresented and other segments underrepresented, leading to a distortion in the sample distribution.

Whatever the situation, the investigator needs to compare the number of cases in a specific portion of the sample distribution with what should be in a sample distribution that is normal. The researcher wants to know how many scores are above or below a given raw score, or how many scores are between two given raw scores of a normal sample distribution. The obtained values are compared with the actual number of scores in these regions to provide some indication of the degree to which the sample distribution approximates the normal distribution in form.

The Normal Curve

The bell-shaped normal curve (or normal distribution) is shown in figure 5.4. Dividing vertical lines are shown in this figure at 1, 2, and 3 standard deviation distances above and below the mean. The mean, marked $\emptyset$ in figure 5.4, is at the center of the curve, where the height of the curve above the base line is greatest. In the region between the mean (M) and 1 standard deviation (S.D.) above the mean, there are approximately 34% of all the scores in the normal distribution. In the region between the mean and 1 S.D. below the mean there are also approximately 34% of the scores. The regions below the mean contain the same percentages as the regions above the mean since the curve is symmetric: the left half is a mirror image of the right half. From the 1 S.D. point to the 2 S.D. point there are about 13.5% of the cases. This same percentage of cases falls in the corresponding region below the mean.

These figures show that in the interval from -1 S.D. to $+1$ S.D. in the normal curve there are approximately $34 + 34$, or 68% of the cases; $M \pm 1$ S.D. covers about 68% of the cases. In the interval from -2 S.D. to $+2$ S.D. there are approximately $13.5 + 34 + 34 + 13.5$, or 95% of the cases; $M \pm 2$ S.D. covers about 95% of the cases. This leaves about 5% of the cases outside the region $M \pm 2$ S.D., 2 1/2% above $+2$ S.D. and 2 1/2% below -2 S.D.

Standard Scores   When raw scores are converted to standard scores, the transformed scores have a mean of zero and a standard deviation of 1.0. The standard-score scale is shown in figure 5.4 just below the percentile ranks row. A person whose raw score is at the mean of the distribution receives a standard score of zero. If the raw score is 1 S.D. above the mean, the standard score is 1.0; if the raw score is 1.2 S.D. below the mean, the standard score is -1.2. Raw scores are also transformed to distributions with means and standard deviations different from zero and 1 (see Problem 14). The bottom two scales in figure 5.4 are examples. The CEEB Score Scale has a mean of 500 and a standard deviation of 100. The McCall *T*-Score Scale has a mean of 50 and a standard deviation of 10. A score of $+3.0$ on the *Z*-score scale is equivalent to an 80 on the McCall *T*-Score Scale and an 800 on the CEEB Score Scale. All these points are at 3 S.D. above the mean. It is just the units of the scale that are changing.

The Normal Curve Table   The proportions of total area under the normal curve for positive standard scores (*Z* values) are given in table B, Appendix B. The table lists values of *Z* by increments of 0.01 between $Z = 0.00$ and $Z = 4.00$. Each entry in column B for a given *Z*-score is the proportion of the total area under the curve that falls below that particular *Z* value. Note that at $Z = 0.00$, the tabled value is 0.5000, or one-half the total area below $Z = 0.00$. This is because $Z = 0.00$ is at the mean of the distribution and the normal distribution is symmetrical about the mean, half above the mean and half below.

As already stated, the mean ± 1 S.D. covers approximately the middle 68% of the cases. This statement can be verified by reference to the normal curve table. Entering table B with $Z = 1.00$ gives an area of 0.8413 in column B, or 84.13% below the $+1$ S.D. point, or 1 S.D. above the mean. Subtracting 0.50 (the half below the mean), from 0.8413, gives 0.3413 as the proportion of cases between the mean and $+1$ S.D. (column C of table B). Since the normal curve is symmetric, there is an equivalent proportion of the total area between the mean and -1 S.D. Adding these two proportions together gives $0.3413 + 0.3413 = 0.6826$, or approximately 68% of the total area. The cases in a distribution are arranged as the area under the curve, so approximately 68% of the cases fall in the interval $M \pm 1$ S.D. provided that the distribution is normal in form.

The interval $M \pm 2$ S.D. contains about 95% of the cases in a normal distribution. Again this can be verified by entering table B with $Z = 2.00$, which gives a value of 0.9772 for the area under the curve below $Z = 2.00$. Subtracting 0.5000, the area below the mean, from 0.9772 gives 0.4772 as the area between the mean and $+2$ S.D. There is an equal area between the mean and -2 S.D., so adding $0.4772 + 0.4772$ gives 0.9544 as the proportion of the total area in the interval $M \pm 2$ S.D. This rounds to a proportion of 0.95, which corresponds to about 95% of the cases falling between -2 S.D. and $+2$ S.D. in a normal distribution.

Table values are given for *Z* by increments of 0.01 up to $Z = 3.99$. The last tabled value of *Z* is 4.00 where the area is 1.0000. This value is the rounded off value, correct to four decimal places. The normal curve actually runs from $-\infty$ to $+\infty$, but the proportion of area beyond ± 3.99 S.D. is so small that it is closer to 0.0000 than it is to 0.0001.

Using the Normal Curve Tables to Find Frequencies

Use of the normal curve tables is best illustrated with examples. These examples generally only use columns A and B of table B. The columns C and D are provided to eliminate many of the calculations of the area under the normal curve. Column B gives the area from $-\infty$ to Z, column C gives the area between zero and Z, and column D gives the area from Z to $+\infty$. Given a distribution of cases for a sample with $N = 300$, which has a mean of 60 and a standard deviation of 15, find the number of scores that fall below a raw score of 55 in this distribution if it is distributed normally.

The first step is to convert the raw score of 55 to a standard score:

$$Z = \frac{X - M}{\text{S.D.}} = \frac{55 - 60}{15} = \frac{-5}{15} = -0.33$$

There is no negative Z-score in table B, the table of the normal curve, but since the normal curve is symmetrical, look at $Z = +0.33$ to get an area proportion equal to 0.6293 falling below the Z-score of $+0.33$. The area above $Z = +0.33$ is obtained as the difference between 0.6293 and the total area, so $1 - 0.6293 = 0.3707$ as the proportion of area above $Z = +0.33$. This proportion is the same as the proportion of area below $Z = -0.33$, since the normal curve is symmetrical (fig. 5.5). The proportion of cases below $Z = -0.33$ is 0.3707. The total number of cases in the sample is 300, so the frequency, the number of cases below 55, is 0.3707 $\times$ 300 = 111.2100 or 111.2 cases, rounding to one decimal place. In a normal distribution of 300 cases with a mean of 60 and a standard deviation of 15 there are 111.2 scores below 55. Compare 111.2 with the actual number of cases below 55 in this sample distribution. When making this comparison be sure to include half the scores of 55 as being below 55.0 and half as being above 55.0.

Example   Another type of problem encountered is to find the number of cases *above* a given raw-score point in a normal distribution rather than *below* that raw-score point. As an example, assume a sample distribution has a mean of 110 and a standard deviation of 20. There are 500 cases in the sample. The problem is to find the number of scores that are above 85 in the sample distribution if it is normal in shape.

As before, convert the raw score to a standard score:

$$Z = \frac{X - M}{\text{S.D.}} = \frac{85 - 110}{20} = \frac{-25}{20} = -1.25$$

Looking up $Z = +1.25$ in table B, since negative values are not tabled, gives 0.8944 as the proportion of area below $Z = +1.25$. Since the normal curve is symmetrical in shape, the proportion of area below $Z = +1.25$ is the same proportion of area above $Z = -1.25$ (fig. 5.6). The proportion of cases above 85 is 0.8944. Multiplying 0.8944 by 500 gives 447.2 cases above 85.

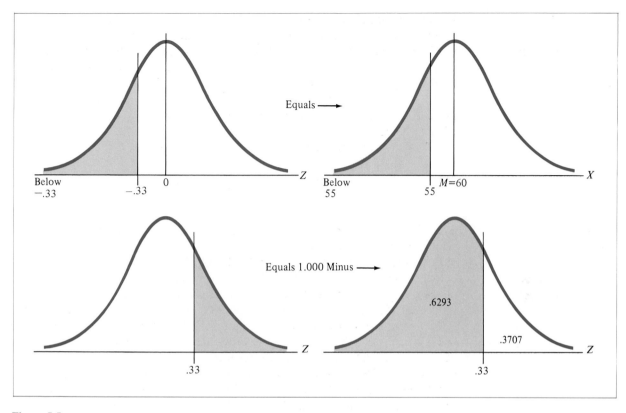

Figure 5.5
Proportions under the normal curve.

Example   There are occasions when the researcher wants to find the number of cases between two raw-score points in a normal distribution. There are three variations of this situation:

1. The two scores are on the right of the mean.
2. The two scores are on the left of the mean.
3. One of the scores is on the left of the mean and the other score is on the right of the mean.

In the first variation, assume a sample distribution of Scholastic Aptitude Test (SAT) scores with a mean of 500 and a standard deviation of 100. There are 1200 cases in the sample. The problem is to find the number of cases between the scores 510 and 600. To do this, 1200 is multiplied by the area under the normal curve between 510 and 600. The first step in finding the area between 510 and 600 is to convert each raw score to a Z-score.

$$Z \text{ (for 510)} = 510 - 500/100 = 10/100 = +0.10$$
$$Z \text{ (for 600)} = 600 - 500/100 = 100/100 = +1.0$$

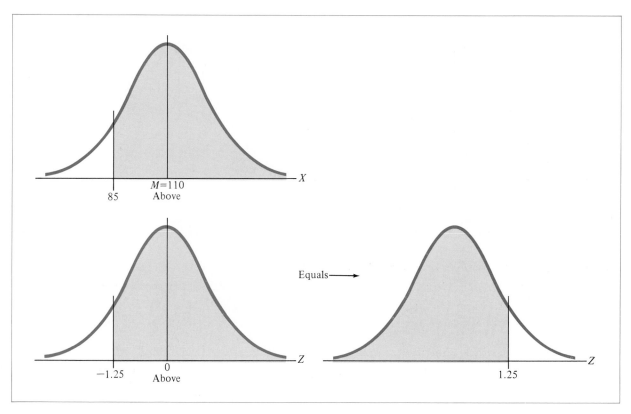

Figure 5.6
Corresponding areas of raw
and Z-scores under the normal
curve.

The area between $Z = 0.10$ and $Z = 1.00$ corresponds to the same area under the normal curve as between $X = 510$ and $X = 600$ (fig. 5.7).

Table B (column B) in Appendix B gives 0.8413 as the proportion of area below 1.00 for $Z = 1.00$ and 0.5398 as the proportion of area below $Z = 0.10$. The area between $Z = 0.10$ and $Z = 1.00$ is found by subtracting 0.5398 from 0.8413 (0.8413 $- 0.5398 = 0.3015$) (fig. 5.8). This area (0.3015) is the proportion of cases between test scores 510 and 600. Multiplying 1200 by 0.3015 gives 361.8 cases between SAT scores of 510 and 600.

The second variation in this example is to find the number of cases between SAT scores of 430 and 480. Note that both scores are on the left side of the mean. Similar to the first variation, Z-scores for each of these two raw scores are computed. Then the area between these two Z-scores is calculated by subtracting the

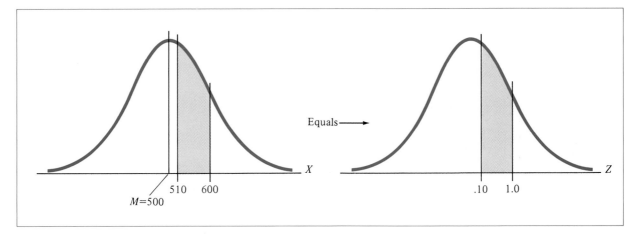

Figure 5.7
Corresponding areas of raw and Z-scores under the normal curve.

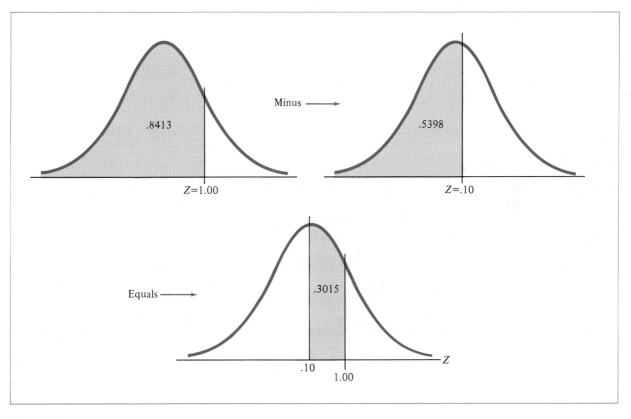

Figure 5.8
Finding the area between two Z-scores on the right side of the mean.

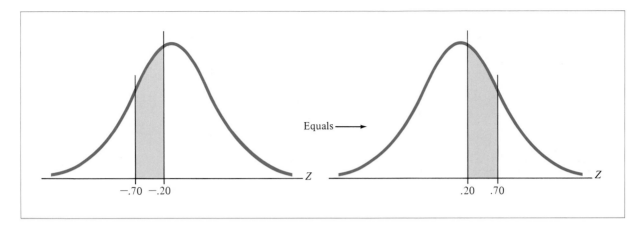

Figure 5.9
Equivalence of areas under the
normal curve.

smaller area from the larger area. The resulting area represents the proportion between 430 and 480. This proportion is multiplied by 1200 to give the number of
cases. The actual steps are as follows:

$$Z \text{ (for 430)} = 430 - 500/100 = -70/100 = -0.7$$

$$Z \text{ (for 480)} = 480 - 500/100 = -20/100 = -0.2$$

The area under the normal curve between $Z = -0.7$ and $Z = -0.2$ is the same
as the area between $Z = +0.7$ and $Z = +0.2$ because the normal curve is symmetrical (fig. 5.9).

Looking up $Z = 0.70$ in table B (column B) gives 0.7580 as the proportion of
area below $Z = 0.70$. The proportion of area below $Z = 0.20$ is 0.5793. Subtracting
0.5793 from 0.7580 gives 0.1787 as the area or proportion of scores between $Z =$
0.20 and $Z = 0.70$. This area is equivalent for the scores between $Z = -0.20$ and
$Z = -0.70$ and between $X = 480$ and $X = 430$. Multiplying this proportion by
1200 gives 214.4 cases between SAT scores of 430 and 480.

For the third variation, the problem is to find the number of cases between SAT
scores of 425 and 615. Note that the 425 occurs on the left of the mean (500) and
615 is to the right of the mean. To solve this problem, 1200 is multiplied by the
proportion of scores between 425 and 615. To find this proportion, the two raw scores
are converted to Z-scores.

$$Z \text{ (for 425)} = 425 - 500/100 = -75/100 = -0.75$$

$$Z \text{ (for 615)} = 615 - 500/100 = 115/100 = 1.15$$

Finding the area or proportion between $Z = -0.75$ and $Z = 1.15$ is equivalent to
finding the area between $X = 425$ and $X = 615$. This is computed by subtracting
the area below $Z = -0.75$ from the area below $Z = 1.15$ (fig. 5.10).

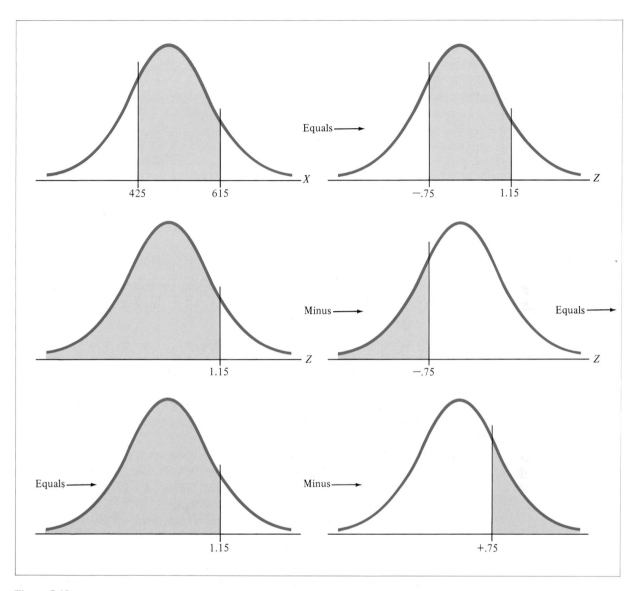

Figure 5.10
Finding the area between two scores with one below and one above the mean.

Checking table B (column B) the area, or proportion, below $Z = 1.15$ is 0.8749. The area *below* $Z = -0.75$ is equivalent to the area *above* $Z = +0.75$. Using table B, that area is found by subtracting from 1.000 (the total area under the curve) the area below $Z = +0.75$. The area below $Z = 0.75$ is 0.7734. Subtracting 0.7734 from 1.000 gives 0.2266, which represents the area below $Z = -0.75$ and above $Z = +0.75$ (fig. 5.11). Subtracting 0.2266 from 0.8749 gives 0.6483, the proportion of cases between $Z = -0.75$ and $Z = 1.15$ and SAT scores 425 and 615. Multiplying 0.6483 by 1200 gives 778 cases between SAT scores 425 and 615.

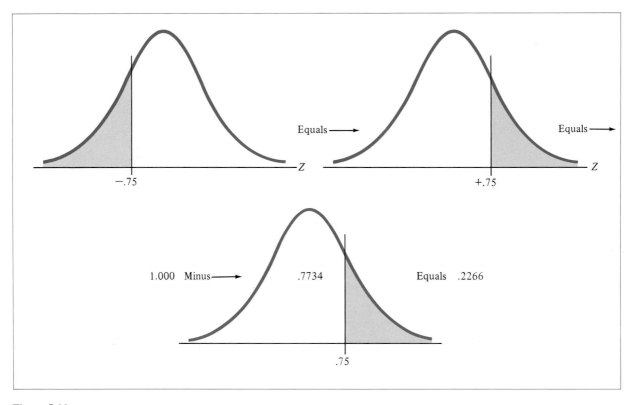

Figure 5.11
Finding areas below negative
Z-score.

In a final example, the value of Z is determined so that $M \pm (Z \times S.D.)$ includes the middle 99% of the cases in a normal distribution. Solving this problem requires using table B in the reverse way, not looking at a Z value to find a proportion, but looking at a proportion to find a Z value.

If the middle 99% is desired, this indicates that there will be one-half of 1% above the Z value and one-half of 1% below the negative Z value at the other end of the distribution (fig. 5.12). The two halves add up to the missing 1%. A Z-score is sought, therefore, that has 0.9950 as the proportion of cases below it. Inspection of table B shows that $Z = 2.57$ has 0.9949 as the proportion below it and $Z = 2.58$ has 0.9951 as the proportion below it. Since 0.9950 is half way between the two tabled proportions, an arbitrary preference for the even value is made. In the normal curve, $M \pm 2.58$ S.D. encompasses the middle 99% of the cases. In the distribution where the mean is 110 and the S.D. is 20, this gives:

$$110 \pm (2.58 \times 20) = 110 \pm 51.6$$
$$= 58.4 \text{ to } 161.6$$

Figure 5.12
Illustration of $Z = 2.58$.

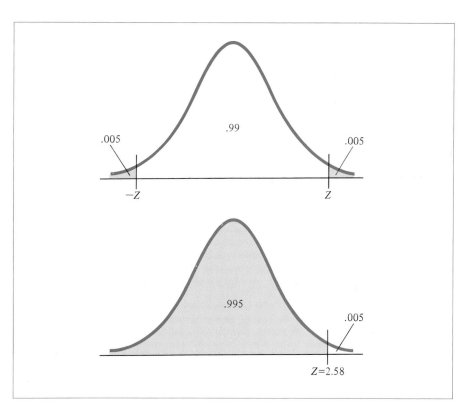

In that sample of 500 cases (observations or scores), then, if the distribution is exactly normal, there are $0.99 \times 500 = 495$ cases between 58.4 and 161.6. Only 5 cases lie outside this interval.

**Finding Normal Curve
Frequencies
Corresponding to
Obtained Frequencies in
a Sample Distribution**

Apply the methods of the last example to determine what the normal curve frequencies are for a distribution with a specified M, S.D., and N. These normal curve frequencies may be compared with the actual obtained frequencies to determine how well the sample distribution approximates the normal distribution in form. The steps are outlined below for finding these normal curve frequencies for the distribution in table 5.3. This is called the "area method" for getting normal curve frequencies.

1. For each class interval, write down the lower limit of the interval to the right of the obtained frequency value, f_o. These values are shown in table 5.3, in the column headed by X_{LL}. The value for the bottom interval is not recorded because it is not used.
2. Convert the lower limit scores to standard scores by the linear transformation $Z = (X - M)/\text{S.D.}$, $(X - 31.3)/9.2$ in this problem. For the interval 50–54 this is $(49.5 - 31.3)/9.2 = 18.2/9.2 = 1.98$. This value is placed in the column headed by Z in table 5.3. Each other Z value is obtained in the same

TABLE 5.3

Computing normal curve frequencies by the area method

X	f_o	X_{LL}	Z	Area $-\infty$ to Z	Interval area	f_e
50–54	1	49.5	1.98	0.9761	0.0239	1.1
45–49	2	44.5	1.44	0.9251	0.0510	2.3
40–44	5	39.5	0.89	0.8133	0.1118	5.0
35–39	9	34.5	0.35	0.6368	0.1765	7.9
30–34	10	29.5	−0.19	0.4247	0.2121	9.5
25–29	8	24.5	−0.74	0.2296	0.1951	8.8
20–24	6	19.5	−1.28	0.1003	0.1293	5.8
15–19	2	14.5	−1.82	0.0344	0.0659	3.0
10–14	1	9.5	−2.37	0.0089	0.0255	1.1
5–9	1				0.0089	0.4
N	45					44.9

$M = 31.3$
S.D. $= 9.2$

way. As a check on the accuracy of these Z values, the difference between adjacent values is checked. They should be $5/9.2 = 0.54$ points apart since they are separated by 1 class interval, which is 5 raw-score points or $5/9.2$ standard-score points. Rounding errors may cause this to be 0.53 or 0.55.

3. From the normal curve table, table B, find the areas from $-\infty$ to Z for each Z-score and place these areas in the next column (table 5.3), headed "Area $-\infty$ to Z."

4. Find the area within each class interval and place it in the column headed by "Interval Area." For the top interval, subtract 0.9761 from 1.00 because all the area above the lower limit of the top interval is included in the top interval. This gives $1.00 - 0.9761 = 0.0239$ as the area for the top interval. For the bottom interval, use the area below the lower limit of the next higher interval since everything below the upper limit of the bottom interval is placed in the bottom interval. Thus, 0.0089 is the area for the bottom interval.

   For the other intervals, the interval area is the difference between the areas from $-\infty$ to Z for adjacent intervals. For the interval 45–49, the area is $0.9761 - 0.9251 = 0.0510$. Since the area below 49.5 is 0.9761 and the area below 44.5 is 0.9251, the area between must be the difference between them, or 0.0510. The remaining areas are computed in a similar manner.

5. The interval areas are multiplied by $N = 45$ to convert areas to frequencies. These normal curve frequencies, often referred to as "theoretical" frequencies, or "expected frequencies," are shown in the last column of table 5.3, headed by "f_e." The failure to have the normal curve frequencies add up to N with the area method is due to rounding error, presuming no computational mistake was made. Typically, one or two of the values is adjusted by 0.1 to 0.2 as necessary to make the total come out to equal N exactly.

In Chapter 9 a statistical test called chi square is described for determining if the differences between the obtained and theoretical frequencies are large enough to reject the hypothesis that the distribution of scores in the population from which the sample was drawn is normal. In the present case, as shown in table 5.3, no normal curve frequency value is different from its corresponding obtained frequency, f_o, by more than 1.1 cases, so this distribution is very close to being normal in form.

| MATHEMATICAL NOTES | The simplest normal curve formula is that for the standard normal curve: |

$$y = \frac{1}{\sqrt{2\pi}} e^{-(Z^2/2)} \tag{5.9}$$

where y = the ordinate, the height of the curve above a point, Z, on the base line,

   π = pi, approximately 3.14159,

   e = the base of the Naperian logarithm system, approximately 2.71828, and

   Z = the point on the standard-score scale to be selected on the base line.

Use Formula 5.9 to find y, the height of the standard normal curve (the ordinate), above any selected point, Z, on the standard-score scale.

To find the height, y, above any given point on the base line, Z, first square Z and divide by 2. Then e is raised to minus this power. If Z is $+2$, then e is raised to the $-(4/2)$ or -2 power. This is the same as $1/e^2$ since raising a number to a negative power is the same as 1 over the number to the positive power. The result of $1/e^2$ is $1/(2.718)^2 = 1/7.388 = 0.1354$. This value must be multiplied by $1/\sqrt{2\pi}$ or $1/\sqrt{2(3.1416)} = 1/\sqrt{6.2832} = 1/2.507 = 0.3989$; thus 0.1354 $\times$ 0.3989 gives 0.05401 as the height of the standard normal curve above the standard-score value of $+2$.

This is a small value, as expected since the curve is very low at $Z = +2.0$. As a standard of comparison, the height of the curve above the point $Z = 0$ is computed since this is the point where the curve is highest.

Squaring $Z = 0$ still gives zero, so e is raised to the minus zero over 2 power, i.e., $e^{-(0/2)}$. This is e^0, but any number to the zero power is equal to 1.0. So 1.0 is multiplied by $1/\sqrt{2\pi}$, which is already computed to be 0.3989. The height, then, of the standard normal curve above the zero point (the mean) on the base line is 0.3989. Note that when computing the value of y for a given value of Z, the value of Z is always squared. This means that y comes out to be the same for a Z value whether or not it has a minus sign on it. The height of the curve (the ordinate) above -3 is the same as the height of the curve above $+3$.

Computing the value of y for other values of Z is more difficult since e must be raised to a power that is a decimal fraction. This is most easily accomplished through the use of logarithms. The process is tedious at best. Fortunately, there are tables available to give the value of y for selected values of Z within the usual normal limits. Most values of Z lie between -3 and $+3$. The normal curve actually runs, however, from $-\infty$ to $+\infty$. If Z is a very large number, positive or negative, e is raised to a large negative power, giving 1 divided by a large number. This makes the value of y, the ordinate, very small when Z is very large. Beyond $+4$, the value of the ordinate is so small that it is ordinarily not even listed in the normal curve tables. The values of the ordinates, or y values, that correspond to the various standard scores are listed in the normal curve table, table B (Column E) in Appendix B. This table is discussed at greater length in a later section of this chapter.

Areas Under the Normal Curve   The ordinate values, or y values, described in the previous section are used to obtain approximations to the areas under the normal curve within given intervals. Suppose, for example, the area under the normal curve from -0.25 S.D. to $+0.25$ S.D. is desired, the mean ± 0.25 S.D. This represents an interval of $(+0.25)$ to (-0.25), or 0.50 on the standard-score scale, one-half point along the base line in figure 5.4. In the last section, it was determined that the height of the standard normal curve above the zero point (the mean) on the base line is 0.3989. If a rectangle of height 0.3989 and base equal to 0.50 standard-score points is superimposed on the standard normal curve as shown in figure 5.13, it is evident that the area of the rectangle is very close to the area under the curve between $Z = -0.25$ to $Z = +0.25$. The area of the rectangle, which equals the base times the height, $0.50 \times 0.3989 = 0.1994$, is just slightly greater than the area under the curve because the ends of the top of the rectangle slightly extend above the curve. The correct area under the curve in this interval is 0.1974, which is 0.002 less than the rectangular approximation to it.

The approximation to the area under the curve by taking the area of a rectangle becomes better as the base of the rectangle becomes smaller. The area obtained is an approximation to the proportion of the cases in the standard normal curve that falls within that interval. When the area under the standard normal curve is found for a particular interval along the base line, that area represents the proportion of the standard scores (Z) that fall in that interval. This applies only for normalized Z-scores (see Problem C), not for linearly transformed Z-scores (Problem 13).

Since the areas under the curve for given intervals represent proportions of cases falling in those intervals, it follows that the total area under the curve between $-\infty$ and $+\infty$ must add up to 1.0. A proportion of 1.0 is necessary to encompass all the cases. By dividing the area under the curve into very narrow strips, using very small intervals along the base line, it is possible to estimate the areas with sufficient accuracy for practical purposes using areas of rectangles. The rectangles have the same base as the narrow strips and heights equal to the ordinates at the center of the strips.

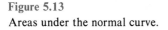

Areas under the normal curve.

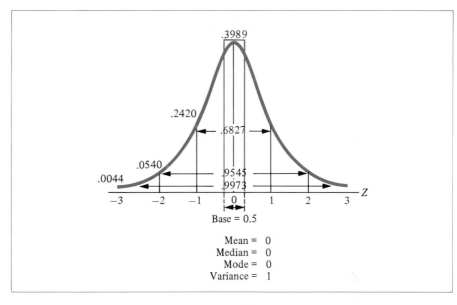

It is impossible to go clear to $-\infty$ and $+\infty$, but if the areas between -4.0 and $+4.0$ are computed, the remaining areas in the tails are quite small. The area between -4.0 and $+4.0$ is subtracted from 1.0 to obtain the area in the two tails, one-half of it below -4.0 and one-half of it above $+4.0$. The area above 4.0 is less than one part in 10,000, or 0.0001.

By adding up the areas in these rectangles with the very small bases, the areas under the standard normal curve between $-\infty$ and any given Z-score is approximated with sufficient accuracy for all practical purposes. Fortunately, these areas have been computed and are available in table B in Appendix B.

Use the previous methods to obtain approximations to the normal curve frequencies corresponding to obtained frequencies in a sample distribution. The midpoint of each class interval is first converted to a Z-score by the formula $Z = (X - M)/\text{S.D.}$ and the y values, or ordinates, are determined from table B. These are the heights of rectangles approximating the areas under the curve for the class intervals. The base of each rectangle is given by $i/\text{S.D.}$ in standard-score units. The base, $i/\text{S.D.}$, is multiplied by the height, y, to obtain the interval area. This area value, a proportion between zero and 1, is multiplied by N to obtain the normal curve frequency estimate for the interval. These frequencies usually add up to less than N because there is some area in the normal curve beyond the ends of the extreme intervals. The difference between the sum of the normal frequencies and N is usually divided equally and added to the end interval normal curve frequencies so the sum of all normal curve frequencies equal N, the total number of cases and the sum of the actual obtained frequencies. Normal curve frequencies computed by this method, called the ordinate method, are shown in table 5.4 together with the corresponding area method normal curve frequencies for the same data.

TABLE 5.4

Normal curve frequencies compared with actual frequencies*

X-score intervals	Actual frequencies	Ordinate method frequencies	Area method frequencies
50–54	1	0.8	1.1
45–49	2	2.3	2.3
40–44	5	4.9	5.0
35–39	9	8.0	7.9
30–34	10	9.7	9.5
25–29	8	8.7	8.8
20–24	6	5.9	5.8
15–19	2	2.9	3.0
10–14	1	1.1	1.1
5–9	1	0.3	0.4
N	45	44.6	44.9

*The sum of the normal curve frequencies computed by the ordinate method do not equal 45 here, because the areas above 54.5 and below 4.5 have been ignored. To include these areas, add 0.2 to the top interval frequency and 0.2 to the bottom interval frequency, making them 1.0 and 0.5, respectively. Then, the frequencies by the ordinate method will add up to N exactly. One of the frequencies for the area method may be adjusted arbitrarily by adding 0.1 so that these frequencies will also add up exactly to equal 45.0.

PROBLEM C

Convert Raw Scores to Normalized Standard Scores

Conversion of raw scores to standard scores by means of a linear transformation, Formula 5.3, does not alter the general shape of the distribution of scores. If the distribution is badly skewed in the negative direction before the conversion, it still has that character after the transformation. If it is suspected that the distribution of scores is normal in form, or bell-shaped, it might be desirable to carry out a transformation of raw scores to standard scores that are normalized. The distribution of the transformed scores will be normal or bell-shaped regardless of the shape of the raw-score distribution. This type of transformation is preferred where there is reason to believe that irregularities in the raw-score distribution reflect inadequacies in the measuring instrument rather than fundamental properties of the underlying trait that the instrument was designed to assess. A transformation that converts skewed raw scores to normalized standard scores is a nonlinear transformation. It is not possible, in such a case, to find values of b and a in Formula 5.2 that convert the raw scores to their corresponding normalized standard scores. This nonlinear transformation is effected using the normal curve.

Converting Raw Scores to Normalized Standard Scores

The areas from $-\infty$ to Z in the normal curve (see table B, Appendix B) give the proportions of cases under the normal curve that correspond to the Z-scores themselves. By locating a specific standard score or Z-score in the table and taking the corresponding area from $-\infty$ to Z, the proportion of cases in the normal curve below that Z-score is immediately known. If this value is multiplied by 100 and rounded off to the nearest whole number from 1 to 99, the centile rank for that Z-score is obtained. With little effort, the centile rank can be obtained for every tabled value of Z in the normal curve table. This process can also be reversed. Given a certain centile rank, it is possible to use the areas from $-\infty$ to Z to find the Z-score corresponding to that centile mark. However, some interpolation is required since the areas from $-\infty$ to Z do not fall on whole numbers in the normal curve table shown in table B, Appendix B.

The basic principle for converting raw scores to normalized standard scores is to find the Z-score that has the same centile rank as the raw score. This is the normalized Z-score corresponding to that raw score. To accomplish this, find the proportion of cases in the raw-score distribution falling below the raw score that is to be converted to a normalized Z-score. The proportion is usually computed to three decimal places, four if greater precision is needed. Once the proportion of cases below the raw score is obtained, this value is used to enter the normal curve table (table B, Appendix B). This value is treated as an area from $-\infty$ to Z and by interpolation the value of Z is found that would correspond to this proportion. If an approximation is sufficient, the Z value for that area closest to the proportion is taken as the normalized standard score corresponding to the raw score that was converted. The interpolation process, by which a more accurate Z value is obtained, is explained in the following example.

Example   As an illustration of the procedure, a raw score of 36 in the frequency distribution shown in table 5.1 (see p. 94) is converted to a normalized standard score, Z. The first step is to find the number of cases in this distribution falling below 36. The computations for this are as follows:

$$\text{Number of cases below } 36.0 = 28 + \frac{36.0 - 34.5}{5} \times 9$$
$$= 28 + (0.3)\,(9)$$
$$= 28 + 2.7 = 30.7$$

There are 28 cases below the lower limit of the interval containing the score of 36. To this add the proportion of the 9 cases in the interval that fall between 34.5, the lower limit of the interval, and 36.0. Divide 30.7 by 45, the number of scores in the distribution, to obtain the proportion of cases in the distribution below 36.0. This value is 0.682.

The proportion of cases below 36.0, 0.682, is used to enter the normal curve table (table B, Appendix B) to find the corresponding normalized standard score, Z. This requires interpolation between tabled values as follows:

Standard score, Z	Area from $-\infty$ to Z		
0.47	0.6808		
	0.6820 $\}$ 0.0012	$\}$ 0.0033	
0.48	0.6844		

The proportion 0.6808 is tabled and corresponds to a Z-score of 0.47; the proportion 0.6844 corresponds to a Z-score of 0.48. These two adjacent values were picked from the table because the proportion of cases below the score of 36 is 0.682, which lies between 0.6808 and 0.6844. This means that the Z-score corresponding to the proportion 0.6820 must lie between 0.47 and 0.48. The unknown Z value is determined by finding that Z lies relatively at the same point between 0.47 and 0.48 as 0.6820 does between 0.6808 and 0.6844. The process used to find this value is called linear interpolation. The distance between 0.6808 and 0.6820 is 0.0012. The distance between 0.6808 and 0.6844 is 0.0036. This means 0.6820 is 0.0012/0.0036 or one-third of the way between 0.6808 and 0.6844. Therefore, Z is one-third of the way between 0.47 and 0.48. The distance between 0.47 and 0.48 is 0.01. Multiplying 0.01 by one-third equals 0.01/3 or 0.0033, which rounds to 0.003 for three decimal places. Adding this to 0.47 gives 0.473 as the normalized standard score, Z, corresponding to the raw score of 36 in the distribution shown in table 5.1.

If this value is rounded to two decimal places, it becomes 0.47. If two decimal place accuracy is sufficient for the Z-score, as it usually is, it is necessary only to find the tabled value in the area from $-\infty$ to the Z column, which is nearest to the look-up value and use the corresponding Z-score without interpolation. Three decimal place accuracy is usually called for only if the Z-score is to be further transformed to some other score scale.

Formula 5.9 is the formula for the standard normal curve where the mean equals zero, the standard deviation equals 1, and the area under the curve equals 1. Actually the normal curve is not a single curve but rather a family of curves. As each value of the area under the curve, the mean, and the standard deviation changes, the particular normal curve that results changes. All of these are bell-shaped curves. They vary in height and width as the values of these constants, or *parameters,* change.

Consider a hypothetical population distribution that is normal in form. The parameters of this population are designated as follows:

σ = the standard deviation of the population (not S.D., the standard deviation of a sample from the population),

μ = the mean of the population (not M, the mean of a sample from the population), and

N = the number of cases in the population. (In a theoretical population this value is normally infinitely large. Practically, populations are finite.)

The μ and σ above are the lowercase Greek letters mu and sigma, which are commonly used to represent the mean and standard deviation of a population. It is convenient to make a distinction between population parameters and sample statistics. For this reason the mean and standard deviation of a sample are labeled in this text as M and S.D., while population values are labeled as μ and σ.

Using these symbols for the population parameters, the general equation for the normal distribution family of curves is given by Formula 5.10 as follows:

$$Y = \frac{N}{\sigma \sqrt{2\pi}} \, e^{-[(X - \mu)^2/2\sigma^2]} \qquad (5.10)$$

where Y = the height of the curve,

N = the number of cases in the distribution and total area under the curve,

π = pi, approximately 3.14159,

e = the base of the Naperian logarithm system, approximately 2.71828,

X = a raw score in the distribution,

μ = mu, the mean of the distribution, and

σ = sigma, the standard deviation of the distribution.

When $N = 1$, $\mu = 0$, and $\sigma = 1$ in Formula 5.10, the equation reduces to the standard normal curve shown in Formula 5.9. This requires exchanging Z for X in the resulting equation. If the X-score distribution consists of X-scores with a mean of zero, a standard deviation of 1, and a normal distribution, the X-scores are normalized standard scores or Z-scores.

Exercises

5-1.   Find the centile rank for the raw score of 89 in the frequency distribution shown in table 3.1 (see page 47).

    a. Find the class interval containing the score.

    b. Find the number of scores, ns, below this class interval.

    c. Find the number of scores within the class interval containing the score, *n.*

    d. Find the lower limit, *LL,* of the class interval containing the score.

    e. Compute the class interval size, *i.*

    f. Compute $\text{ns} + \dfrac{\text{raw score} - LL}{i} \times n$

    g. Divide the result of f by the total number of scores, round to two decimal places, and multiply by 100.

5–2.   Find the centile rank for the raw score of 44 in the frequency distribution shown in table 3.2 (see page 47).

    a. Find the class interval containing the score.

    b. Find the number of scores, ns, below this class interval.

    c. Find the number of scores within the class interval containing the score, *n*.

    d. Find the lower limit, *LL*, of the class interval containing the score.

    e. Compute the class interval size, *i*.

    f. Compute $\text{ns} + \dfrac{\text{raw score} - LL}{i} \times n$

    g. Divide the result of f by the total number of scores, round to two decimal places, and multiply by 100.

5–3.   Find the centile rank for the raw score of 42 in the frequency distribution shown in table 3.5 (see page 64).

    a. Find the class interval containing the score.

    b. Find the number of scores, ns, below this class interval.

    c. Find the number of scores within the class interval containing the score, *n*.

    d. Find the lower limit, *LL*, of the class interval containing the score.

    e. Compute the class interval size, *i*.

    f. Compute $\text{ns} + \dfrac{\text{raw score} - LL}{i} \times n$

    g. Divide the result of f by the total number of scores, round to two decimal places, and multiply by 100.

5–4.   Find the centile rank for the raw score of 8 in the frequency distribution shown in table 3.6 (see page 64).

    a. Find the class interval containing the score.

    b. Find the number of scores, ns, below this class interval.

    c. Find the number of scores within the class interval containing the score, *n*.

    d. Find the lower limit, *LL*, of the class interval containing the score.

    e. Compute the class interval size, *i*.

    f. Compute $\text{ns} + \dfrac{\text{raw score} - LL}{i} \times n$

    g. Divide the result of f by the total number of scores, round to two decimal places, and multiply by 100.

5–5.   Find the raw score that corresponds to the centile rank of 68 in the frequency distribution shown in table 3.5 (see page 64).

    a. Multiply the centile rank by the total number of scores and divide by 100, *X*.

    b. Find the class interval where the total cumulative scores found in (a) lies (use cumulative frequency).

    c. Determine the lower limit, *LL*, of the class interval found in (b).

    d. Determine the number of scores, ns, below the class interval found in (b).

    e. Determine the number of scores, n, in the class interval found in (b).

    f. Determine the class interval size, i.

    g. Compute $LL + \dfrac{(X - ns)\,(i)}{n}$

    h. Round to two decimal places.

5–6. Find the raw score that corresponds to the centile rank of 16 in the frequency distribution shown in table 3.6 (see page 64).

    a. Multiply the centile rank by the total number of scores and divide by 100, X.

    b. Find the class interval where the total cumulative scores found in (a) lies (use cumulative frequency).

    c. Determine the lower limit, LL, of the class interval found in (b).

    d. Determine the number of scores, ns, below the class interval found in (b).

    e. Determine the number of scores, n, in the class interval found in (b).

    f. Determine the class interval size, i.

    g. Compute $LL + \dfrac{(X - ns)\,(i)}{n}$

    h. Round to two decimal places.

5–7. Given a distribution with a mean of 26.1 and a standard deviation of 5.7, convert a raw score of 21 to a standard score, using a linear transformation. Compute Z to two decimal places. Use Formula 5.3:

$$Z = \frac{X - M}{\text{S.D.}}$$

5–8. Given a distribution with a mean of 99.9 and a standard deviation of 15.1, find the standard score corresponding to a raw score of 121, using a linear transformation. Compute Z to two decimal places. Use Formula 5.3.

5–9. Given a distribution with a mean of 69.1 and a standard deviation of 18.9, find the standard score corresponding to a raw score of 52, using a linear transformation. Compute the score to two decimal places. Use Formula 5.3.

5–10. Given a distribution with a mean of 10.2 and a standard deviation of 4.0, find the standard score corresponding to a raw score of 9, using a linear transformation. Compute the score to two decimal places. Use Formula 5.3.

5–11. The McCall T-Score distribution has a mean of 50 and a standard deviation of 10. Convert a standard score of -1.2 to this scale. Use Formula 5.8.

5–12. Convert a standard score of 2.5 to a McCall T-Score. Use Formula 5.8.

5–13.   Convert a standard score of -3.12 to a scale with $M = 500$ and S.D. $= 100$. Use Formula 5.8.

5–14.   Convert a standard score of $+1.68$ to a scale with $M = 500$ and S.D. $= 100$. Use Formula 5.8.

5–15.   Using the area method, find the normal curve frequencies for the distribution below ($M = 69.1$, S.D. $= 19.9$). Hint: fill in the table (the first line has been completed), and read pages 115 to 117.

Frequency distribution

X	f	cf	X_{LL}	Z	Area $-\infty$ to Z	Interval area	f_e
110–119	3	161	109.5	2.03	0.9788	0.0212	3.41
100–109	8	158					
90–99	12	150					
80–89	20	138					
70–79	36	118					
60–69	33	82					
50–59	24	49					
40–49	16	25					
30–39	7	9					
20–29	2	2					
N = 161							

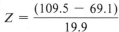

$$Z = \frac{(109.5 - 69.1)}{19.9}$$

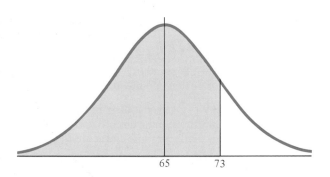

65   73

5–16.   Given a distribution with $M = 65$, S.D. $= 20$, and $N = 200$, find the number of cases that fall below a score of 73 if the distribution is normal.

    a. Convert 73 to a Z-score.

    b. Find the area under the curve for $Z = -\infty$ to Z found in (a) (use table B).

    c. Multiply the area times N.

5-17.   Given a distribution with $M = 48$, S.D. $= 10$, and $N = 300$, find the number of cases that fall below a score of 44 if the distribution is normal.
   a. Convert 44 to a Z-score.
   b. Find the absolute value of the Z-score.
   c. Find the area under the curve for $Z = -\infty$ to Z found in (b) (use table B).
   d. Subtract the area found in (c) from 1.00.
   e. Multiply the result of (d) by N.

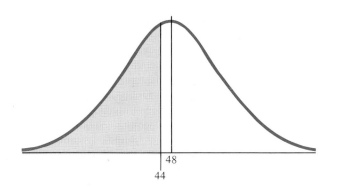

5-18.   Given a sample distribution with $M = 80$, S.D. $= 15$, and $N = 150$, find the number of cases that are above 98 if the distribution is normal.
   a. Convert 98 to a Z-score.
   b. Find the absolute value of the Z-score.
   c. Find the area under the curve for $Z = -\infty$ to Z found in (b) (use table B).
   d. Subtract the area found in (c) from 1.00.
   e. Multiply the result of (d) by N.

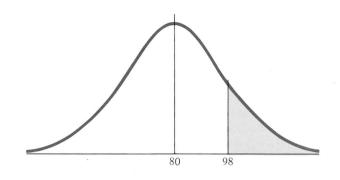

5–19.   Given a sample distribution with $M = 100$ and S.D. $= 16$, find the percentage of cases that are below 132 if the distribution is normal.

 a. Convert 132 to a Z-score.
 b. Find the area under the curve for $Z = -\infty$ to Z found in (a).
 c. Multiply the number found in (b) by 100.

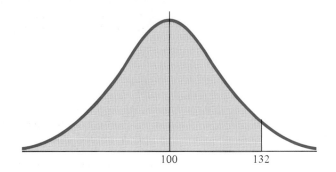

5–20.   Using the ordinate method, find the normal curve frequencies for the distribution with a mean of 69.1 and a standard deviation of 19.9 (see table 3.5, p. 64). Make the f_e values add up to $N = 161$ by slight adjustments to the computed values (see Mathematical Notes for Problem 15).

Frequency distribution

X	f	cf	MdPt	Z_M	y	i/S.D.	Y(i/S.D.)	f_e $N \cdot Y(i/\text{S.D.})$
110–119	3	161	115	2.31	0.0277	10/19.9	0.014	2.25
100–109	8	158						
90–99	12	150						
80–89	20	138						
70–79	36	118						
60–69	33	82						
50–59	24	49						
40–49	16	25						
30–39	7	9						
20–29	2	2						

 a. Find the midpoint for each class interval.
 b. Using M and S.D., convert each midpoint to a Z-score.
 c. Using table B of Appendix B, find the y value for each Z-score.
 d. Divide the class interval size, i, by S.D.
 e. Multiply the results in (d) by the ys found in (c).
 f. Multiply the results found in (e) by the total number of scores.

5–21.   Given the distribution in table 3.5, convert a raw score of 52 to a normalized standard score. Report the score to two decimal places.

a. Find the number of scores falling below the raw score, n.
b. Find the class interval where the raw score falls.
c. Compute the lower limit, LL, of the class interval.
d. Find the class interval size, i.
e. Find the number of scores, ns, falling in that class interval.
f. Use the following formula:

$$n + \frac{\text{raw score} - LL}{i} \times \text{ns}$$

g. Divide the result of (f) by the total number of scores.
h. Using table B of Appendix B find the Z-score corresponding to the proportion found in (g).

5–22.   Given the distribution in table 3.6, convert a raw score of 9 to a normalized standard score. Report the score to two decimal places.

a. Find the number of scores falling below the raw score, n.
b. Find the class interval where the raw score falls.
c. Compute the lower limit, LL, of the class interval.
d. Find the class interval size, i.
e. Find the number of scores, ns, falling in that class interval.
f. Use the following formula:

$$n + \frac{\text{raw score} - LL}{i} \times \text{ns}$$

g. Divide the result of (f) by the total number of scores.
h. Using table B of Appendix B find the Z-score corresponding to the proportion found in (g).

5–23.   Given the distribution in table 3.5, convert a raw score of 86 to a normalized standard score. Report the score to two decimal places.

a. Find the number of scores falling below the raw score, n.
b. Find the class interval where the raw score falls.
c. Compute the lower limit, LL, of the class interval.
d. Find the class interval size, i.
e. Find the number of scores, ns, falling in that class interval.
f. Use the following formula:

$$n + \frac{\text{raw score} - LL}{i} \times \text{ns}$$

g. Divide the result of (f) by the total number of scores.
h. Using table B of Appendix B find the Z-score corresponding to the proportion found in (g).

5–24.   Given the distribution in table 3.6, convert a raw score of 18 into a normalized standard score. Report the score to two decimal places.

    a. Find the number of scores falling below the raw score, n.
    b. Find the class interval where the raw score falls.
    c. Compute the lower limit, LL, of the class interval.
    d. Find the class interval size, i.
    e. Find the number of scores, ns, falling in that class interval.
    f. Use the following formula:

$$n + \frac{\text{raw score} - LL}{i} \times \text{ns}$$

    g. Divide the result of (f) by the total number of scores.
    h. Using table B of Appendix B find the Z-score corresponding to the proportion found in (g).

5–25.   A normal distribution of AGCT scores is given. Find the percentage of cases that are between a AGCT score of 89 and 124.

    a. Convert 89 and 124 to Z-scores.
    b. Find the area under the curve from $-\infty$ to $|Z_{89}|$.
    c. Subtract the area found in (b) from 1.000.
    d. Find the area under the curve from $-\infty$ to Z_{124}.
    e. Subtract the area found in (c) from the area found in (d).
    f. Multiply (e) by 100.

5–26.   Given a normal distribution of scores from the Social Conformity versus Rebelliousness Scale of the Comrey Personality Scales, find the proportion of scores that are between 53 and 61.

    a. Convert the scores 53 and 62 to Z-scores.
    b. Find the area under the curve from $-\infty$ to Z_{53}.
    c. Find the area under the curve from $-\infty$ to Z_{61}.
    d. Subtract the area found in (b) from the area found in (c).

Chapter 6

Regression Analysis and Correlation

I
n any scientific field, a very important activity is the study of the relationships between major variables in that domain. In the physical sciences, it is
sometimes possible to express these relationships in terms of rather exact
mathematical (deterministic) equations. For example, the distance that a body falls
in a vacuum in a given time period is expressed by the following equation:

$$S = 1/2 \ gt^2 \tag{6.1}$$

where   S = the distance fallen in feet,

g = a gravitational constant = 32 ft./sec.², and

t = the elapsed time in seconds.

abscissa The *x*-axis or
horizontal axis in a 2-
dimensional plot.

ordinate The *y*-axis or
vertical axis in a 2-dimensional
plot.

The relationship expressed in Formula 6.1 is represented graphically in figure 6.1.
By selecting a time (*t*) along the base line (**abscissa** or *X*-axis) and drawing a vertical line from that point up to where it intersects the curve and then drawing a
horizontal line over the **ordinate** (*Y*-axis), the number of feet that a body falls in a
vacuum in *t* seconds can be read off. Actual measured values of *t* and S are expected
to give points falling directly on this line, within the limits of measuring error and
the approximate nature of Formula 6.1. The relationship expressed by Formula 6.1
and graphically represented in figure 6.1 is an exact mathematical (deterministic)
relationship relating the variables S and *t* (g is a constant). The measured value of
S is an exact function of *t,* at least within very close limits under proper measurement conditions. S does not depend on other variables under the specified conditions
where this equation holds. Examples of deterministic mathematical models are found
in physics, chemistry, and engineering books. These exact mathematical models are
used in explaining the relationship between physical phenomena.

In the social and behavioral sciences models that describe relationships among
variables are rarely expressible as exact mathematical relationships. If two variables
are related at all in the social sciences, the nature of the relationship is apt to be
statistical or probabilistic rather than *mathematical* or deterministic. While exact
mathematical models make no allowance for error or random factors, the statistical
model contains one or more random elements, where each element has a specific
probability distribution.

This is illustrated in figure 6.2, which graphically represents the relationship
between Intelligence Quotients (I.Q.s) and scores on a reading comprehension test
for a fictitious sample of high school students. Each point in figure 6.2 represents
the pair of scores for one student. The students' I.Q. scores are located on the *X*-
axis (abscissa) and their Reading Comprehension scores are located on the *Y*-axis
(ordinate). The intersection of lines through these points parallel to the coordinated
axes locates the point on the graph for one particular student.

Inspection of the points readily reveals that it is impossible to draw any smooth,
regular curve that will go through or even near all these points. The best that can
be done here is to draw a straight line that comes as close to as many of the points
as possible. This is done by placing a straight edge on the graph and moving it around

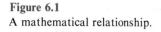

Figure 6.1
A mathematical relationship.

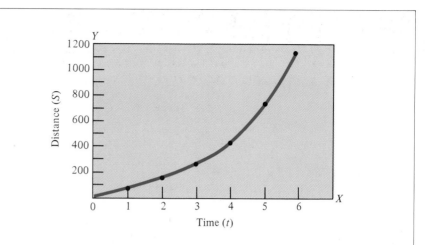

Figure 6.2
A statistical relationship.

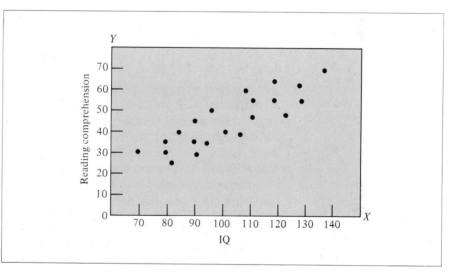

until the edge of the ruler seems to pass through the points, thereby providing a line that "best fits" the data. The relationship between these variables is statistical rather than mathematical in character because the points in the graph scatter about the line on either side of it at some distance.

Thus, a person who has an I.Q. of 95 might have a Reading Comprehension score somewhere between 30 and 50, according to the appearance of this graph. It will not necessarily be 40, as would be expected if all points fell right on a line running through the middle of these points. Figure 6.3 shows the same points with such a "best-fitting" line running through them. The points scatter about the line because the relationship between I.Q. scores and Reading Comprehension scores is an imperfect one. One student with a modest I.Q. might have had intensive training

Figure 6.3
A regression line, or line-of-
best-fit.

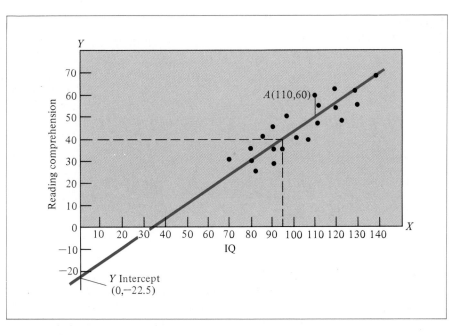

and experience in reading difficult material. Another person of moderate I.Q. might have had less than the usual exposure to reading. Other persons of very high I.Q. might resent reading because their parents had pushed them too hard, developing resistance on their part that decreased performance. These and other factors might be at work to keep the relationship between these two variables from approaching a mathematical type of relationship in precision.

The relationship shown in figures 6.2 and 6.3 is still substantial, since people with high I.Q. scores generally have high Reading Comprehension scores. It is just not possible to determine *exactly* what the Reading Comprehension score will be, knowing only the student's I.Q. Most of the relationships between variables encountered in the social sciences will not be as strong as the one shown in figures 6.2 and 6.3.

Another difference to be noted between the relationships shown in figures 6.1 and 6.3 is that figure 6.1 shows a *curvilinear* relationship and figure 6.3 shows a *linear* relationship. The best-fitting line through the points in figure 6.1 is a curve, whereas the best-fitting line through the points in figure 6.3 is a straight line. Of course it is possible to snake a very curved line through the points in figure 6.3 so that all the points are contacted, however, not only is this inadmissible, but it would look like a plate of spaghetti! It is not possible to create a mathematical equation for a straight line to fit all the points in figure 6.3. Most of the relationships that the social sciences deal with involve linear relationships. A straight line fits the points

Figure 6.4
Typical error in fitting points
with a line.

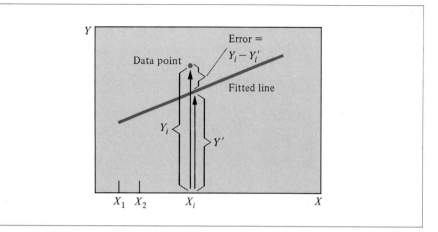

relating the two variables better than any other regular smooth curve. For this reason, only methods of expressing the degree of linear relationship are considered in this text. Methods for expressing the degree of curvilinear relationship are infrequently used in the social sciences in comparison with linear methods.

line-of-best-fit   A straight line fitted to the paired data either visually or mathematically to best represent the data linearly.

The visual approach to fitting the **line-of-best-fit** is not necessarily the most desirable method, largely because it requires an enormous amount of work. The researcher must plot each data point and then "guesstimate" where the line falls. For a large data set it is very time-consuming to plot all the points before visually fitting the line. If the data are quite dispersed, but a relationship exists between the two variables, the placement of the line might vary from researcher to researcher. A more precise and faster method is to define a mathematical procedure that captures the same effect of plotting the data and visually fitting the line. This mathematical procedure acts as the researcher's eyes in minimizing the differences (error or deviation) of the points from the prospective line. This minimization procedure is shown in figure 6.4 where, for a given data point, there is defined a value for X_i and Y_i. The Y_i's are the observed Y values. Seldom do these Y_i's fall on the fitted line. The Y values falling on the line are denoted as Y_i'. The difference between Y_i and Y_i' is called an error or deviation.

method of least squares   The criterion that measures which straight line best fits the given data; refers to the minimum possible value of the sum of the squared residuals of predicted y-values from observed y-values.

The goal of the desired mathematical procedure is to minimize the error or deviation. A number of criteria are proposed to accomplish this minimization, including minimizing the sum of errors, the sum of the absolute value of the errors, and the sum of the squared errors. Under certain circumstances, one criterion works better than others, but the minimization of the sum of squared errors is generally the superior method. This mathematical method, which defines a line such that the sum of the squared errors $\Sigma(Y_i - Y_i')$ is a minimum, is called the **method of least squares.** Reference materials at the end of the chapter give the derivation of the method of least squares and discuss why it is superior to other methods.

EXPRESSING LINEAR RELATIONSHIPS QUANTITATIVELY

The general equation for a straight line, from high school algebra, is shown in Formula 6.2:

$$Y = bX + a \qquad (6.2)$$

where Y = a raw score in the Y variable,

X = a raw score in the X variable,

b = the slope of the line-of-best-fit, and

a = the Y-intercept.

Y-intercept The point where the regression line (line-of-best-fit) intersects the *y*-axis. Usually considered as the additive constant in the regression equation.

By reference to the plot of Reading Comprehension scores against I.Q. scores in figure 6.3 it is apparent that the approximate line of best fit, as drawn by eye, passes through the Y-axis at about the point $(0,-22.5)$, where zero is the X-axis coordinate of the point and -22.5 is the Y-axis coordinate. The value -22.5 is the **Y-intercept** in this plot. This value is substituted for the constant a in Formula 6.2 in obtaining the equation of the line shown in figure 6.3.

slope The coefficient of *x* in the regression equation; representing the degree of change in the dependent variable as a function of the independent variable

The value b in Formula 6.2 represents the **slope** of the line. In figure 6.3 the slope of the line is determined by dropping a perpendicular from any point on the line to the X-axis and taking the ratio of this height to the distance from the base of this perpendicular to the point where the line crosses the X-axis (the X-intercept). For example, a triangle is formed by the line-of-best-fit to the point $(95,0)$ on the X-axis. The slope of the line is the height of this triangle divided by the base. This approximately equals 0.658.

Any other perpendicular dropped from the line-of-best-fit to the X-axis also forms a triangle for which the height to base ratio approximately is 0.658, the slope of the line-of-best-fit. The equation for the line-of-best-fit in figure 6.3 is:

$$Y' = 0.658X - 22.5 \qquad (6.3)$$

Substituting $X = 95$ in Formula 6.3 gives

$$Y' = 0.658\,(95) - 22.5 = 62.5 - 22.5 = 40$$

The point $(95,40)$ indeed falls on the line-of-best-fit. Any number substituted for X in Formula 6.3 produces a value of Y such that the point (X,Y) falls on the line-of-best-fit. In algebra, this equation is determined mathematically without error. In using this equation in statistics, there is an error term, e in a theoretical equation, added to take care of the unmeasurable factors, $y = ax + b + e$.

dependent variable The variable whose value is determined when the value of the independent variable(s) is(are) known. Usually the variable of the vertical axis, or *y*-axis.

independent variable The variable over which the experimenter has control, normally the variable of the horizontal axis, or *x*-axis.

The use of the regression line is more than just a description of the relationship between two variables. Through the proper development and use of the equation for the regression line, a value for one of the variables can be predicted (estimated) by obtaining a value for the other variable. When used in this fashion, the regression equation is a probabilistic prediction equation. The variable whose value is being predicted is referred to as the **dependent variable** (Y), and the variable whose value is known and used to predict the value of the dependent variable is called the **independent variable** (X).

Figure 6.5
Representation of a and b in
the equation $Y' = a + bX$.

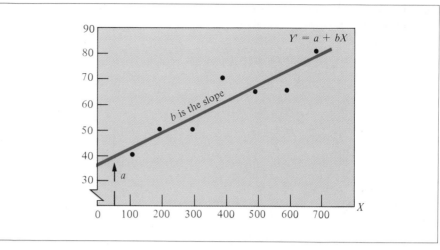

For example, predicting first-year college GPA (Y') using only SAT scores (X)
of high school students is a typical application. In this example, college GPA is the
dependent variable and the SAT score is the independent variable. Figure 6.5 yields
another representation showing the physical meaning of a and b in the equation
$Y' = a + bX$.

The Regression Line for Deviation Scores

regression line   The straight
line that best fits the given
data.

The equation of the line-of-best-fit, which will also be called a **regression line,** becomes simplified when scores are expressed in terms of deviation scores instead of
raw scores:

$$y' = bx \qquad (6.4)$$

where y' = a predicted deviation score in y ($Y' - M_Y$),

b = the slope of the regression line, and

x = the deviation score in x ($X - M_X$)

Formula 6.2 simplifies to Formula 6.4 for deviation scores because the mean of deviation scores is zero for both variables and the regression line, or line-of-best-fit,
passes through the origin in every case, the point (0,0) (fig. 6.4). The regression line
passes through the origin because the point on the Y-axis most appropriately associated with the mean point on the X-axis is the mean of the Y-scores, zero. In
other words, if it is known that one person's I.Q. is at the mean, giving them a
deviation score of zero, the best prediction of their score on Reading Comprehension
will be the mean of all the Reading Comprehension scores, a reading comprehension
deviation score of zero. If the regression line, or line-of-best-fit, passes through the
origin, it follows that the Y-intercept will be zero. Hence, Formula 6.2 reduces to
Formula 6.4 for plots involving (regression lines) deviation scores.

Figure 6.6
A deviation-score regression plot.

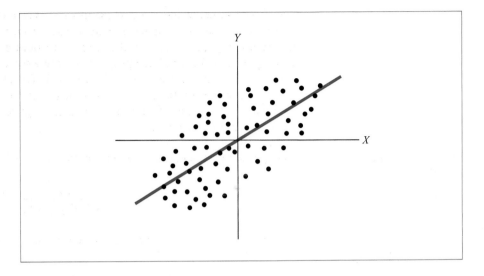

It is possible to prepare a graph of the points showing the relationship between any two variables of interest, draw a best-fitting straight line by eye, and then determine from the placement of the line what the slope and Y-intercept values are. This makes it possible to obtain approximate equations, such as that in Formula 6.3, relating raw scores in Y to raw scores in X. By using deviation scores to make the plots instead of raw scores, only the slope needs to be determined and substituted for b in Formula 6.4.

Determining regression lines by visual approximation, however, is seldom carried out in practice. It is demonstrated in the Mathematical Notes section on pages 145–47 that a least-squares determination of the value of b in Formula 6.4 leads to the following equation for the regression line for predicting deviation scores in Y given the deviation scores in X:

$$y' = r \cdot \frac{\text{S.D.}_y}{\text{S.D.}_x} \cdot x \tag{6.5}$$

where $y' =$ a predicted deviation score in Y,

$x =$ a deviation score in X,

$r =$ the correlation coefficient relating Y- and X-scores (see page 139 for definition),

$\text{S.D.}_y =$ the standard deviation of the Y-scores, and

$\text{S.D.}_x =$ the standard deviation of the X-scores.

A prime (') is placed on the y at the left in Formula 6.5 to designate that it is a *predicted* score rather than an actual score. Some researchers use the notation "hat" (ˆ) to designate predicted scores, so $\hat{Y}$ is an alternative notation for Y'. If the relationship between y and x was mathematically exact, rather than statistical, there would be only one value of y associated with a given value of x. With a statistical relationship, however, where the points scatter about the line of best fit rather than falling directly on it, there might be several values of y corresponding to a given value of x. That is, for all the people who have a score in x of 10, the values might vary considerably, such as $y = -1, 2, 4, 5, 8$. The obtained value of y, predicted by y', will ordinarily be near the mean of the actual y values for that particular value of x. The values of (x,y') are all on the regression line. The values of (x,y) scatter over the entire plot.

The Correlation Coefficient

The correlation coefficient, r, is defined by the following formula:

$$r = \frac{\Sigma xy}{N(\text{S.D.}_x)(\text{S.D.}_y)} \tag{6.6}$$

where

r = the correlation coefficient, which varies between -1.0 and $+1.0$,

x = a deviation score in X,

y = a deviation score in Y,

N = the number of cases,

S.D._x = the standard deviation of the X-scores, and

S.D._y = the standard deviation of the Y-scores.

The proof of this formula is given in the Mathematical Notes section on pages 149–50. It states that the correlation coefficient might be computed as follows:

1. Multiply each person's deviation score in X by their deviation score in Y. Do this for each person in the sample.
2. Add these products for all N cases to get the numerator of Formula 6.6.
3. Multiply N, the number of cases, by the product of the two standard deviations for variables X and Y (standard deviations for raw scores and deviation scores are identical) to obtain the denominator of Formula 6.6.
4. Divide the numerator from Step 2 by the denominator from Step 3 to obtain r.

correlation coefficient   A statistic or parameter that measures the degree of correlation.

positive correlation   A relationship in which two variables increase or decrease together.

Once the correlation coefficient is determined, it is multiplied by the ratio of the standard deviation of the Y-scores to the standard deviation of the X-scores, as called for in Formula 6.5, to produce the equation relating predicted deviation scores in Y (y'-scores), to actual deviation scores in X.

The **correlation coefficient** gives a numerical index of the degree of linear relationship or association between two variables. For example, if $r = +1.0$, there is a perfect positive relationship, or a **positive correlation.** This is a mathematical type relationship in which all the observed points fall right on the regression line rather than scattering on either side of the line. The highest score in Y is held by the same person who has the highest score in X. The lowest score in X belongs to the person who also has the lowest score in Y.

negative correlation A relationship in which one variable decreases when the other variable increases.

If $r = -1.0$, the relationship is still a perfect mathematical relationship, but is a **negative correlation.** That is, the highest score in Y is associated with the lowest score in X; the lowest score in Y belongs to the person who has the highest score in X. Scatter plots for several types of correlation situations are shown in figure 6.7. The more closely the points are clustered about the line of best fit, the higher the correlation. The larger the standard deviation in X, relative to Y, the more the scatter plot is stretched in the horizontal direction. The larger the standard deviation in Y, relative to X, the more the scatter plot is stretched in the vertical direction. When the correlation is zero, there is no apparent linear relationship between the two variables. The graphs in figure 6.5 are for deviation scores, since the origin is at the center of every plot.

When the standard deviations of the two variables are equal, Formula 6.5 reduces to $y' = r \cdot x$. Since standard scores all have $M = 0$ and S.D. $= 1$, Formula 6.5 for standard scores reduces to:

$$Z'_y = r \cdot Z_x \tag{6.7}$$

where $Z'_y =$ a predicted standard score in Y,

$Z_x =$ a standard score in X, and

$r =$ the correlation between Y and X.

Formula 6.7 shows most clearly how the correlation coefficient expresses the relationship between Y and X. For example, if the standard score in X is increased 1 point, the increase in the predicted standard score of Y is r points. If the correlation coefficient is 1.0, perfect, then an increase of 1 point in the standard score in X leads to a 1 point increase in the predicted standard score in Y. It is also clear that for standard scores, or even scores with the same standard deviations, the slope of the line of best fit equals the correlation coefficient. This slope is 1.0 for a perfect correlation and the regression line is at 45° to the X-axis. As the correlation decreases from 1.0 toward zero, the angle that the regression line makes with the X-axis declines from 45° to 0° and becomes coincident with the X-axis when $r = 0$. With $r = -1$, the regression line is at $-45°$ and moves toward 0° as r moves from -1.0 toward zero.

A Second Regression Line   By convention, in this text, Y' is used to designate the variable to be predicted (the dependent variable), and X is used to designate the variable from which the predictions are made (the independent variable). For example, grade point average can be predicted from knowledge of Scholastic Aptitude Test score. GPA is the dependent variable to be predicted, Y, and the aptitude score is the independent variable,

Figure 6.7
Regression plots for different correlations.

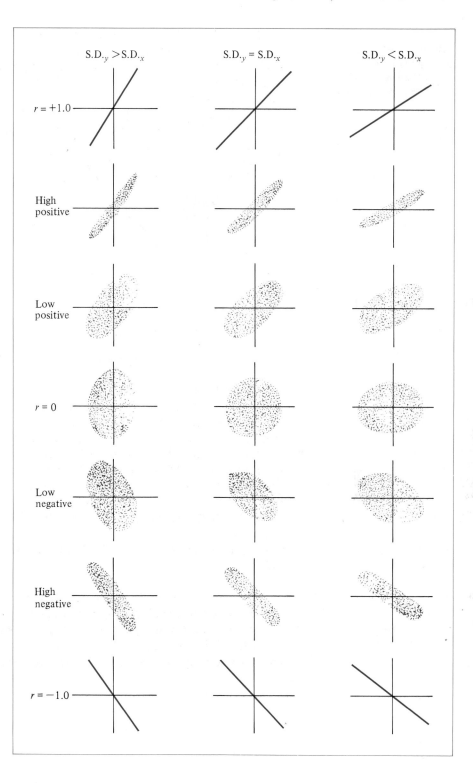

Figure 6.8
Two deviation-score regression lines.

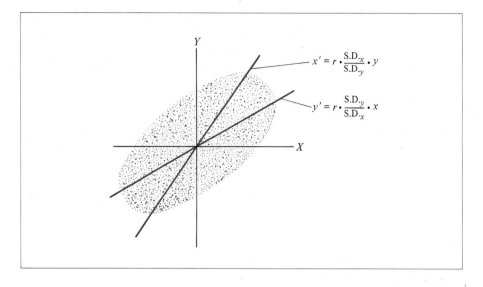

X. It makes little sense to predict aptitude scores from GPA values. It is always possible to specify that the predicted score be placed on the ordinate, or the Y-axis, making Formula 6.5 appropriate. If it is desired to obtain an equation, however, predicting x-scores from a knowledge of y-scores, the corresponding equation is

$$x' = r \cdot \frac{S.D._x}{S.D._y} \cdot y \tag{6.8}$$

where the terms are defined as in Formula 6.5. There are, therefore, two distinct regression lines. They coincide and become one line only when $r = 1.0$ or $r = -1.0$. These two lines are illustrated in figure 6.8. Prediction is almost always in one direction only, making it possible to use only Formula 6.5 rather than requiring the determination of two regression lines using both Formulas 6.5 and 6.8.

INTERPRETATION OF THE CORRELATION COEFFICIENT

coefficient of correlation A statistical measure of the degree, or strength, of the association between two variables.

The **coefficient of correlation** provides a numerical index of the extent to which two variables co-vary, but it does not shed any light on the reasons for this covariation. Suppose a study reveals a substantial positive correlation between a person's age at the time of marriage and the number of years that he or she remains married. This correlation indicates that people who get married at a younger age do not stay married as long as those who get married at a later age.

An uninformed or overzealous social reformer observing this correlation might be tempted to begin a program to prevent early marriages, assuming that this will reduce the divorce rate. This is an example of confusing correlation with causation. It *might* be true that getting married at a younger age tends to cause divorce, but the correlation does not establish this conclusion as fact. Instead it might be true that the same traits that lead a person to an early marriage, such as impulsiveness and/or irresponsibility, also lead the person to divorce. Then postponing marriage until a later age would not necessarily affect the rate of divorce.

In another example, higher incidence of crime is associated with lower-income residential areas in urban communities. Just because this correlation exists, however, we cannot infer that the crime rate can be lowered by raising income level. To assume this is true on the basis of the correlation alone is to mistake correlation for causation.

Size of the Correlation Coefficient

The magnitude of the correlation coefficient must be interpreted with caution, because it is not on an absolute scale like height or weight. For example, a correlation of 0.25 is not half as large as a correlation of 0.50, and a correlation of 0.50 is not half as large as a correlation of 1.00. The relationship is more complicated than that. A better indication of the magnitude of the correlation relative to other correlation coefficients is obtained by taking r^2, the coefficient of determination. A correlation of 0.707 gives $r^2 = 0.50$, and an r of 0.50 gives $r^2 = 0.25$. Thus, an r of 0.707 is in a sense approximately twice as great as a correlation of 0.50, since their **coefficients of determination** are on the order of two to one.

coefficient of determination A statistical measure of the proportion of the variation in the dependent variable that is explained by the regression equation.

The coefficient of determination is the amount of total variation shared by the two variables. Thus, r^2 appears to be a more meaningful interpretation of the strength of the relation between the two variables. The sample correlation coefficient is an estimate of the population correlation coefficient. As mentioned previously, the convention in statistics is to use Greek letters for population coefficients. The Greek letter used for the correlation coefficient is ρ (rho). A test of an hypothesis covering the value of ρ is given in Chapter 8.

Figure 6.9 is a pictorial representation of r^2. If the two circles represent the variation in both variables, the area where they intersect is the amount (proportion) of variation shared between them. This shaded area is r^2.

Pearson correlation A correlation coefficient computed on two variables measured on a continuous scale.

Factors Affecting the Size of the Correlation Coefficient   The correlation coefficient presented here is often referred to as the **Pearson product-moment correlation,** which is designed to measure the degree of linear (straight line) relationship between two variables measured on a continuous scale. If the data are nonlinear, then the Pearson r will *underestimate* the true degree of relationship. The data and graph in figure 6.10 demonstrates that the relationship between X and Y is a perfect nonlinear one. However, fitting a straight line to these nonlinear data produces a correlation coefficient below perfect ($r = 0.57$).

The size of r is also dependent on the range of the X and Y variables. If the range of one of the variables is limited or truncated, then the size of r is reduced. Recently, it was determined that there is only a 0.48 correlation between GRE (Graduate Record Examination) scores and success in graduate school (professor evaluations and grades). The correlation is low because the range of one of the variables is restricted. Only those students who have achieved a high score on the GRE are normally admitted to graduate school. Hence, only those who are admitted are evaluated for success in graduate school. Those who scored low are usually not admitted to graduate school, and as a result, no evaluation of their graduate work is available. By exercising this screening procedure, a relatively homogeneous group exists in graduate school.

Figure 6.9
A representation of r^2.

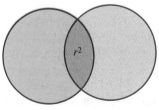

Variance of Y

Variance of X

Figure 6.10
Fitting a straight line to
nonlinear data.

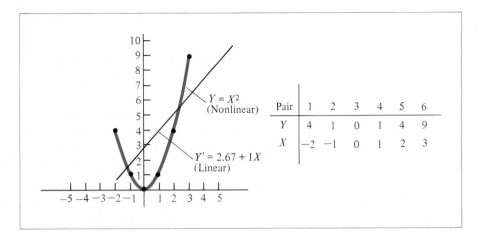

Figure 6.11
Effects on r when sample
contains different subgroups
(means).

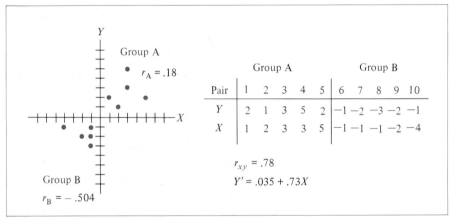

Figure 6.12
Effects on r when sample
contains different subgroups
(standard deviation).

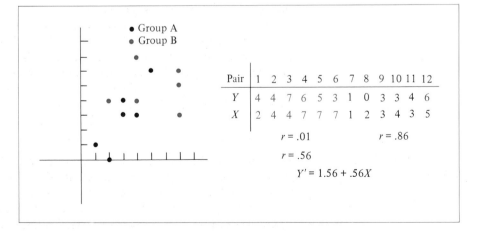

STATISTICS IN THE WORLD AROUND YOU

A team of doctors recently reported that the amount an 11-year-old girl's blood pressure rises during exercise can probably predict whether she or a family member will die of heart disease before reaching the age of 55. The report is based on a study involving 300 pairs of 11-year-old twins and their parents. Extensive statistical analysis was performed on the collected data. A comparison of blood pressures during exercise and family medical histories revealed a direct relationship (correlation) between high blood pressure and a family history of heart disease.

"How depressing" you say, to know with some certainty at age 11 that you or a family member will die of heart disease by the age of 55. But this is an example of how statistics provide an early warning to help improve and extend life. The early warning provides the girls and their families with the chance to be counseled about proper diet, avoidance of smoking, stress, and other factors associated with early death from cardiovascular disease.

By the way, those other factors associated with heart disease were "associated" through the use of statistical analysis of morbidity data.

The correlation coefficient can be spuriously elevated by collecting a sample of subjects that contains two or more subgroups with means and/or standard deviations that differ on both variables. Examine the data given in figure 6.11. There are two groups, A and B, which differ on the means for the two variables X and Y. The correlation of X and Y within each group is relatively low ($r_A = 0.18$, $r_B = -0.50$). However, when the two groups are combined to form one sample, the correlation between X and Y increases substantially ($r = 0.78$). Figure 6.12 shows a similar occurrence when the standard deviations of the two groups are different.

If the distributions of the X and Y variables are highly skewed, the value of the Pearson r is lower than the r's computed from X's and Y's that are nearly normal in shape (bell-shaped). The skewness causes a greater dispersion around the line of best fit. In general, the use of the Pearson r does not require that X and Y be normal. However, with skewed distributions of X and Y, the value of r will be closer to zero.

| MATHEMATICAL NOTES | The derivation of Formula 6.4 is based on the principle of least squares. Consider point A (110,60) in figure 6.3. The actual value of Y for $X = 110$ is 60, whereas the value of Y on the regression line for $X = 60$ is $Y = 50$. Thus, there is a discrepancy between the actual Y and the predicted Y' of $60 - 50$, or $+10$. The principle of least squares states that the regression line should be placed in such a way |

as to minimize the sum of squares of these discrepancies over all cases. The sum of squares of these discrepancies to be minimized is expressed mathematically as follows for deviation scores:

$$\Sigma(y - y')^2 = \text{the sum of squared discrepancies between} \qquad (6.9)$$
$$\text{predicted and obtained deviation scores.}$$

By Formula 6.4, $y' = bx$. Substituting this in Formula 6.9 gives:

$$\Sigma(y - bx)^2 = \text{the sum of squared discrepancies.} \qquad (6.10)$$

The expression for the sum of squared discrepancies between obtained and predicted deviation scores in Y, as shown in Formula 6.10, is to be minimized by the appropriate choice of b. That is, only the slope of the regression line determines the location of the line when scores are expressed as deviation scores since the line always goes through the origin, giving a zero Y-intercept. Thus, the task is to pick a value of b that minimizes the expression in Formula 6.10.

Finding a value of b that minimizes the sum of squared discrepancies between predicted and obtained deviation scores in Y is a calculus problem. Students without training in calculus cannot be expected to understand fully the rest of the proof. The expression in Formula 6.10 is differentiated with respect to b and set equal to zero to give:

$$2\Sigma(y - bx)(-x) = 0$$

$$\Sigma(-xy + bx^2) = 0$$

$$-\Sigma xy + b\Sigma x^2 = 0$$

$$b = \frac{\Sigma xy}{\Sigma x^2}$$

$$b = \frac{\Sigma xy}{N \cdot \dfrac{\Sigma x^2}{N}}$$

$$b = \frac{\Sigma xy}{N \cdot (\text{S.D.}_x)^2} \cdot \frac{\text{S.D.}_y}{\text{S.D.}_y}$$

$$b = \left(\frac{\Sigma xy}{N \cdot \text{S.D.}_x \text{S.D.}_y}\right) \cdot \frac{\text{S.D.}_y}{\text{S.D.}_x} \qquad (6.11)$$

$$b = r \cdot \frac{\text{S.D.}_y}{\text{S.D.}_x} \qquad (6.12)$$

The value of b in Formula 6.4 is given by Formula 6.12 above. Making this substitution in Formula 6.4 gives Formula 6.5. The correlation coefficient is by definition that part of the formula for b, which is in parentheses in Formula 6.11.

Derivation of Formula 6.12 can be accomplished working with raw scores but it is more difficult. Since the raw-score regression line does not pass through the origin, the line is determined by two constants, the slope, b, and the Y-intercept, a. It is necessary to differentiate the expression for the sum of the squared discrepancies partially with respect to b and then a, set both expressions equal to zero, and then solve them simultaneously for the values of b and a.

PROBLEM 16

Compute the Correlation Coefficient from Raw Scores Using the Raw-Score Formula

Formula 6.6 gave a formula for the correlation coefficient in terms of deviation scores. That formula is seldom used in actual calculation of the correlation coefficient because it requires the use of deviation scores. Deviation scores are obtained by subtracting the mean from raw scores ($x = X - M_x$, $y = Y - M_y$). In addition to being a computational chore in itself, this process yields decimal fractions that are difficult to cross multiply to obtain the xy cross-product terms in Formula 6.6.

Raw-Score Formula for the Correlation Coefficient

A more convenient formula than Formula 6.6 for computation of the correlation coefficient directly from raw scores, rather than from deviation scores, is given by:

$$r = \frac{N\Sigma XY - (\Sigma X)(\Sigma Y)}{\sqrt{N\Sigma X^2 - (\Sigma X)^2} \; \sqrt{N\Sigma Y^2 - (\Sigma Y)^2}} \tag{6.13}$$

where r = the correlation coefficient,

X = a raw score in the X variable,

Y = a raw score in the Y variable.

The derivation of Formula 6.13 from Formula 6.6 is given in the Mathematical Notes section on pages 149–50. Formula 6.13 requires ΣX, ΣY, ΣX^2, ΣY^2, and ΣXY as well as N. All but ΣXY are terms that are used in computing the standard deviations of these two variables by the raw-score method. In fact, multiplying the left-hand radical in the denominator of Formula 6.13 by $1/N$ gives the raw-score formula for the standard deviation in X. Multiplying the radical on the right in the denominator of Formula 6.13 by $1/N$ gives the raw-score formula for the standard deviation of the Y variable. The term ΣXY requires that each person's raw scores in X and Y are multiplied together. The products are then added up for all N cases in the sample to get ΣXY. The XY term for subject number 1 in table 6.1 is found by multiplying $X = 5$ times $Y = 10$.

TABLE 6.1

Computation of the correlation coefficient using the raw-score formula

Subject number	X	Y	X^2	Y^2	XY
1	5	10	25	100	50
2	7	13	49	169	91
3	10	2	100	4	20
4	10	8	100	64	80
5	11	6	121	36	66
6	15	4	225	16	60
7	9	7	81	49	63
8	9	7	81	49	63
9	8	9	64	81	72
10	6	9	36	81	54
11	9	8	81	64	72
12	12	5	144	25	60
Sum	111	88	1107	738	751

$$r = \frac{N\Sigma XY - (\Sigma X)(\Sigma Y)}{\sqrt{N\Sigma X^2 - (\Sigma X)^2}\ \sqrt{N\Sigma Y^2 - (\Sigma Y)^2}}$$

$$r = \frac{12(751) - (111)(88)}{\sqrt{12(1107) - (111)^2}\ \sqrt{12(738) - (88)^2}}$$

$$r = \frac{9012 - 9768}{\sqrt{13284 - 12321}\ \sqrt{8856 - 7744}}$$

$$r = \frac{-756}{\sqrt{963}\ \sqrt{1112}} = \frac{-756}{(31.03)(33.35)}$$

$$r = \frac{-756}{1034.85} = -0.731 = -0.73$$

Computation of the Correlation Coefficient by the Raw-Score Formula

Computation of the correlation coefficient, sometimes called the *Pearson r* or the *product-moment r*, by Formula 6.13 is illustrated in table 6.1. X represents the reported average number of hours studied per week for students in a statistics class, and Y represents the number of problems missed on the final examination for the course. The first person in the group of 12 students reported studying an average of 5 hours per week and missed 10 problems on the final examination. The squared raw scores in X are placed in the column headed by X^2, the squared raw scores in Y are placed in the column headed by Y^2, and the cross-product terms are placed in the last column. Substituting $N = 12$ and the sums of the columns in table 6.1 in Formula 6.13 results in a correlation of -0.73 between hours of study and number of errors on the final exam, as shown by the computations at the bottom of table 6.1. The negative correlation means that greater numbers of hours spent studying tend to be associated with smaller numbers of errors on the final examination.

This method of computing the correlation coefficient is particularly useful when hand calculators or computers are available. If it is necessary to do the computations without such aids, multiplying and adding large raw scores can be laborious, especially for large numbers of cases. To shorten labor, subtracting a constant or dividing by a constant reduces the size of the numbers. The labor on smaller numbers is less and the arithmetic operation of subtraction or division *does not* affect the magnitude or direction of the correlation coefficient.

SUMMARY OF STEPS

Computing the Correlation Coefficient from Raw Scores Using the Raw-Score Formula

Step 1   Obtain the sum of raw scores for the X-score variable, ΣX.

Step 2   Obtain the sum of the squared raw scores for the X-score variable, ΣX^2.

Step 3   Obtain the sum of raw scores for the Y-score variable, ΣY.

Step 4   Obtain the sum of the squared raw scores for the Y-score variable, ΣY^2.

Step 5   Obtain the sum of the cross-product terms, ΣXY, by multiplying each X-score by the corresponding Y-score and adding up these values over all cases from 1 to N.

Step 6   Substitute the sums obtained in Steps 1 through 5 along with the value of N, into Formula 6.13 and solve for r, the correlation coefficient.

MATHEMATICAL
NOTES

The derivation of Formula 6.13 proceeds from Formula 6.6 as follows:

$$r = \frac{\Sigma xy}{N(\text{S.D.}_x)(\text{S.D.}_y)} \tag{6.6}$$

Substituting the raw-score formula for the standard deviation and the raw-score equivalent to the deviation score in Formula 6.6 gives:

$$r = \frac{\Sigma(X - M_x)(Y - M_y)}{N \cdot \dfrac{1}{N}\sqrt{N\Sigma X^2 - (\Sigma X)^2} \cdot \dfrac{1}{N}\sqrt{N\Sigma Y^2 - (\Sigma Y)^2}}$$

Multiplying numerator and denominator by N and multiplying out the two terms in parentheses gives:

$$r = \frac{N\Sigma(XY - M_x Y - M_y X + M_x M_y)}{\sqrt{N\Sigma X^2 - (\Sigma X)^2}\ \sqrt{N\Sigma Y^2 - (\Sigma Y)^2}}$$

b. Sum the X (ΣX).

c. Square each X and place result in the X^2 column.

d. Sum the X^2 (ΣX^2).

e. Sum the Y (ΣY).

f. Square each Y and place result in the Y^2 column.

g. Sum the Y^2 (ΣY^2).

h. Multiply each X by the corresponding paired Y and place result in column XY.

i. Sum XY (ΣXY).

j. Determine N, the number of pairs.

k. Enter the appropriate sums into Formula 6.13.

6–4.   Using the raw-score formula, compute the correlation coefficient for the second 20 cases in the data for question 6–13 (subjects 21–40). A calculator is needed for this exercise.

a. Place the first number of each pair in the X column and the second number in the Y column.

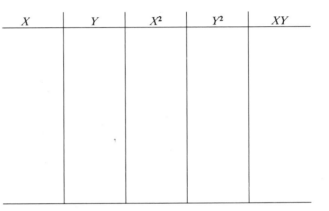

X	Y	X^2	Y^2	XY

b. Sum the X (ΣX).

c. Square each X and place result in the X^2 column.

d. Sum the X^2 (ΣX^2).

e. Sum the Y (ΣY).

f. Square each Y and place result in the Y^2 column.

g. Sum the Y^2 (ΣY^2).

h. Multiply each X by the corresponding paired Y and place result in column XY.

i. Sum XY (ΣXY).

j. Determine N, the number of pairs.

k. Enter the appropriate sums into Formula 6.13.

6–5.   Using the raw-score formula, compute the correlation coefficient for all 60 cases in the data for question 6–13. (Use the totals from questions 6–3 and 6–4 if they have been obtained already.) A calculator is needed for this exercise.

a. Place the first number of each pair in the X column and the second number in the Y column.

X	Y	X^2	Y^2	XY

b. Sum the X (ΣX).
c. Square each X and place result in the X^2 column.
d. Sum the X^2 (ΣX^2).
e. Sum the Y (ΣY).
f. Square each Y and place result in the Y^2 column.
g. Sum the Y^2 (ΣY^2).
h. Multiply each X by the corresponding paired Y and place result in column XY.
i. Sum XY (ΣXY).
j. Determine N, the number of pairs.
k. Enter the appropriate sums into Formula 6.13.

6–6.   Using the raw-score formula, compute the correlation coefficient for all 40 cases in the data for question 6–14. A calculator is needed for this exercise.

a. Place the first number of each pair in the X column and the second number in the Y column.

X	Y	X^2	Y^2	XY

b. Sum the X (ΣX).
c. Square each X and place result in the X^2 column.
d. Sum the X^2 (ΣX^2).
e. Sum the Y (ΣY).
f. Square each Y and place result in the Y^2 column.
g. Sum the Y^2 (ΣY^2).
h. Multiply each X by the corresponding paired Y and place result in column XY.
i. Sum XY (ΣXY).
j. Determine N, the number of pairs.
k. Enter the appropriate sums into Formula 6.13.

6–7. Given the following pairs of error scores on successive administrations of a given test, compute the correlation between the sets of scores using the raw-score formula: (9,10); (7,2); (1,5); (12,14); (3,0); (6,8); (9,7); (10,11); (10,5); (2,1).

a. Place the first number of each pair in the X column and the second number in the Y column.

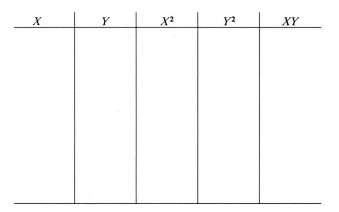

X	Y	X^2	Y^2	XY

b. Sum the X (ΣX).
c. Square each X and place result in the X^2 column.
d. Sum the X^2 (ΣX^2).
e. Sum the Y (ΣY).
f. Square each Y and place result in Y^2 column.
g. Sum the Y^2 (ΣY^2).
h. Multiply each X by the corresponding paired Y and place result in column XY.
i. Sum XY (ΣXY).
j. Determine N, the number of pairs.
k. Enter the appropriate sums into Formula 6.13.

6–8.   Given the following age and error scores on an ability test for a small sample of boys, find the correlation using the raw-score formula: (8,8); (6,12); (12,5); (15,1); (10,10); (5,15); (12,5); (9,10); (13,3); (10,9).

    a. Place the first number of each pair in the X column and the second number in the Y column.

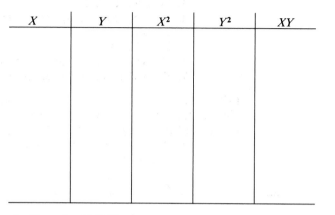

X	Y	X^2	Y^2	XY

    b. Sum the X (ΣX).
    c. Square each X and place result in the X^2 column.
    d. Sum the X^2 (ΣX^2).
    e. Sum the Y (ΣY).
    f. Square each Y and place result in Y^2 column.
    g. Sum the Y^2 (ΣY^2).
    h. Multiply each X by the corresponding paired Y and place result in column XY.
    i. Sum XY (ΣXY).
    j. Determine N, the number of pairs.
    k. Enter the appropriate sums into Formula 6.13.

6–9.   In a certain industrial plant, the average number of items produced per day per worker is 50 with a standard deviation of 10. A manual dexterity test correlates 0.50 with this criterion of production. Develop a deviation-score regression equation to predict performance from test scores. The test score mean is 80 and the S.D. is 15. For a person with a test score of 60, what is the predicted deviation production score?

    a. Determine r.
    b. Determine S.D. for items produced (S.D.$_y$).
    c. Determine S.D. for test score (S.D.$_x$).
    d. Determine deviation test score x by subtracting the person's test score from the test score mean.
    e. Insert the information in a–d into Formula 6.5 to find y'.

6–10.   Given two sets of test scores, develop two deviation-score regression equations to predict either score from the other where $r_{12} = 0.6$, $M_1 = 20$, S.D.$_1 = 3$, $M_2 = 40$, S.D.$_2 = 5$. What raw score in variable 2 results in a predicted raw score of 18 in variable 1?

    a. Determine r.
    b. Determine S.D. for each variable.

c. Determine M for each variable.

d. Insert appropriate information in (a–c) into formula

$$y_1' = r \cdot \frac{\text{S.D.}_1}{\text{S.D.}_2} \cdot y_2$$

Note: No substitution is required for y_1 and y_2.

e. Insert information in (a–c) into formula

$$y = r \cdot \frac{\text{S.D.}_2}{\text{S.D.}_1} \cdot y_1$$

f. Substitute values from (a–c) and solve for y_2.

$$18 - M_1 = r \cdot \frac{\text{S.D.}_1}{\text{S.D.}_2} \cdot (y_2 - M_2)$$

6–11.   Express the regression equations in question 6–9 in both standard-score form and raw-score form. Use Formula 6.15 to predict the raw score in production.

    a. Determine r.

    b. Insert the value of r into Formula 6.7 to get standard-score regression equation.

    c. Use Formula 6.15.

    d. Insert test score 60 into equation found in (c).

6–12.   Express the regression equations in question 6–10 in both standard-score form and raw-score form. Use these formulas to solve for the standard score and raw score, respectively, in variable 2 that would result in a predicted score equivalent to a raw score of 18 in variable 1.

    a. Use Formula 6.7 for standard-score regression equation.

    b. Use Formula 6.15 for raw-score regression equation.

    c. Change test score of 18 to standard score and insert into equation as variable 1 in (a) and solve for variable Z.

    d. Insert test score of 18 into equation as variable 1 found in (b) and solve for variable Z.

6–13.   Given the following scores for two personality factors O, Orderliness vs Lack of Compulsion, and C, Social Conformity vs Rebelliousness, for a sample of 60 male subjects, find the regression equation predicting O using C.

    a. Find ΣO.

    b. Find ΣO^2.

    c. Find ΣC.

    d. Find ΣC^2, ΣOC.

    e. Find r using Formula 6.14.

    f. Find S.D.$_O$ and S.D.$_C$.

    g. Find M_O and M_C.

    h. Substitute information in (e–g) into Formula 6.15, where Y is O and X is C.

Subject	O	C	Subject	O	C	Subject	O	C
1	72	59	21	93	87	41	77	69
2	81	87	22	111	83	42	56	53
3	54	36	23	70	53	43	97	83
4	91	61	24	103	81	44	71	50
5	74	58	25	77	82	45	71	84
6	88	86	26	97	93	46	88	97
7	78	70	27	91	58	47	103	84
8	90	39	28	53	95	48	81	92
9	77	104	29	81	67	49	56	56
10	88	84	30	73	101	50	53	76
11	90	70	31	93	72	51	77	77
12	69	55	32	70	49	52	104	73
13	94	89	33	82	66	53	83	60
14	100	94	34	86	92	54	81	48
15	74	51	35	66	57	55	70	44
16	76	54	36	81	90	56	71	66
17	96	101	37	101	95	57	75	105
18	50	46	38	94	68	58	76	62
19	103	80	39	108	86	59	94	73
20	93	91	40	69	51	60	94	85

6–14.   Consider the following scores to represent weights and number of minutes played in a randomly selected football game for all team players. Find the regression equation(s) predicting minutes played as a function of weight. Find the raw score, deviation score, and standard score.

Player	Weight	Minutes	Player	Weight	Minutes	Player	Weight	Minutes
1	180	15	16	215	0	31	205	19
2	220	20	17	170	0	32	200	17
3	200	29	18	185	50	33	195	15
4	175	8	19	155	2	34	220	49
5	195	0	20	230	46	35	185	11
6	200	38	21	225	38	36	190	5
7	240	6	22	180	25	37	200	18
8	190	21	23	175	10	38	175	12
9	205	33	24	200	25	39	190	20
10	230	40	25	195	31	40	200	17
11	160	8	26	210	35			
12	180	0	27	190	21			
13	185	13	28	200	26			
14	200	19	29	195	16			
15	210	28	30	185	12			

Note: (a–k) were done for question 6–6.
  a. Place the first number of each pair in the X column and the second number in the Y column.

X	Y	X^2	Y^2	XY

  b. Sum the X (ΣX).
  c. Square each X and place result in the X^2 column.
  d. Sum the X^2 (ΣX^2).
  e. Sum the Y (ΣY).
  f. Square each Y and place result in the Y^2 column.
  g. Sum the Y^2 (ΣY^2).
  h. Multiply each X by the corresponding paired Y and place result in column XY.
  i. Sum XY (ΣXY).
  j. Determine N, the number of pairs.
  k. Enter the appropriate sums into Formula 6.13 to find r.
  l. Find S.D.$_y$, S.D.$_x$, M_y, M_x.
  m. Enter appropriate values into Formula 6.15 for raw-score equation.
  n. Enter appropriate values into Formula 6.5 for deviation-score equation.
  o. Enter appropriate values into Formula 6.7 for standard-score equation.

6–15.   A personnel researcher wanted to determine if there was any relationship between age and the number of sick days used by employees. The data are given below. For this study the information was gathered over a 12-month period.

Employee	Sick days	Age
1	5	35
2	0	24
3	1	55
4	2	30
5	7	29
6	2	33
7	2	43
8	0	44
9	3	47
10	1	26
11	4	37
12	0	23

Compute the correlation coefficient and compute the coefficient of determination.

a. Place the first number of each pair in the X column and the second number in the Y column.

X	Y	X^2	Y^2	XY

b. Sum the X (ΣX).
c. Square each X and place result in the X^2 column.
d. Sum the X^2 (ΣX^2).
e. Sum the Y (ΣY).
f. Square each Y and place result in the Y^2 column.
g. Sum the Y^2 (ΣY^2).
h. Multiply each X by the corresponding paired Y and place result in column XY.
i. Sum XY (ΣXY).
j. Determine N, the number of pairs.
k. Enter the appropriate sums into Formula 6.13.

6–16.   In a university job placement office, the dean is attempting to determine whether there is any relationship between undergraduate grade point average (GPA) and starting salaries received. The dean obtained the data of 15 recent graduates.

GPA	Starting salary (in $)
2.56	11,235
3.14	17,190
3.95	25,000
2.75	20,000
3.75	24,500
2.98	26,000
3.00	14,000
2.11	12,000
2.31	11,200
3.49	18,900
2.66	16,700
3.06	17,900
2.89	19,010
3.82	26,600
2.74	16,950

Develop the linear regression equation predicting starting salaries using GPA. (GPA is the independent variable).

    a. Find ΣGPA.
    b. Find ΣSalary.
    c. Find ΣGPA^2.
    d. Find ΣSalary^2.
    e. Find Σ(GPA)(Salary).
    f. Find r using Formula 6.13.
    g. Find S.D.$_{GPA}$, S.D.$_{SALARY}$.
    h. Find M_{GPA}, M_{SALARY}.
    i. Substitute information in (f–h) into Formula 6.15 where Y is salary and X is GPA.

6–17. Compute the correlation coefficient and the coefficient of determination using the data in question 6–16.
    a. Use value in (f) of question 6–16.
    b. Square r.

6–18. What is the estimated average starting salary for graduates in the data for question 6–16 with a grade point average of 2.69? Of 3.18 using the data in question 6–16?
    a. Insert GPA $= 2.69$ into equation found in question 6–16.
    b. Insert GPA $= 3.18$ into equation found in question 6–16.

6–19. A market research study was conducted at a supermarket to determine the relationship between the amount of shelf-space allotted to a brand of cereal (Brand X) and its weekly sales. The amount of space allocated to Brand X was varied over 3, 6, and 9 linear feet displays in a random fashion over 12 weeks. The competitors' brands were maintained at a constant 3 linear feet for the duration of the study. The data recorded are given below for the weekly sales of the entire cereal category.

Week	1	2	3	4	5	6	7	8	9	10	11	12
Space	3	3	6	9	6	6	3	9	3	9	9	6
Sales	10	12	16	20	18	17	8	12	9	15	24	19

Compute the correlation coefficient, r.

6–20. Develop a regression equation that predicts sales using display space. Use the data in question 6–19.
    a. Use r found in question 6–19.
    b. Find S.D.$_{SPACE}$, S.D.$_{SALES}$.
    c. Find M_{SPACE}, M_{SALES}.
    d. Substitute information in (a–c) into Formula 6.15.

Chapter 7

Testing Hypotheses with Frequency Data Using the Binomial Distribution

U p to this point in the text, the emphasis has been on "descriptive" statistics. That is, the various statistical features of a body of data are described, much as a biographical sketch describes a person. Batting averages, number of votes, and charts and graphs of numbers are also classified as descriptive statistics. The remainder of the text will be concerned with "inferential" statistics. While descriptive statistics are concerned with collecting and summarizing a set of data and can be used in assisting in a decision-making process, inferential statistics can be interpreted in a much broader sense as being concerned with the problem of drawing conclusions about a large collection of things (populations) based on measuring a subset of the collection, or a sample. Figure 7.1 shows this inferential process.

Suppose, for example, that it is important to determine what proportion of males in the United States smoke cigarettes. A large random sample is drawn from the population of U.S. males and the proportion of them who are smokers determined by interview or questionnaire. This proportion is a descriptive statistic for the sample and provides an estimate of the proportion of smokers in the total population of U.S. males. The question, How accurate is this estimate? raises a problem in statistical inference. The sample proportion is not apt to be exactly equal to the population proportion and in some cases may depart from it to an appreciable degree. Through the methods of statistical inference it is possible to estimate the **probability** or likelihood, that the sample proportion will deviate from the population proportion by more than a predetermined amount. This helps to fix the location of the true population proportion within certain limits rather than just to estimate its value without giving any idea about how far off the estimate could be.

This chapter is concerned with statistical inferences that are made by applying the binomial distribution to frequency data. Chapter 8 will treat problems of statistical inference with continuous variable data. The last chapters, on Chi Square, Analysis of Variance, and Nonparametric Methods are also concerned with problems in statistical inference. In certain statistical definitions, the population proportion or any population characteristic is usually referred to as a **parameter,** while the sample proportion or any sample characteristic is referred to as a **statistic.** Greek letters are usually used to represent population parameters.

probability   The proportion or percentage of the time that an event occurs. The likelihood that something will happen.

parameter   A number derived from knowledge of the entire population, such as the minimum of the population, or the mean of the population.

statistic   A number derived from observing a sample from a population of numbers. A value computed from a sample.

Figure 7.1
The inference process.

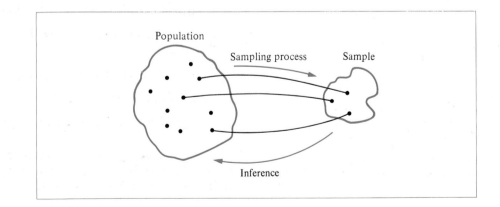

PROBLEM **19**

Given a Situation with Two Possible Outcomes, Determine the Probability of a Particular Combination of Outcomes in *n* Trials

One of the major tools for making statistical inferences with frequency data is the binomial distribution. Problem 19 represents an application of this distribution, which is not an example of statistical inference. It is, however, a necessary preliminary to more complicated applications of the binomial, which do represent examples of statistical inference.

It is also necessary at this time to present, in an elementary way, the nature of probability. Probability can be viewed as a numerical measure of uncertainty, or a numerical measure of the likelihood an uncertain event will occur. Zero is on one end of the numerical scale used in probability, 1 is at the other end. The value zero corresponds to the impossibility of an event occurring, and the value 1 corresponds to the absolute certainty of an event occurring. Any probability between zero and 1 expresses a relative likelihood, or a ratio or proportion. For example, if an event is just as likely to occur as it is not to occur, then the probability of that event is 1/2 or 0.5. Probabilities are often expressed as percentages. A probability of 0.5 that it will rain is usually expressed by the weather person as "a 50% chance of rain."

The Binomial Distribution

coin toss   The result of an imaginary experiment of tossing a fair coin and allowing it to fall to rest so that the outcome of the coin facing heads or tails is due to chance and cannot be predicted in advance.

Consider the classical example of **tossing a coin** *n* times where the coin might fall heads or tails on any one trial. If *p* is the probability of a head coming up on any one trial and *q* is the probability of a tail coming up on any one trial, where $p + q = 1$, then the probability of getting exactly *x* heads in *n* trials is given by:

$$P_x = \frac{n!}{x!(n - x)!} p^x q^{n - x} \tag{7.1}$$

where P_x = the probability of exactly *x* heads showing up in *n* coin tosses,

p = the probability of a head on any given trial (usually $p = 1/2$ in this situation),

q = the probability of a tail (usually 1/2) where $q = 1 - p$,

n = the number of coin tosses, and

x = number of heads being tested for the probability value.

factorial   The product of the numbers 1, 2, 3, and so on, up to the given number. The given number must be a positive integer or 0. By definition, 0 factorial is 1. It is written as "*n*!".

The "!" symbol is used to indicate a mathematical concept called **factorial.** Here, a number, let's say *n,* followed by "!" says that all integers between 1 and *n* are multiplied by each other ($6! = 6 \cdot 5 \cdot 4 \cdot 3 \cdot 2 \cdot 1 = 720$).

TABLE 7.1

Probabilities for x heads in ten coin tosses

x	Probability
0	1 : 1024
1	10 : 1024
2	45 : 1024
3	120 : 1024
4	210 : 1024
5	252 : 1024
6	210 : 1024
7	120 : 1024
8	45 : 1024
9	10 : 1024
10	1 : 1024
Sum	1024 : 1024

If 10 coins are tossed, or 1 coin is tossed 10 times, Formula 7.1 is used to determine the probability that the 10 tosses will result in any particular combination of heads and tails totaling 10, 4 heads and 6 tails. Substituting $x = 4$, $n = 10$, and $p = 1/2$ (remember $q = 1 - p$) in Formula 7.1 gives:

$$P_4 = \frac{10!}{4!(10 - 4)!} \left(\frac{1}{2}\right)^4 \left(\frac{1}{2}\right)^6 \tag{7.2}$$

The definition of 10! (read as 10 factorial) is $10 \cdot 9 \cdot 8 \cdot 7 \cdot 6 \cdot 5 \cdot 4 \cdot 3 \cdot 2 \cdot 1$, and that $(1/2)^p (1/2)^q = (1/2)^{p+q}$, Formula 7.2 becomes:

$$P_4 = \frac{10 \cdot 9 \cdot 8 \cdot 7 \cdot 6 \cdot 5 \cdot 4 \cdot 3 \cdot 2 \cdot 1}{(4 \cdot 3 \cdot 2 \cdot 1)(6 \cdot 5 \cdot 4 \cdot 3 \cdot 2 \cdot 1)} \left(\frac{1}{2}\right)^{10}$$

$$= \frac{10 \cdot 9 \cdot 8 \cdot 7}{4 \cdot 3 \cdot 2 \cdot 1} \cdot \frac{1}{1024}$$

$$= \frac{210}{1024}$$

If the probability of no heads is called for, then Formula 7.2 gives:

$$P_0 = \frac{10!}{0!(10 - 0)!} \left(\frac{1}{2}\right)^0 \left(\frac{1}{2}\right)^{10}$$

$$P_0 = \frac{1}{1024}$$

This is due to the fact that 0! is defined to be equal to 1 and any quantity to the zero power is defined to be 1.

Applying Formula 7.2 successively for all the values of x from 0 to 10 with $n = 10$ and $p = 1/2$ gives the probabilities shown in table 7.1. Note that the sum of all the probability values adds up to $1024/1024 = 1$; since these outcomes exhaust all the possibilities, the sum of their probabilities must equal 1.0. Also notice that the distribution in table 7.1 is symmetrical about $x = 5$, the central value, 4 and 6 have the same probability, 3 and 7 have the same probability, and so on. The distribution is symmetrical in this way when $p = q$.

Formula 7.1 then, can be used to determine the probability of a particular combination of outcomes (numbers of heads and tails) in n trials, or equivalently, the probability of a particular value of x, such as the number of heads, in tossing n coins. It is assumed that p, the probability of a head, for instance, remains constant from trial to trial and it is assumed that each trial is independent of each other trial. That is, what happened on previous trials does not affect the probabilities for the current trial. A derivation for Formula 7.1 is given in the Mathematical Notes on page 172.

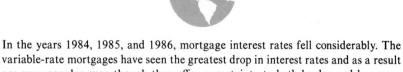

STATISTICS IN THE WORLD AROUND YOU

In the years 1984, 1985, and 1986, mortgage interest rates fell considerably. The variable-rate mortgages have seen the greatest drop in interest rates and as a result are very popular even though they offer uncertainty to both lender and borrower during unstable periods. However, many borrowers prefer the fixed-rate interest loans, which have a higher rate, because they remain fixed at the agreed rate no matter how the other interest rates fluctuate. Initially the variable-rate loans might be lower but as the rate increases, the borrower might ultimately pay more than the rate set on a fixed-rate loan.

To determine if consumers favor a fixed-rate or the variable-rate loans, seventeen applicants at a large savings and loan were surveyed. The data showed that five favored the variable-rate loan, ten favored the fixed-rate loan, and two showed no preference. Using the binomial formula in this chapter, the preference of the borrowers can be determined. After eliminating the two applicants who showed no preference, fifteen eligible data points remain. The probability of observing five or less out of fifteen given that there is a 0.5 probability of selecting one type of rate over the other is 0.1509. This probability is not less than 0.05 (level of significance, see page 174), so the conclusion is that there is insufficient evidence to claim a difference in preference between variable-rate and fixed-rate loans.

binomial distribution   The "histogram" of possible outcomes from a binomial process of some specified number of trials, with a fixed probability of "success" on each.

The Binomial Expansion   The general term of the **binomial distribution** is expressed by Formula 7.1. That formula can be used to generate each and every term for all the values of $x = 0, 1, 2, \ldots n$. It is sometimes more convenient, however, to use the following scheme for generating these terms:

$$(p + q)^n = p^n + \frac{n}{1} p^{n-1} q^1 + \frac{n(n-1)}{2 \cdot 1} p^{n-2} q^2$$
$$+ \frac{n(n-1)(n-2)}{3 \cdot 2 \cdot 1} p^{n-3} q^3$$
$$+ \frac{n(n-1)(n-2)(n-3)}{4 \cdot 3 \cdot 2 \cdot 1} p^{n-4} q^4 \qquad (7.3)$$
$$+ \cdots + \frac{n!}{n!} p^{n-n} q^n$$

The sequence of terms in Formula 7.3 provides a device for producing all the terms of the binomial distribution. The terms follow a regular progression from the first term. The first term is always p^n. The exponent of the first term is divided by 1 and multiplied by the second term. The exponent of p is reduced by 1 in the second term and the exponent of q is increased by 1. It was q^0 in the first term, which is just 1, hence it does not show there. The third term adds an additional multiplier, $(n-1)$, in the numerator and 2 in the denominator. The numerator of the multiplier with each successive term has an additional element multiplied in which there is 1 less than that multiplied in for previous terms. The denominator of the term multiplier keeps multiplying in an extra element with each term that is 1 larger than the one for the previous term. This process terminates (the last term will be reached), when the term multiplier is $n!/n! = 1$. The p exponent decreases by 1 each time until, with the last term, it becomes p^0 and drops out since $p^0 = 1$. The exponent of q increases by 1 each term until it becomes q^n for the last term.

Formula 7.3 yields $n+1$ terms in the binomial expansion. The first term gives the probability of all heads and no tails; the second term gives the probability of $n-1$ heads and 1 tail; and so on. The last term gives the probability of no heads and all tails. The sum of all the terms in a particular expansion equals 1.0.

The terms for Formula 7.3 give the results that would be obtained by successively substituting $x = n, n-1, n-2, \ldots 1, 0$ in Formula 7.1 for the general term of the binomial distribution. Letting $x = 0$, for example, gives:

$$P_0 = \frac{n!}{0!(n-0)!} p^0 q^{n-0}$$

$$= \frac{n!}{n!} q^n$$

$$P_0 = q^n \tag{7.4}$$

The value of P_0 is q^n, the last term of Formula 7.3.

Substituting $x = n$ in Formula 7.1 gives:

$$P_n = \frac{n!}{n!(n-n)!} p^n q^{n-n}$$

$$= \frac{n!}{n!(0!)} p^n q^0$$

$$P_n = p^n \tag{7.5}$$

The value of P_n in Formula 7.5 is the first term of Formula 7.3.

To get all the terms of the binomial expansion, or the first few terms of the series, Formula 7.3 is probably most convenient. To obtain a single term, it is usually more convenient to use Formula 7.1, particularly if it is not an early term in the series.

Application of the Binomial Distribution to Research Problems

There are many situations that can result in two possible outcomes and where many trials can be run. The binomial distribution can be applied to such situations to estimate the probability of obtaining a particular outcome, assuming $p = 1/2$, or some other specified value, n is known, and the trials are independent.

Consider the case of the T-maze in experimental psychological research. A rat placed on the runway proceeds to the point in the maze where it has to make a choice to go either left or right. No other possibilities exist. Assuming that rats are making random choices at the choice point ($p = 1/2$), the probability of obtaining a specified number of left and right turns when n rats are run in the maze is determined using the binomial distribution.

Example   Suppose 15 rats are run in a T-maze. It turns out that 10 rats go left and 5 go right. What is the probability of obtaining this particular result if the rats are making random choices (if the probability of a right turn or a left turn is $1/2$)?

Since this is a single term of the binomial not near the beginning, it is more convenient to use Formula 7.1. Substituting $x = 5$ and $n = 15$ in Formula 7.1 gives:

$$P_5 = \frac{15!}{5!(15 - 5)!} \left(\frac{1}{2}\right)^5 \left(\frac{1}{2}\right)^{15 - 5}$$

$$= \frac{15 \cdot 14 \cdot 13 \cdot 12 \cdot 11}{5 \cdot 4 \cdot 3 \cdot 2 \cdot 1} \left(\frac{1}{2}\right)^{15}$$

$$P_5 = \frac{3003}{32768} = 0.09$$

Since $p = q = 1/2$ in this problem, it does not matter whether $x = 5$ or $x = 10$ is substituted in Formula 7.1. When p is not equal to q, however, the value of x must be related to the event for which p is the probability on a given trial, not q.

For example, consider the probability of getting a "snake eye," or 1, on exactly 4 out of 6 tosses of a die. There are six faces on a die, only one with a single dot, so if we denote p as the probability of getting a 1 on a single throw of the die, $p = 1/6$ and $q = 1 - 1/6 = 5/6$. Here, $5/6$ is the probability of not getting a 1 but instead a 2, 3, 4, 5, or 6. To obtain the probability of getting a 1 exactly four times out of six, substitute $x = 4$ and $p = 1/6$ in Formula 7.1:

$$P_4 = \frac{6!}{4!(6 - 4)!} \left(\frac{1}{6}\right)^4 \left(\frac{5}{6}\right)^2$$

$$= \frac{6 \cdot 5}{2 \cdot 1} \cdot \frac{1}{36 \cdot 36} \cdot \frac{25}{36}$$

$$P_4 = 15 \cdot \frac{25}{46656} = \frac{375}{46656} = 0.008$$

Inserting $x = 2$ (i.e., $n - 4$) in Formula 7.1 does not give the correct answer, however, unless p is designated as $5/6$ and q as $1/6$. Putting $p = 1/6$ and $x = 2$ in Formula 7.1 gives the probability of two 1s out of 6 instead of 4.

Most applications of the binomial distribution in statistical inference require addition of the probabilities for several terms rather than the use of the probability of a single term. Problem 19, therefore, is concerned with the solution of a part of a statistical inference problem more often than it represents a solution to the entire problem. In Problems 20, 21, and 22 the techniques used in Problem 19 are applied to the solution of actual statistical inference problems.

MATHEMATICAL
NOTES

This section presents a derivation for Formula 7.1. Let's assume there are only two possible outcomes of an event, A with probability p, and B with probability q, where $p + q = 1$ but p does not necessarily equal q. There is a given set of n trials resulting in x A's and $n - x$ B's. The obtained results and their associated probabilities are schematized as follows:

$$\text{A B A A B} \cdots \text{B A A A} \tag{7.6}$$
$$p \cdot q \cdot p \cdot p \cdot q \cdots q \cdot p \cdot p \cdot p$$

where there are x A's interspersed with $n - x$ B's. The product of the probabilities is commutative, so the p's and q's above can be rearranged to give:

$$(p \cdot p \cdot p \cdots p)(q \cdot q \cdot q \cdots q) \tag{7.7}$$

where there are x p's in the first parentheses and $n - x$ q's in the second parentheses. This then gives $p^x q^{n-x}$ as the probability of getting the precise order of A's and B's schematized above in Formula 7.6.

Any other sequence of A's and B's that have x A's and $n - x$ B's can be schematized and the p's and q's rearranged to give Formula 7.7. Every ordering of x A's and $n - x$ B's has the probability of $p^x q^{n-x}$. Since each of these is an independent event, the probability of any one of them occurring is the sum of their probabilities because there are no duplicate orders in the list of possible orders. This is equivalent to multiplying $p^x q^{n-x}$ by the number of such possible arrangements of x A's and $n - x$ B's since each has the same probability of occurring.

If each A and each B are different in some minor way so that each movement of a given A (or B) to the place of another A (or B) creates a different order, the total number of arrangements of the A's and B's is $n(n - 1)(n - 2) \cdots 1 = n!$ since n different ones are picked for the first spot, $n - 1$ for the second, and so on until only 1 is left.

A great many of these $n!$ different orders become indistinguishable from each other if the A's cannot be distinguished from each other and the B's cannot be distinguished from each other. Each different placement of the A's in a given dispersal arrangement among the B's gives $x(x-1)(x-2) \cdots 1 = x!$ different arrangements when the distinguishing marks on the A's are left on. Thus, $x!$ different orders collapse to 1 when the distinguishing marks on the A's are removed. This is true for every one of the possible placements of the A's. The total number of orders with distinguishing marks ($n!$) reduces to $n!/x!$ when the distinguishing marks are removed from the A's. Similarly, every different placement of the B's without distinguishing marks on them gives $(n - x)!$ different orders when the marks are distinguishable. Thus $(n - x)!$ orders reduce to 1 when the marks on the B's are removed. The total number of different orders of A's and B's when the marks are removed from both is $n!/x! (n - x)!$. When this number is multiplied by $p^x q^{n-x}$, the probability of such order, the result is Formula 7.1.

$$P_x = \frac{n!}{x!(n - x)!} p^x q^{n-x} \tag{7.1}$$

PROBLEM **20**

Using the Binomial Distribution, Make a Statistical Test of a Hypothesis, Selecting the Appropriate One- or Two-Tailed Test

The methods of Problem 19 provide a way of determining the probability of a particular distribution of two alternate outcomes under the proper conditions, for example, the probability of getting 12 heads and 4 tails in 16 coin tosses. The methods of Problem 19 are not sufficient, however, to answer the question of whether or not this particular outcome is reasonable.

In the long run, a coin should turn up heads 50% of the time if it is properly balanced. On any given sequence of 16 tosses do not expect that there will be 8 heads and 8 tails. Even if the coin is properly balanced, the number of heads will sometimes be more, sometimes less than the number of tails in a given finite sequence of trials.

If a coin is not balanced, the number of heads might consistently exceed the number of tails. How unequal does the split have to be to raise a question about the balance or fairness of the coin? Is a 12–4 split enough to render the coin suspect?

For example, in a given series of trials, suspicions are aroused about a coin because it turns up more heads than tails and a test is conducted to see if the coin is biased in favor of giving too many heads. A series of 16 tosses is decided upon as the test and the results are 12 heads and 4 tails. Is the coin biased?

The solution to this problem requires the use of Formula 7.3 to calculate the first five terms of the binomial corresponding to $x = 16, 15, 14, 13$, and 12 heads. These terms give, respectively, the probabilities for splits of 16–0, 15–1, 14–2, 13–3, and 12–4 for the division of the 16 trial results into heads and tails. These terms are as follows:

$$\left(\frac{1}{2} + \frac{1}{2}\right)^{16} = \left(\frac{1}{2}\right)^{16} + \frac{16}{1}\left(\frac{1}{2}\right)^{15}\left(\frac{1}{2}\right) + \frac{16 \cdot 15}{2 \cdot 1}\left(\frac{1}{2}\right)^{14}\left(\frac{1}{2}\right)^{2}$$

$$+ \frac{16 \cdot 15 \cdot 14}{3 \cdot 2 \cdot 1}\left(\frac{1}{2}\right)^{13}\left(\frac{1}{2}\right)^{3} + \frac{16 \cdot 15 \cdot 14 \cdot 13}{4 \cdot 3 \cdot 2 \cdot 1}\left(\frac{1}{2}\right)^{12}\left(\frac{1}{2}\right)^{4} + \cdots$$

Evaluation of these terms gives the following probabilities for 16, 15, 14, 13, and 12 heads, respectively, out of 16 trials:

$$\frac{1}{65,536} + \frac{16}{65,536} + \frac{120}{65,536} + \frac{560}{65,536} + \frac{1820}{65,536}$$

When trying to answer the question of whether a 12–4 split of heads and tails is extreme enough to cast doubt on the balance of the coin, it is not enough to consider only the probability of the 12–4 split. It is necessary to add the probabilities for those splits that are even more extreme to the probability of the 12–4 split (all the terms above), giving 2517/65536 or 0.038. The probability for the 12–4 split (exactly 12 heads and 4 tails) is 1820/65536 or 0.028.

The reason it is necessary to add the probabilities for the more extreme heads–tails splits to the probability of the 12–4 split is that any one of these events might have occurred instead of the 12–4 split and if so, the result might have been used as evidence against the fairness of the coin. Since any of these events might have occurred, the probability of getting one or the other of them is the sum of their separate probabilities. This sum, 0.038, gives the probability that at least 12, and possibly *more* heads will occur in 16 tosses of a balanced coin.

Consequently, the probability of getting at least 12 heads in 16 trials with a fair coin is 0.038. On the basis of this probability, is the coin biased in favor of heads when 12 out of 16 tosses result in heads?

In about 4 series out of 100 where a fair coin is tossed 16 times, 12 or more heads are obtained. Based on this, it is impossible to state categorically that a coin is biased in favor of heads if on a test series it gives 12 heads in 16 trials, since in about 4 series out of 100 this happens even with a balanced coin. Yet, it is disturbing to have a coin give such a disproportionate number of heads when it is specifically being tested because it is suspected of yielding too many heads, which happens only about 4 times in 100 with a fair coin. Is this particular test 1 of the 4 out of 100 in which such a result occurs with a fair coin and not 1 of the 96 times it does not occur? Or is it that a probability of 0.038, or about 4 times in 100, is too unlikely for it to have happened on this particular test?

Levels of Confidence

confidence level   The percentage of time that a statement about the location of some parameter will be correct.

When it is asked whether the probability of 0.038 for 12 or more heads in 16 trials is too small to believe that the coin is fair, a little thought makes it clear that any answer must be based on an arbitrary standard. This is because there is always a finite probability of getting any possible result, no matter how extreme, even with a fair coin. There is 1 chance in 65,536 of getting 16 heads in 16 tosses with a fair coin. There is about 1 chance in 4.3 billion of getting 32 heads in 32 tosses. It is certainly unreasonable, however, to believe the coin is fair if it gives 32 heads in 32 tosses even if it does happen on the average of 1 time in 4.3 billion attempts! At what point does it become unreasonable to believe the coin is fair even though it gives a disproportionate number of heads?

significance level   The proportion of the time that a hypothesis test will lead to a Type I error, usually 0.10, 0.05, or 0.01. The probability at which an observed deviation can no longer be attributed to chance.

Scientific custom has established two arbitrary **confidence levels** that are used more than any others to provide an answer to such questions. These are the 5% level of confidence and the 1% level of confidence. In practice, the probability, p, for a given result, such as 12 or more heads out of 16 coin tosses, is compared to either 0.05 or 0.01, with the following conclusions: (a) if $p > 0.05$, the result is "not significant"; (b) if $0.05 \geq p > 0.01$, the result is significant at the 5% level of confidence; and (c) if $p \leq 0.01$, the result is significant at the 1% level of confidence. The coin is not fair since there is less than a 5% chance of this number of heads occurring for a fair coin. The level of confidence expressed as a decimal fraction is sometimes referred to as the probability of a Type I error, which is described later in this chapter. Another interchangeable term is **significance level.**

General practice tends to conflict with what is theoretically appropriate in this situation. Before making the test, the investigator should decide what level of confidence to use. If p for the resulting outcome is less than or equal to the level selected, the result is "significant" at that level. If p for the resulting outcome of the test is greater than the confidence level selected, the result is "not significant." A real, or statistically significant difference states that the difference between what is observed (in a sample) and what is expected is not attributable to chance. A large discrepancy between the sample statistic and the population parameter lends evidence that the sample came from a population with a different parameter value. The investigator should not wait to see the result before selecting the level of confidence. To do so introduces the temptation to set the confidence level at a less stringent value (0.05 instead of 0.01) if a "significant" result is desired and the outcome fails to satisfy a more stringent criterion. On the other hand, if the obtained result yields $p \leq 0.01$, the researcher might be tempted to declare significance at the 0.01 level to increase the apparent clarity of the experiment when, in fact, there was an unwillingness to declare the result "not significant" if p had been less than 0.05 but greater than 0.01. The need for presetting the confidence level is widely ignored, however, despite the lack of theoretical justification for this practice.

There is nothing to prevent a researcher from selecting some value other than 0.05 or 0.01 for the predetermined standard to use in deciding whether the results are to be labeled as "significant" or "not significant." In fact, it is not uncommon for researchers in some areas to use the 10% level of confidence (0.10) as a criterion for identifying potentially meaningful results. It is also not uncommon to report the significance level as equivalent to the highest level of significance the results can justify. For example, if the value p for the obtained results is 0.019, the results are reported as "significant at the 0.02 level" or equivalently, "significant at the 2% level." If p is 0.0009, the results are reported as "significant at the 0.001 level." Although common, this practice is not strictly appropriate from the standpoint of the theory of statistics.

Depending on certain research or practical reasons, different levels of significance can be set. The 0.05 and 0.01 levels are conventional figures, usually used as criteria to judge whether the study is publishable or worth reporting. Despite the attack on using these conventional figures (see Barber, 1976 and Lykken, 1968), their use persists.

hypothesis testing   The procedure of forming a hypothesis and testing it by the use of some statistic.

Testing Hypotheses   In the coin tossing experiment, a hypothesis was tested, namely, that the coin was balanced. This hypothesis is stated as follows:

$$h_0 : p = 1/2$$

where h_0 = the hypothesis to be tested, and p = the true probability of a head.

The hypothesis to be tested, h_0, states that p, the true probability of getting a head on any given trial, is $1/2$. If h_0 is true, the coin is indeed balanced. Of course, in practice the number of heads obtained with an unbiased coin cannot be guaranteed

stringent criterion of significance, such as 0.01. On the other hand, if there is more concern to not miss something, use a less stringent level, such as 0.05. Table 7.2 and figure 7.2 show the relationship between the Type I and Type II errors. Propst (1988) presents an absorbing discussion of the trade-offs between Type I and Type II errors in some real world situations.

In examining figure 7.2, the diagonally cross-hatched area indicates the probability of a Type I error. The critical decision point is the point that divides the "h_0 is true" distribution so 0.05 or 0.01 of the area lies to the right of the point. By setting the probability of a Type I error, the probability of a Type II error is set. By moving the critical decision point, the Type I error becomes smaller or larger, and in return, the Type II error becomes larger or smaller.

The size of the sample is related to both types of errors. With a fixed value of α and a fixed sample size n, the value of β is predetermined. If β is too large, it can be reduced by either raising the level of α for fixed n, or by increasing n for a fixed

TABLE 7.2

Type I and Type II decision errors

		Real world	
		No difference; the null hypothesis, h_0, is correct.	A real difference; the experimental hypothesis, h_1, is correct.
Experimenter's decision	The data support the null hypothesis, h_0.	Correct decision (No error) $1 - \alpha$	Type II error β
	The data support the alternative hypothesis, h_1 (reject h_0).	Type I error (Probability $= \alpha$)	Correct decision (No error) $1 - \beta$

Figure 7.2
Relationship between h_0, h_1, Type I, and Type II errors.

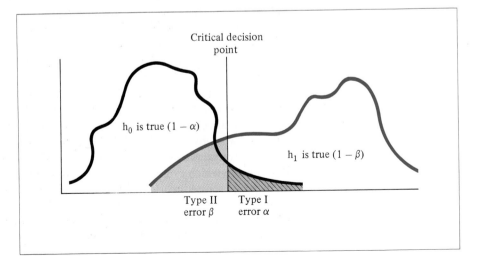

Critical decision point

h_0 is true $(1 - \alpha)$

h_1 is true $(1 - \beta)$

Type II error β Type I error α

level of α. Although β is seldom determined in an experiment, researchers can be assured that it is reasonably small by collecting a large sample.

power The probability that a test will reject the null hypothesis when it is false.

The concept of **power** of a test arises from the Type II error, β. In fact, the power of a test is defined as $1 - \beta$. The power of a test is the probability of rejecting a false null hypothesis. A test that is termed to be more powerful than another is defined as a test that is more likely to discover significant differences than another. These tests with different power levels can be further compared with an index of power efficiency. The power efficiency index usually ranges from 0.63 to 1.00. When a test has a power efficiency of 0.75 in comparison to another test, it indicates that the weaker test requires a sample size of 100 to achieve the same power level as the stronger one does with a sample size of 75. Power of test is usually not computed. Tables are available to estimate the power of a test. A more complete treatment of this material may be found in Cohen, 1988.

One-Tailed and Two-Tailed Tests

In the coin tossing experiment previously described, 12 out of 16 tosses yielded heads and 4 yielded tails. The first five terms of the binomial expansion added up to 2517/65536 or 0.038 as the probability of getting 12 or more heads in 16 tosses. This result was termed "significant at the 5% level" because 0.038 is less than 0.05, indicating that the obtained result occurs less than 5% of the time in a large number of such experiments with an unbiased or balanced coin in which the true probability of getting a head on any one trial is 1/2. Since the result was significant, h_0, the hypothesis that p, the true probability of getting a head on any one trial, equals 1/2 was rejected at the 5% level of confidence.

one-tailed test A hypothesis test in which the entire critical or rejection region is located in only one tail of the sampling distribution of the test statistic.

The test just described constitutes a **one-tailed test** of the hypothesis, h_0. It is a one-tailed test because the coin is suspected of giving too many heads and only that particular type of bias is being considered. The null hypothesis states that the coin is fair. The alternate hypothesis, h_1, states that the probability of heads is greater than 0.50. Only the probability for 12 or more heads is computed. There is no consideration of the probabilities of events involving an excess number of tails. When $p = q = 1/2$, the binomial distribution is symmetric, so the computed probabilities for 16, 15, 14, 13, and 12 *tails* out of 16 trials is exactly the same as those for the corresponding number of heads. To compute the probability of getting a division of heads and tails as extreme or more extreme than 12–4 in either direction requires adding the probabilities for 12 or more heads and the probabilities for 12 or more tails. Since the sum of the probabilities for 12 through 16 heads is approximately 0.038, twice this figure, 0.076, is the sum for the terms representing 12 or more heads *and* 12 or more tails.

If the question is merely whether the coin is biased or not, without reference to the direction of bias, then it is necessary to add the probabilities from *both* ends of the binomial expansion, taking into account all terms for outcomes as extreme or more extreme than the outcome actually obtained. It does not matter whether heads or tails comes up more often; both outcomes are considered anyway. The reason for this is that the coin is presumed to be biased if it gives an extreme result in *either* direction, too many heads *or* too many tails. Since any of these events has a finite possibility of occurring with an unbiased coin, their probabilities must all be included in the total representing the probability of getting a result as extreme or more extreme than the obtained result under the assumption that h_0 is true.

two-tailed test A test for which values can fall in either tail of the distribution and be cause for rejection of the null hypothesis. The critical or rejection region for the test is divided between the two tails of the sampling distribution of the test statistic.

When both sides of the binomial are considered in this way, the test of h_0 is called a two-tailed test. The **two-tailed test** is more commonly used than the one-tailed test. There must be special justification to use the one-tailed test. In this particular situation, if a two-tailed test is made, the value of p for the obtained outcome is 0.076. Since this probability is greater than 0.05, the result is found to be "not significant" and hence the investigator must "fail to reject" h_0. The hypothesis, h_0, remains tenable according to the 5% level of significance. Of course, if the 10% level had been preset as the criterion of significance, then h_0 would have been rejected at the 10% level of confidence since 0.076 is less than 0.10. Most researchers, however, are reluctant to use such a lenient test as the 10% level.

It should be emphasized here that using a one-tailed test would have resulted in significance at the 5% level of confidence while using a two-tailed test would not. Use of a one-tailed test, therefore, effectively permits an apparently less extreme result to achieve significance in certain cases. Although this appears to be getting "something for nothing," this is not the case due to the following reasons: (a) the use of a one-tailed test is justified only in certain cases, (b) it must be labeled as such, and (c) it places a limit on the interpretations that can be made.

In the test of the coin suspected of giving too many heads, the experiment was specifically set up ahead of time to test *only* that notion, thereby permitting the use of a one-tailed test. In such a case, h_0 is rejected *only* if the outcome shows too many heads. It is not appropriate to reject h_0 if too many tails come up, even if all 16 tosses are tails. If the experimenter had any intention to reject h_0 if an extreme number of tails appeared as well as if an extreme number of heads appeared, a two-tailed test would be obligatory since either type of outcome increases the chances of getting an extreme result, either too many heads or too many tails.

Normally, the appropriate thing to do is to make a two-tailed test rather than a one-tailed test. Certainly a coin is biased if it gives an extreme result in either direction. In any event, it is most explicitly improper to wait to see how the experiment comes out to decide whether a one-tailed or a two-tailed test is to be made and then to choose in such a way as to make the results significant with a one-tailed test when they would not have been significant with a two-tailed test. There are situations where one-tailed tests are appropriate, but these are primarily situations in which only a result in one direction has any importance or meaning. The following situation justifies the use of a one-tailed test.

A stranger comes to town and cleans everybody out betting on heads using his own coin. The sheriff seizes the stranger, impounds the coin, and sets up an experiment to see if the stranger's coin is honest. Since the stranger did not win betting on tails, it does not matter if too many tails show up in the test, only if too many heads show up. At a given level of confidence (5% level), it takes a less extreme result to incriminate the stranger. With a 5% level one-tailed test, the stranger is incriminated if 12 heads out of 16 trials result. With a two-tailed test, 13 or more heads out of 16 trials is required to yield a significant result at the 5% level. The probability of 13 or more heads plus the probability of 13 or more tails in 16 trials adds up to 0.04, which is less than 0.05, leading to the rejection of h_0 at the 5% level of confidence with a two-tailed test.

A surefire method of determining whether a one- or two-tailed test is needed is to examine how the alternative hypothesis, h_1, is stated. If it is stated indicating less than or greater than, the test is a one-tailed test. If the alternative hypothesis makes a statement that there is a difference, but no direction is given, the test is two-tailed.

Symbolically, the null hypothesis about a population parameter is

$$h_0 : \theta = \theta_0$$

As mentioned before, h_0 means null hypothesis. A null hypothesis states that the population parameter θ (lowercase Greek letter theta) is equal to the value θ_0. In the coin tossing example, the null hypothesis can be stated as:

$$h_0 : P_H = 0.5$$

where P_H = the proportion of heads.

Using symbols, this is saying the coin is fair, the proportion of heads equals the proportion of tails. The alternative hypothesis, h_1, can take the following forms:

1. $h_1 : P_H \neq 0.5$

2. $h_1 : P_H < 0.5$

3. $h_1 : P_H > 0.5$

The first form of the alternative hypothesis is nondirectional. It states that the proportion of heads is not equal to 0.5. It may be more or less, but it is not equal to 0.5. This alternative hypothesis is a statement of a two-tailed test. The second form is a one-tailed test, but a special one called a *left-tailed* test where the proportion of heads is definitely less than 0.50. In real world terms, this translates to a statement that the coin is unfair and biased toward producing too few heads. The third form is also a one-tailed test, but a *right-tailed* one. This states that the coin is unfair and biased toward producing too many heads.

Using the normal curve for illustration, figure 7.3 shows a two-tailed test of the null hypothesis using a level of significance of 0.05. Figure 7.4 is a left-tailed test, and figure 7.5 is a right-tailed test.

Application of the Binomial to a "Crucial" Experiment

After a careful study of two competing theories of animal behavior, an independent experimenter deduces consequences from the two theories that contradict each other. Theory A implies that after a specified type of training rats will jump from a jumping stand through a door with a black circle painted on a white background, preferring it to the other alternative, a door with a black cross painted on a white background. Theory B implies that with the exact same type of training, the rats prefer the door with the cross. The investigator, therefore, plans an experiment that will provide a crucial test of the two theories: 14 rats are put on the jumping stand, one at a time, and it is observed which door they prefer, having each door half the time on the left and half the time on the right to control for position effects. If the rats choose the circle, the results favor Theory A. If they choose the cross, the results are more consistent with Theory B.

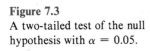

Figure 7.3
A two-tailed test of the null
hypothesis with $\alpha = 0.05$.

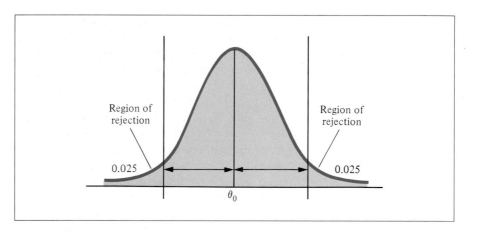

Figure 7.4
A one-tailed test (left) of the
null hypothesis with $\alpha = 0.05$.

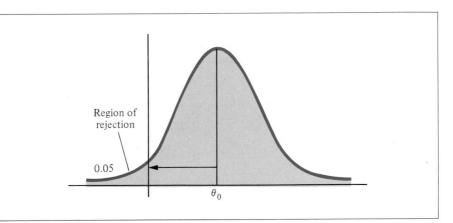

Figure 7.5
A one-tailed test (right) of the
null hypothesis with $\alpha = 0.05$.

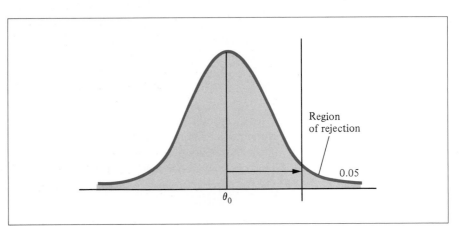

In either case, a two-tailed test is clearly demanded since an extreme result in either direction will lead to rejection of h_0 (the selection of each door will have equal probability). The 5% level of confidence is selected as the criterion (instead of the 1% level) since the number of rats is not large and this is an exploratory investigation.

The investigator runs the experiment and finds out that 12 of the 14 rats jump through the door with a circle on it and 2 rats prefer the door with the cross. The hypothesis to be tested here, h_0, is that $p = 1/2$ where p is the true probability of jumping through the door with a circle on it (or a cross). The probabilities associated with outcomes as extreme as the obtained result or more extreme include the first three terms ($x = 0,1,2$) and the last three terms ($x = 12,13,14$) of the binomial expansion for $(1/2 + 1/2)^{14}$. Since the distribution is symmetrical with $p = q$, the sum for the first three terms may be computed and doubled to give the sum for the first three *and* the last three terms together. The first three terms may be obtained using Formula 7.3 as follows:

$$\left(\frac{1}{2} + \frac{1}{2}\right)^{14} = \left(\frac{1}{2}\right)^{14} + \frac{14}{1}\left(\frac{1}{2}\right)^{13}\left(\frac{1}{2}\right) + \frac{14 \cdot 13}{2 \cdot 1}\left(\frac{1}{2}\right)^{12}\left(\frac{1}{2}\right)^2 + \ldots$$

$$\left(\frac{1}{2} + \frac{1}{2}\right)^{14} = \frac{1}{16384} + \frac{14}{16384} + \frac{91}{16384} + \ldots$$

The sum of the first three terms is 106/16384. Doubling this value to include the last three terms gives 212/16384 or 0.013. This value is less than 0.05 but not less than 0.01 and therefore the result is significant at the 5% level of confidence, but not at the 1% level.

Although h_0 cannot quite be rejected at the 1% level of significance (probability $= 0.013$), the outcome of the experiment would encourage the investigator to repeat the experiment using a larger number of rats if the 1% level had been preset. If the rats continue to prefer the door with the circle on it in the same proportion, an experiment with a larger number of rats would certainly yield a result significant at the 1% level. In no case, however, would it be appropriate for the investigator to lower the criterion of significance to the 5% level or to use a one-tailed test so that the results of the present experiment turn out to be significant if the 1% level had been preset.

SUMMARY OF STEPS

Using the Binomial Distribution to Make a Statistical Test of a Hypothesis

Step 1   Determine h_0, the hypothesis to be tested. Usually h_0 states that the true probability of a given outcome (a left or right turn in a maze) is $1/2$ where there are two possible outcomes.

Step 2   Decide whether h_0 is to be tested by means of a two-tailed test or a one-tailed test. A one-tailed test can be used only if it has been decided in advance of the experiment that h_0 will be rejected by an extreme result in one direction only, not by an extreme result in either direction.

Step 3   Determine n, the number of trials, and x, the number of trials in which a "success" occurred. A "success" is defined as an outcome in the expected direction (in the direction that results in a rejection of h_0 if enough successes occur in n trials). In a two-tailed test, either of the two possible outcomes may be designated as a "success," but usually the outcome with the greater frequency is chosen.

Step 4   Using either Formula 7.1 or Formula 7.3, determine the probability of x successes under the hypothesis that h_0 is true. This is the probability of getting exactly x successes.

Step 5   Using the same formula, determine the probability of getting $x + 1$ successes, $x + 2$ successes, and so on up to and including n successes.

Step 6   Add all the probabilities together from Steps 4 and 5 to get the probability of getting x or more successes in n trials.

Step 7   If a one-tailed test is being conducted, let p equal the probability obtained in Step 6. If a two-tailed test is being conducted, let p be *twice* the probability obtained in Step 6 (assumes two equally likely outcomes, $p = q$ in Formulas 7.1 and 7.3).

Step 8   If p is less than 0.05, reject h_0 at the 5% level of confidence. If p is less than 0.01, reject h_0 at the 1% level of confidence. If p is greater than 0.05, do *not* reject h_0. To be technically correct, the decision should be made in advance as to whether the 0.05 or the 0.01 level will be used to decide whether h_0 is to be rejected or not. In this case, h_0 is merely rejected or not rejected at the given level of confidence that has been preset as the criterion. If a one-tailed test is made, this fact should be indicated in giving the results of the test.

PROBLEM 21

Using the Normal Curve Approximation to the Binomial Distribution, Make a Statistical Test of a Hypothesis

Although it is beyond the scope of this text to do so, it is possible to prove that the binomial distribution approaches the normal distribution in form as the number of trials approaches infinity. The plausibility of this theorem can be grasped intuitively by inspection of figure 7.6, which gives the expected frequencies in 1,048,576 series for different combinations of heads and tails when tossing an unbiased coin 20 times per series. These expected frequencies are the coefficients of the terms in the binomial expansion by Formula 7.3 of $(1/2 + 1/2)^{20}$. Each of these terms has a denominator of 1,048,576. Thus, the probability in one 20-trial experiment of getting

Figure 7.6

Normal curve approximation to the binomial for $p = 1/2$ and $n = 20$.

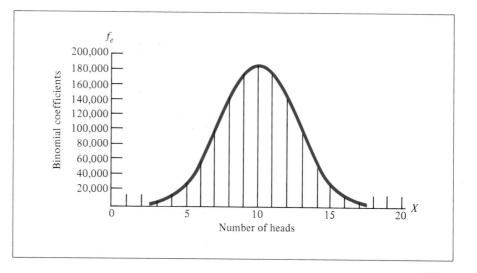

20 heads is 1/1,048,576. In 1,048,576 repetitions of this experiment (20 tosses in each experiment) it is expected that in one experiment there will be 20 heads out of 20. The sequence of terms in the binomial is represented as follows:

$$\left(\frac{1}{2} + \frac{1}{2}\right)^{20} = \left(\frac{1}{2}\right)^{20} + \frac{20}{1}\left(\frac{1}{2}\right)^{19}\left(\frac{1}{2}\right)^{1} + \frac{20 \cdot 19}{2 \cdot 1}\left(\frac{1}{2}\right)^{18}\left(\frac{1}{2}\right)^{2}$$

$$+ \frac{20 \cdot 19 \cdot 18}{3 \cdot 2 \cdot 1}\left(\frac{1}{2}\right)^{17}\left(\frac{1}{2}\right)^{3} + \ldots + \frac{20!}{20!}\left(\frac{1}{2}\right)^{0}\left(\frac{1}{2}\right)^{20}$$

There will be $n + 1$ or 21 of these terms, the first 11 of which have the following coefficients: 1, 20, 190, 1140, 4845, 15,504, 38,760, 77,520, 125,970, 167,960, and 184,756. The last 10 will be the same as the first 10 in reverse order: 184,756, 167,960, . . ., 1. Each of these coefficients is multiplied by 1/1,048,576. The sum of these coefficients must equal 1,048,576 since the sum of the probabilities for all possible events must equal 1.

Figure 7.6 shows the plot of the binomial coefficients, or expected frequencies, f_e, in 1,048,576 sets of trials for the different possible numbers of heads ($x = 0, 1, 2, 3, \ldots, 20$ when $p = 1/2$). The vertical lines above the possible values of x extend as high as the expected frequency in 1,048,576 repetitions of the experiment. These are the binomial coefficients when $n = 20$ and $p = 1/2$. The end points of the vertical lines are connected with a smooth curve. The closeness in appearance of this curve to the normal curve is obvious. Figure 7.7 shows how well the normal curve fits the binomial distribution for certain values of n and p. The larger the n, the closer the approximation.

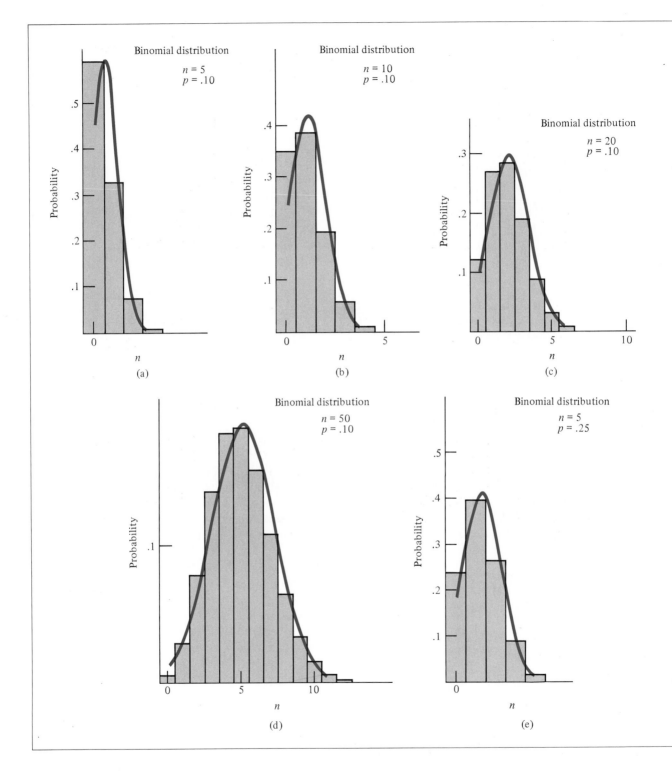

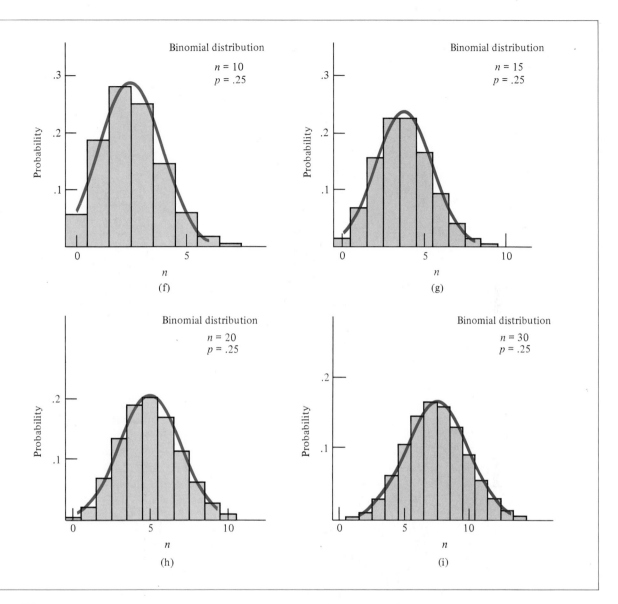

Figure 7.7
(a–i) The fit of the normal
curve and the binomial
distribution for several values
of *n* and *p*.

FORMULAS FOR THE MEAN AND STANDARD DEVIATION OF THE BINOMIAL

The mean and standard deviation of the binomial are given by the following formulas:

$$M_b = np \tag{7.8}$$

$$\text{S.D.}_b = \sqrt{npq} \tag{7.9}$$

where M_b = the mean of the binomial x scale,

S.D._b = the standard deviation of the binomial x scale,

n = the number of trials,

p = the probability of a specified outcome on any one trial (a head), and

$q = 1 - p$, the probability of the alternative outcome (a tail).

Formula 7.8 and 7.9 are derived in the Mathematical Notes section on page 193. Applying Formula 7.8 to the case where $n = 20$ gives $M_b = 20 \cdot 1/2 = 10$ and in figure 7.6, it can be observed that $X = 10$ is the middle point in the distribution where the maximum expected frequency occurs. The standard deviation for $n = 20$ is $\sqrt{20 \cdot 1/2 \cdot 1/2} = \sqrt{5} = 2.24$.

USING THE NORMAL CURVE APPROXIMATION TO DETERMINE THE PROBABILITY OF A PARTICULAR BINOMIAL OUTCOME

With $n = 20$ and $p = 1/2$, the normal curve approximates the binomial distribution with sufficient accuracy so that it can be used to obtain estimates of the probabilities of particular binomial outcomes. In the first example of Problem 19, the probability of obtaining a result of 10 rats turning left out of 15 under the assumption (h_0) that the probability of a left turn is $1/2$ was computed by Formula 7.1 to be 0.09. Using the same formula, or the binomial coefficients given above for $n = 20$, the probability for exactly 15 rats out of 20 turning left is $15504/1048576 = 0.0148$.

The area under the normal curve is now used to obtain an estimate of the probability just computed. The area under the normal curve involved is that portion of the area under the curve between 14.5 and 15.5 since the score 15 covers an entire point (15.0 ± 0.5) (fig. 7.6). There is no area associated with $X = 15.0$ in the binomial, of course, since it is a discrete distribution. The scores all occur at integral values of X, with nothing in between. The sum of the probabilities for the binomial and the total area under the standard normal curve are both 1.0, so adding up the area under the curve is equivalent to adding probabilities for discrete outcomes on the binomial.

To use the standard normal curve everything must be converted to standard scores. Remembering that the standard score Z is given by $(X - M)/\text{S.D.}$, $X = 14.5$ and 15.5 must be converted to standard scores by subtracting

M_b and then dividing by S.D.$_b$. For $n = 20$, it has already been determined that $M_b = 10$ and S.D.$_b = 2.24$. The standard scores for the two end points of the desired interval, 14.5 to 15.5, are:

$$Z_{15.5} = \frac{15.5 - 10}{2.24} = \frac{5.5}{2.24} = 2.455$$

$$Z_{14.5} = \frac{14.5 - 10}{2.24} = \frac{4.5}{2.24} = 2.009$$

If these values are used to enter the table of normal curve values (table B, Appendix B) interpolation will give the following areas from $-\infty$ to Z (column B):

for $Z = 2.455$	Area $= 0.9930$
for $Z = 2.009$	Area $= \underline{0.9777}$
	Difference $= 0.0153$

This normal curve approximation value agrees closely with the 0.0148 value computed by the exact Formula 7.1. Both values round to 0.015. With $n = 20$, these approximations are sufficiently close for most practical purposes. As n increases, the approximations improve. With n less than 20, it is safer to use Formula 7.1. A general rule for applying the normal curve to approximate the binomial is when np or $n(1 - p)$, whichever is less, is at least equal to 10.

USING THE NORMAL CURVE APPROXIMATION TO THE BINOMIAL TO MAKE A STATISTICAL TEST OF A HYPOTHESIS

Since the normal curve approximation to the binomial may be used to give approximate probabilities for particular binomial outcomes, like Problem 19, it follows that normal curve approximation can also be used to test hypotheses, like Problem 20, provided n is 20 or more.

Suppose the investigator performing the crucial experiment in Problem 20 (where 12 rats out of 14 jumped through the door with the circle) decided to repeat the experiment using 20 rats. The first experiment just failed to achieve significance at the 0.01 level. When the experiment is repeated with 20 rats, 17 out of 20 jump through the door with the circle and 3 jump through the door with the cross. Again, a two-tailed test must be run since h_0 will be rejected with an extreme result in either direction.

The statistical test conducted with the binomial requires adding the probabilities for all terms representing 17 or more choices of either alternative. This requires adding the first four terms in the binomial expansion of $(1/2 + 1/2)^{20}$ and then doubling this figure to take into account the last four terms as well. The first four binomial coefficients given above for $n = 20$ are: 1, 20, 190, and 1140. These sum to 1351. Dividing 1351 by 1,048,576, which is 2^{20}, gives 0.0013. Doubling this figure gives 0.0026 as the probability of getting a result as extreme or more extreme than 17 out of 20 rats preferring one door or the other.

Solving this same problem using the normal curve approximation to the binomial distribution gives a probability reasonably close to the 0.0026 obtained using the binomial itself. Since 17 out of 20 rats jumped through the door with the circle on it, the area under the normal curve is needed for outcomes of 17 through 20, to take into account the probability of the particular outcome *plus* those outcomes that are more extreme. This indicates that the area under the curve to the right of a score of 16.5 must be obtained since the discrete score of 17 in the binomial is represented by the area from 16.5 to 17.5 under the continuous normal curve.

The standard score corresponding to a raw score of 16.5 must be determined before entering the normal curve table to find the area to the right of 16.5. These computations are shown as follows:

$$Z_{16.5} = \frac{16.5 - M_b}{S.D._b}$$

$$= \frac{16.5 - np}{\sqrt{npq}}$$

$$= \frac{16.5 - 20 \cdot \dfrac{1}{2}}{\sqrt{20 \cdot \dfrac{1}{2} \cdot \dfrac{1}{2}}}$$

$$= \frac{16.5 - 10}{\sqrt{5}}$$

$$Z_{16.5} = \frac{6.5}{2.24} = 2.902$$

The standard score corresponding to 16.5 is 2.902. Entering the normal curve table (Appendix B, table B) with this value of Z gives an area from $-\infty$ to Z of 0.9981 or 0.0019 to the right of 16.5. Doubling this figure to 0.0038 gives the area to the right of 16.5 *plus* the corresponding area in the left tail of the distribution (below 3.5). The area below 3.5 approximates the probability for 17 or more rats preferring the door with the cross and the area above 16.5 approximates the probability for 17 or more rats preferring the door with the circle. Adding these together (by doubling one of them) gives an approximation to the probability of getting a result as extreme or more extreme than 17 rats preferring one or the other of the two possible choices.

The value of 0.0038 for the probability of 17 or more preferences in either one direction or the other computed from the normal curve approximation to the binomial compares favorably with the value of 0.0026 obtained by adding probabilities from the terms of the binomial itself. Since the obtained probability is less than 0.01, h_0 can be rejected at the 1% level of confidence.

Repetition of the experiment with a larger number of trials resulted in a statistically significant outcome even though the proportion of rats preferring the door with the circle on it actually dropped from $12/14 = 0.86$ to $17/20 = 0.85$. This comparison illustrates the fact that as the size of n increases, rejection of h_0 can be achieved with a smaller and smaller departure from 50–50 in the choices between two alternatives.

Example   Suppose 100 randomly selected respondents are asked to state a preference for candidate A or candidate B in a forthcoming election. Of the 100 respondents, 57 favored candidate A and 43 favored candidate B. Can it be concluded that there is a preference for candidate A in the population from which this sample was selected?

The hypothesis to be tested, h_0, is that the probability of a choice of A is 1/2 on any given trial. The normal curve approximation to the binomial is applied because n is well over 20. A Z-score is calculated for the point 56.5 since the area under the normal curve is to be obtained for all scores of 57 or more out of 100.

$$Z = \frac{56.5 - np}{\sqrt{npq}}$$

$$= \frac{56.5 - 50}{\sqrt{100 \cdot \frac{1}{2} \cdot \frac{1}{2}}}$$

$$Z = \frac{6.5}{5} = 1.30$$

From the normal curve table, the area from $-\infty$ to Z for $Z = 1.30$ is 0.9032, which leaves 0.0968 as the area to the right of $Z = 1.30$. Taking into account the area in the other tail by doubling 0.0968 gives 0.1936 as the total area and the probability of obtaining a result as extreme as the one obtained or more extreme. This value is larger than all the possible significance level values, indicating that there is almost one chance in five of obtaining a result as extreme as this or more extreme when the true proportion equals 1/2. Thus, even though 57 to 43 seems like a substantial difference in favor of candidate A, it does not prove to be even nearly statistically significant.

Critical Values of Z

In the previous examples using the normal curve approximation to the binomial, the probability associated with a particular outcome was determined. This probability was compared with such criterion probability values as 0.05 and 0.01. An alternate way of determining whether the result is significant or not is to compare the obtained value of Z with the critical values of Z associated with the criterion probabilities. For example, in the normal curve, the area to the right of $Z = 1.96$ is approximately 0.025. Doubling this gives 0.05. Hence, the critical value of Z for the 5% level of significance with a two-tailed test is 1.96 (and -1.96). If the obtained value of Z is 1.96 or greater or -1.96 or smaller, the result is significant and h_0 is rejected at the 5% level of confidence. If the obtained value of Z is less than

TABLE 7.3

Critical values of Z

Level of confidence	Type of test	Critical value of Z
0.10	Two-tailed	1.65, −1.65
0.05	Two-tailed	1.96, −1.96
0.05	One-tailed	1.65 (Right)
0.02	Two-tailed	2.33, −2.33
0.01	Two-tailed	2.58, −2.58
0.01	One-tailed	2.33 (Right)
0.002	Two-tailed	3.08, −3.08
0.001	Two-tailed	3.30, −3.30
0.001	One-tailed	3.08 (Right)

1.96 but greater than −1.96, the result is not significant and h_0 is not rejected. Table 7.3 gives the critical values of Z for various levels and types of tests. The test of h_0 can be carried out either by computing the probability for the given outcome or by comparison of Z with critical values of Z when the normal curve approximation to the binomial is being used. When the binomial itself is being employed this alternative way of making the test is not available.

SUMMARY OF STEPS

Using the Normal Curve Approximation to the Binomial to Make a Statistical Test of a Hypothesis

Step 1   Determine h_0, the hypothesis to be tested.

Step 2   Decide whether a one-tailed test or a two-tailed test is to be used.

Step 3   Determine n, the number of trials (usually 20 or more) and x, the number of trials in which a "success" occurs. The outcome labeled a "success" is chosen in such a way that x is greater than $n/2$.

Step 4   Determine the mean of the distribution by Formula 7.8.

Step 5   Determine the standard deviation of the distribution by Formula 7.9.

Step 6   Subtract $1/2$ point from x and convert the resulting score to a standard score (Z-score). This is done by subtracting the mean to get a deviation score and then dividing the deviation score by the standard deviation.

Step 7   Enter the normal curve table (Appendix B, table B) with the Z-score obtained in Step 6 and find the proportion of area under the normal curve beyond this value.

Step 8   If a one-tailed test is being conducted, let p equal the probability obtained in Step 7. If a two-tailed test is being conducted, let p be twice the probability obtained in Step 7.

Step 9   If p is less than 0.05, reject h_0 at the 5% level of confidence. If p is less than 0.01, reject h_0 at the 1% level of confidence. If p is greater than 0.05, do not reject h_0.

MATHEMATICAL
NOTES

In this section, Formulas 7.8 and 7.9 for the mean and standard deviation of the binomial are developed from the definitions of these statistical concepts.

Mean of the Binomial

The mean of the binomial is defined as follows:

$$M = \frac{\sum_{i=1}^{k} f_i X_i}{N} = \sum_{i=1}^{k} X_i \left(\frac{f_i}{N}\right) \tag{7.10}$$

where

k = the number of different scores,

X_i = a raw score,

f_i = the frequency for score i, and

N = the number of cases.

In Formula 7.10, the value f_i/N is the relative frequency of score X (the ratio of the number of X_i scores to the total number of scores). This can be taken as the probability of getting that particular score, X_i. Substituting the general term of the binomial, Formula 7.1, which gives the probability of a given score, X, for (f_i/N) and modifying the notation to conform with the possible scores in the binomial, Formula 7.10 may be rewritten to give:

$$M_b = \sum_{x=0}^{n} x \cdot \left(\frac{n!}{x!(n-x)!} p^x q^{n-x}\right) \tag{7.11}$$

Since the values of X in the binomial are 0, 1, 2, . . . , n the notation in Formula 7.11 is more convenient and conforms to the notation of Formula 7.1. Formula 7.11 involves $n + 1$ terms the first of which equals zero since when $x = 0$ is substituted in the expression, the result is zero. This permits elimination of the first term in the summation, making the limits $x = 1$ to $x = n$ instead of $x = 0$ to $x = n$. Also, the value of x can be cancelled with the x in $x!$ to give $(x - 1)!$ making Formula 7.11 change to:

$$M_b = \sum_{x=1}^{n} \frac{n!}{(x-1)!(n-x)!} p^x q^{n-x}$$

Remove the constant np outside the summation sign to give:

$$M_b = np \sum_{x=1}^{n} \frac{(n-1)!}{(x-1)!(n-x)!} p^{x-1} q^{n-x} \tag{7.12}$$

Let $y = x - 1$; then $x = y + 1$ and substituting for x in Formula 7.12 gives:

$$M_b = np \left(\sum_{y=0}^{n-1} \frac{(n-1)!}{y!(n-1-y)!} p^y q^{n-1-y} \right) \quad (7.13)$$

The expression inside parentheses in Formula 7.13, however, is just the sum of the binomial terms for $(p + q)^{n-1}$. These terms must add up to 1, so this gives:

$$M_b = np \quad (7.8)$$

Standard Deviation of the Binomial

The raw-score formula for the standard deviation is given by:

$$\text{S.D.} = \frac{1}{N} \sqrt{N\Sigma X^2 - (\Sigma X)^2} \quad (4.3)$$

Taking into account that some scores are repeated and squaring both sides of Formula 4.3 leads to:

$$(\text{S.D.})^2 = \frac{\sum_{i=1}^{k} f_i X_i^2}{N} - \left(\frac{\sum_{i=1}^{k} f_i X_i}{N} \right)^2 \quad (7.14)$$

As with the derivation for the mean of the binomial, the probability of a given score can be substituted for f_i/N, the relative frequency of score X_i. Also, the last term on the right of Formula 7.14 is the square of the mean, which for the binomial is $n^2 p^2$. Taking these facts into account, substituting x as the binomial score, and remembering that the possible binomial scores are $x = 0, 1, 2, \ldots, n$, Formula 7.14 for the binomial may be rewritten as follows:

$$(\text{S.D.}_b)^2 = \left(\sum_{x=0}^{n} x^2 \cdot \frac{n!}{x!(n-x)!} p^x q^{n-x} \right) - n^2 p^2 \quad (7.15)$$

The first term of the summation, for $x = 0$, will equal zero since zero times anything is still zero. Adjusting the summation limits for this fact, moving np outside the summation sign, and cancelling one x from x^2 with the x of $x!$ in the denominator, Formula 7.15 becomes:

$$(\text{S.D.}_b)^2 = \left(np \sum_{x=1}^{n} x \cdot \frac{(n-1)!}{(x-1)!(n-x)!} p^{x-1} q^{n-x} \right) - n^2 p^2 \quad (7.16)$$

Let $y = x - 1$; then $x = y + 1$. Substituting for x in Formula 7.16 gives:

$$(\text{S.D.}_b)^2 = \left(np \sum_{y=0}^{n-1} (y + 1) \cdot \frac{(n - 1)!}{y!(n - 1 - y)!} p^y q^{n-1-y} \right) - n^2 p^2$$

$$(\text{S.D.}_b)^2 = \left(np \sum_{y=0}^{n-1} y \cdot \frac{(n - 1)!}{y!(n - 1 - y)!} p^y q^{n-1-y} \right)$$

$$+ \left(np \sum_{y=0}^{n-1} \frac{(n - 1)!}{y!(n - 1 - y)!} p^y q^{n-1-y} \right) - n^2 p^2 \qquad (7.17)$$

The first of the three terms to the right of the equal sign in Formula 7.17 is like Formula 7.11 for the mean of the binomial except that np appears in front of summation sign, y is the variable symbol instead of x, and it is the sum of the binomial where there are $n - 1$ trials instead of n trials. The mean of the binomial for $n - 1$ trials is $(n - 1)p$ since with n trials it is np. The first term on the right of Formula 7.17, therefore, becomes $np(n - 1)p$. The second term on the right is np multiplied by the sum of the binomial terms for $n - 1$ trials, which equals 1, so this term is just np. Formula 7.17 reduces, therefore, to:

$$(\text{S.D.}_b)^2 = np(n - 1)p + np \cdot 1 - n^2 p^2$$

$$(\text{S.D.}_b)^2 = n^2 p^2 - np^2 + np - n^2 p^2$$

$$(\text{S.D.}_b)^2 = -np^2 + np = np(1 - p) = npq$$

$$\text{S.D.}_b = \sqrt{npq} \qquad (7.9)$$

PROBLEM 22

Using the Binomial Distribution, Test the Significance of the Difference between Correlated Frequencies

In one of the examples for Problem 21, a poll of 100 randomly selected respondents revealed that 57 favored candidate A and 43 favored candidate B. Although this did not prove to be a statistically significant difference in favor of candidate A, it would be a cause for concern among those supporting candidate B and a basis for guarded optimism on the part of supporters of candidate A.

On the basis of this set of results, supporters of candidate B might decide to repeat the poll with a larger number of cases and also to take a third poll two weeks later among the same individuals to see if there is any trend developing, either for or against their candidate. The results of this hypothetical study are shown in a **fourfold table,** table 7.4.

fourfold table   A square table consisting of two rows and two columns (four cells).

TABLE 7.4			
Results of a repeated poll with the same 200 subjects			
	Choices		
	Candidate A	Candidate B	Total
Second poll	115	85	200
Third poll	105	95	200

Two hundred subjects are randomly selected from the defined population and are asked on two different occasions, two weeks apart, to express their preference for candidate A or candidate B. The first occasion of the second poll (the poll of 100 subjects was the first poll) gives 115 out of 200 in favor of candidate A, which is 57.5%, very close to the 57% favoring A in the first poll. When the poll is repeated two weeks later, however, only 105 out of 200 favor A, a drop to 52.5%. This result might lead candidate B supporters to take hope if the difference can be shown to represent a statistically significant change as opposed to a difference that could reasonably be attributed to mere random sampling variations.

This problem can be reduced to a problem of applying the binomial to a two-choice situation as follows. Consider the behavior of the respondents on the two polls of 200 respondents. Examination of individual votes reveal that 103 cases chose A both times and 83 chose B both times. There is certainly no significant change in those cases. Two people, however, shifted from B on the first poll to A on the second and 12 shifted from A on the first poll to B on the second. This gives a 12–2 split in favor of shifting from A to B for the 14 people who changed from preferring one candidate to another.

If there is no trend in opinion, it could be assumed that a change from candidate A to B would be no more likely than a change from B to A. This would yield for h_0, the hypothesis to be tested, that the probability of a shift from A to B is 1/2. The binomial can be applied for n trials, where n is the number of changed votes, to see if the split is unbalanced enough to warrant rejection of h_0. Either the binomial itself can be used or the normal curve approximation to the binomial if n is 20 or more.

In this case, n is 14, far less than 20, so the normal curve approximation is not used. The first three terms of the binomial for $(1/2 + 1/2)^{14}$ are as follows:

$$\left(\frac{1}{2} + \frac{1}{2}\right)^{14} = \left(\frac{1}{2}\right)^{14} + \frac{14}{1}\left(\frac{1}{2}\right)^{13}\left(\frac{1}{2}\right)^{1} + \frac{14 \cdot 13}{2 \cdot 1}\left(\frac{1}{2}\right)^{12}\left(\frac{1}{2}\right)^{2} + \cdots$$

$$\left(\frac{1}{2} + \frac{1}{2}\right)^{14} = \frac{1}{16384} + \frac{14}{16384} + \frac{91}{16384} + \cdots$$

Summing the first three terms gives 106/16384. This is doubled to 212/16384 to take into account the terms in the other tail, since a two-tailed test will be made here. This gives 0.013, which is not significant at the 0.01 level, being slightly larger than 0.01.

If a one-tailed test had been performed here, the result would have been significant at the 0.01 level, since only the terms on one side would have been considered, giving a probability of approximately 0.006, which is less than 0.01. A case could be made for using a one-tailed test here since the poll was being conducted by forces favorable to candidate B who are only interested in a trend favoring their candidate. If the question is being asked, Is there a trend of any kind?, however, a two-tailed test would be called for. On the basis of the findings, supporters of candidate B might take heart that a trend may well be developing in favor of their candidate although it fails to reach significance at the 1% level for a two-tailed test.

This same type of analysis can be applied to a wide variety of situations where changes in opinion or response are being evaluated. The requirements are that the data be two-choice categorical, either "yes" or "no," as opposed to continuous, and the same subjects must be evaluated twice. This method of analysis does not apply to the situation where two different samples of people are taken on the two occasions.

statistically significant   The likelihood that an event has occurred by chance is less than some experimenter-established limit. Rejection of h_0 at some α level.

The approach described here in certain cases can lead to rather counterintuitive results. For example, if 2000 subjects are tested instead of 200 and only 14 change, giving the same split of 12–2 in favor of an A to B change, the level of significance is the same as it was with the 200 cases. The subjects who do not change do not affect the outcome in any way. It is difficult to consider a change as significant or approaching significance when only 14 out of 2000 subjects register a change. Nevertheless, if 13 out of these 14 subjects change in one particular direction, the trend is **statistically significant** at the 1% level of confidence with a two-tailed test.

SUMMARY OF STEPS

Using the Binomial Distribution to Test the Significance of the Difference between Correlated Frequencies

Step 1   Prepare a fourfold table that contains the frequencies for the following: (a) those who gave response one on both occasions; (b) those who gave response two on both occasions; (c) those who gave response one on the first occasion and response two on the second occasion; and (d) those who gave response two on the first occasion and response one on the second occasion.

Step 2   Ignore frequencies *a* and *b* in Step 1, and consider only frequencies *c* and *d*. Let $c + d = n$, and let the larger of the values *c* and *d* be *x*.

Step 3   Using the values of *n* and *x* obtained in Step 2, treat the data as a test of a hypothesis, h_0, using the binomial distribution, use the exact binomial test, as in Problem 20, or if *n* is 20 or more, use the normal curve approximation (Problem 21).

Step 4   If h_0 is rejected at a given level of significance by the test in Step 3, the difference between the frequencies is also significant at the same level. If h_0 is not rejected by the test in Step 3, the difference between frequencies is not significant.

References

Barber, T. X. (1976). *Pitfalls in human research*. New York: Pergamon Press.

Cohen, J. (1988). *Statistical power analysis for the behavioral sciences* (2d ed.). Hillsdale, N.J.: Lawrence Erlbaum Associates.

Kenney, J. M. (1985). Hypothesis testing: An analogy with the criminal justice system. *Statistics Division Newsletter.* American Society for Quality Control, 6(2), 8–9.

Lykken, D. T. (1968). Statistical Significance in Psychological Research. *Psychological Bulletin, 70,* 15–159.

Propst, A. L. (1988). The alpha-beta wars: Which risk are you willing to live with? *Statistics Division Newsletter.* American Society for Quality Control, 8(11), 7–9.

Skipper, J. K., Guenther, A. L. & Nass, G. (1967). The sacredness of .05: A note concerning the uses of statistical levels of significance in social science. *The American Sociologist, 2,* 16–18.

Exercises

7-1. According to a new visual theory developed by Professor X, placing a black dot at the center of a circle will make the circle appear larger. To test this hypothesis, an experiment was conducted in which 17 subjects were asked to state which of two circles appeared larger to them. The circles were actually equal in size but one had a black dot in it. Of the 17 subjects, 13 stated that the circle with the black dot appeared larger. Select an appropriate type of test, give h_0, carry out the test, state the result, and draw conclusions.

    a. State the null hypothesis.

    b. Is it a one-tailed or two-tailed test?

    c. Determine n and x, where n is number of trials and x is the number of subjects that stated that the circle with the black dot did not appear larger.

    d. Use Formula 7.1 for $x = 0,1,2,3,4$.

    e. Add up all the P's from (d). If two-tailed, double the sum.

    f. If the sum in (e) is less than 0.05 reject h_0.

7-2. Eighteen randomly selected congressmen are polled concerning whether they intend to vote for a particular bill coming before the house. Of these, 14 say they will vote no. State h_0 in terms of a vote outcome, select an appropriate test, make the test, and state your conclusions.

    a. State the null hypothesis.

    b. Is it a one-tailed or two-tailed test?

    c. Determine n and x, where n is number of trials and x is the number of congressmen voting "yes" for the bill.

     d. Use Formula 7.1 for $x = 0,1,2,3,4$.
     e. Add up all the P's from (d). If two-tailed, double the sum.
     f. If the sum in (e) is less than 0.05 reject h_0.

7–3.   A given manufacturing process is supposed to turn out plastic bags that can withstand a specified air pressure without breaking with only a 10% failure rate. Periodic tests are made to monitor the quality of production. The last sample of 10 bags yielded 3 bags that broke under pressure. Is the process out of control? Devise and carry out a test and draw conclusions.
     a. State the null hypothesis.
     b. Is it a one-tailed or two-tailed test?
     c. Find n and x, where n is the number of trials and x is the number of bags that broke.
     d. Using information in (c), substitute into Formula 7.3 to obtain first 3 terms.
     e. Sum the probabilities and subtract sum from 1.00.
     f. If probability in (e) is less than 0.05 reject the null hypothesis.

7–4.   A pair of dice is suspected of giving too many "snake eyes" (two ones). In a four-trial test, "snake eyes" turn up twice. Select a test, make it, state results and conclusions.
     a. State the null hypothesis.
     b. Is it a one-tailed or two-tailed test?
     c. Find n and x, where n is the number of trials and x is the number of "snake-eyes."
     d. Using information in (c), substitute into Formula 7.3 to obtain first 3 terms.
     e. Sum the probabilities and subtract sum from 1.00.
     f. If probability in (e) is less than 0.05 reject the null hypothesis.

7–5.   Although the normal curve approximation is not very good for $n = 17$, apply it to the specific outcome of a 13–4 split in question 7–1 and compare with the probability obtained from the general term of the binomial.
     a. State null hypothesis.
     b. Is the test one-tailed or two-tailed?
     c. Find n, the number of trials and x is the larger split.
     d. Use Formula 7.8 to find the mean (M_b).
     e. Use Formula 7.9 to find the standard deviation (S.D.$_b$).
     f. Subtract 1/2 unit from x to get X_0.
     g. Calculate $Z = \dfrac{X_0 - M_b}{\text{S.D.}}$.
     h. Find the probability of Z to $+\infty$.
     i. If the probability is less than 0.05 reject h_0.

7–6.   Using the normal curve approximation, even though n is only 18, find the probability for the 14–4 split in question 7–2 and compare it with the corresponding value as computed by the general term of the binomial.
     a. State null hypothesis.
     b. Is the test one-tailed or two-tailed?
     c. Find n, the number of trials and x is the larger split.
     d. Use Formula 7.8 to find the mean (M_b).

e. Use Formula 7.9 to find the standard deviation (S.D.$_b$).

f. Subtract $1/2$ unit from x to get X_0.

g. Calculate $Z = \dfrac{X_0 - M_b}{\text{S.D.}}$.

h. Find the probability of Z to $+\infty$.

i. If the probability is less than 0.05 reject h_0.

7–7.   Determine the probability in question 7–5 using the ordinate method and the normal curve. (See Mathematical Notes for Problem 15 on page 117.)

   a. Find Z for 12.5.

   b. Find Z for 13.5.

   c. Find the mean of Z's in (a) and (b).

   d. Find the ordinate (height) corresponding to Z in (c). Use the normal curve table (Appendix B, table B).

   e. Compute base $= 1/\text{S.D.}_b$.

   f. Compute area $=$ base $\times$ height.

7–8.   Determine the probability in question 7–6 using the ordinate method and the normal curve. (See Mathematical Notes for Problem 15 on page 117.)

   a. Find Z for 12.5.

   b. Find Z for 13.5.

   c. Find the mean of Z's in (a) and (b).

   d. Find the ordinate (height) corresponding to Z in (c). Use the normal curve table (Appendix B, table B).

   e. Compute base $= 1/\text{S.D.}_b$.

   f. Compute area $=$ base $\times$ height.

7–9.   An audience fills out a questionnaire exploring their attitudes about India and then sees an informational film on India followed by a repetition of the original questionnaire to check for changes in attitude. On one particular question, the breakdown of responses was as follows:

	Yes I	No I
Yes II	72	9
No II	16	48

Select a test, make it, state results, and give conclusions.

   a. Determine a, b, c, and d.

   b. Compute $c + d$ and set equal to n.

   c. Determine the larger of c and d and set equal to X.

   d. Subtract 0.5 unit from X, set equal to X_0.

   e. Determine M and S.D. (Formulas 7.8, 7.9).

   f. Find $Z = \dfrac{X_0 - M_b}{\text{S.D.}_b}$.

   g. Find the probability of Z to $+\infty$.

   h. If the probability is less than 0.05 reject h_0.

7–10. A physical educationist claims to be able to improve people's golf scores through hypnosis. To test this, 100 golfers play one game, receive hypnotic suggestion to do better and then play a second game. The results were as follows: 55 did better the second time, 8 had the same score, and 37 did worse. Evaluate the educationist's claim.

a. Determine a, b, c, and d.
b. Compute $c + d$ and set equal to n.
c. Determine the larger of c and d and set equal to X.
d. Subtract 0.5 unit from X, set equal to X_0.
e. Determine the mean and the standard deviation (Formulas 7.8, 7.9).
f. Find $Z = \dfrac{X_0 - M_b}{\text{S.D.}_b}$.
g. Find the probability of Z to $+\infty$.
h. If the probability is less than 0.05 reject h_0.

7–11. You are interested in seeing if dogs prefer to eat brand X over brand Y. You test 100 dogs and see which brand each dog prefers. How many dogs would have to prefer brand X for you to get significant results at the 0.01 level?

a. Determine n, the number of trials, and p, the proportion of dogs preferring X over Y under the null hypothesis.
b. Find Z corresponding to 0.01 level.
c. Find M_b, S.D.$_b$ (use Formulas 7.8, 7.9).
d. Substitute values into the formula $X = Z\,(\text{S.D.}_b) + M_b + 0.5$.

7–12. Airlines and hotels often grant reservations in excess of capacity to minimize losses due to no-shows. Suppose the records of an airline show that on the average 12% of their prospective riders will not claim their flight reservation. If the airline accepts 320 reservations and there are only 300 seats in the airplane, what is the probability that all riders who arrive to claim a seat on the plane will receive one? (Use normal approximation to binomial.)

a. Determine n, the number of trials.
b. Determine p and q.
c. Find the mean.
d. Find the standard deviation S.D.$_b = \sqrt{npq}$.
e. Find the difference between the number of reservations and the number of seats.
f. Determine the lower limit of differences found in (e) $X_{LL} =$.
g. Substitute values into the formula $Z = \dfrac{X_{LL} - M_b}{\text{S.D.}_b}$.
h. Find probability of $-\infty$ to $|Z|$. ($|Z| =$ absolute value of Z, ignoring the sign.)

7–13. For $n = 20$, $p = 0.6$ in a binomial experiment.

a. $P\,(x = 6)$
b. $P\,(x < 5)$
c. $P\,(8 < x < 10)$
    1. Use Formula 7.1 $P = [\ \ !/\ !(\ -\ )!]\ (.\ \ )\,(.\ \ ) = $.
    2. Use Formula 7.1 five (5) times for $X = 0,1,2,3,4$.
    3. Add up the probabilities $P_0 + P_1 + P_2 + P_3 + P_4 =$.
    4. Use Formula 7.1 for $X = 9$.

7–14.   What is meant by the term "levels of confidence"? (See pages 174–77 this chapter.)

7–15.   Which test, one-tailed or two-tailed, for a fixed α, will yield a greater number of null hypothesis rejections? Why?

7–16.   Expand $(1/6 + 5/6)^3$. Use Formula 7.3.

7–17.   What is the sum of the first 3 terms with the 3 last terms of the binomial expansion $(1/3 + 2/3)^9$? Use Formula 7.3.

7–18.   In a binomial experiment, a coin is tossed 26 times. The coin is biased toward tails, i.e., tails 0.62, heads 0.38. Find the following probabilities.
      a. $P(x = 12)$
      b. $P(x > 14)$
         1. Find n.
         2. Establish p and q.
         3. Use Formula 7.1.
         4. Use the normal approximation to find M, S.D., and Z.

7–19.   Compute the following (see pages 167–70).
      a. 6!
      b. 11!/8!
      c. 20!/(17!)(3!)
      d. 0!/0!

7–20.   Give an example that demonstrates the statement that "fail to reject h_0" does not mean we accept h_0. (See pages 175–77.)

Chapter 8

Testing Hypotheses with Continuously Measured Variables

continuous variable   A variable that can assume any value within some range of values. Observations on a continuous variable are the result of measurements.

When there are many possible outcomes from a particular experimental trial instead of just two, the binomial distribution cannot be employed. For example, if a group of students randomly selected from a large population obtains a certain mean score on an achievement test, the binomial cannot be used to determine how well that sample mean approximates the mean of the population from which the sample was drawn. This chapter considers the methods that are adapted to the problems of statistical inference with **continuously measured variables.** A full understanding of the theory of statistics upon which these methods are based requires considerably more mathematical background than has been assumed for this text. The formulas that are used, therefore, will in most cases have to be taken on faith by the student.

Many of the more theoretical topics, such as the mathematical development of unbiased estimates, are not covered in this text. Interested students with sufficient mathematical background will be able to consult one or more of the many available books on mathematical statistics to satisfy their curiosity about the origins of these methods. See Hoel (1971), or Brunk (1965).

In the section of Chapter 7 describing the binomial distribution, the population proportion is referred to as a *parameter,* and the sample proportion is called a *statistic.* In this chapter, all population entities are referred to as parameters of the population and all sample entities are called statistics. In problems concerning means, the population mean (a parameter) is symbolized as μ (lowercase Greek letter mu). The symbol M, introduced in Chapter 3, is the symbol for the sample mean. For standard deviations, the population standard deviation (a parameter) is written as σ (lowercase Greek letter sigma). The sample standard deviation (a statistic) is either S.D. or s, depending on whether N or $N - 1$ is used in the denominator of the formula. Lowercase sigma squared (σ^2) is used to represent the population variance. S.D.^{2} or s^2 are the symbols for the sample variance. For correlations, ρ (lowercase Greek letter rho) is the population parameter, while r is the sample statistic.

THE SAMPLING DISTRIBUTION

One of the most important concepts in the theory of statistics is the notion of the sampling distribution. Each statistic, M, S.D., and others, has its own sampling distribution. A hypothetically correct, but impractical, method of constructing a sampling distribution for the mean is described by the following example. Consider an infinitely large population (or at least a very, very large one) with a true mean of μ

Figure 8.1
Sampling distribution of M
formed by repeated sampling.

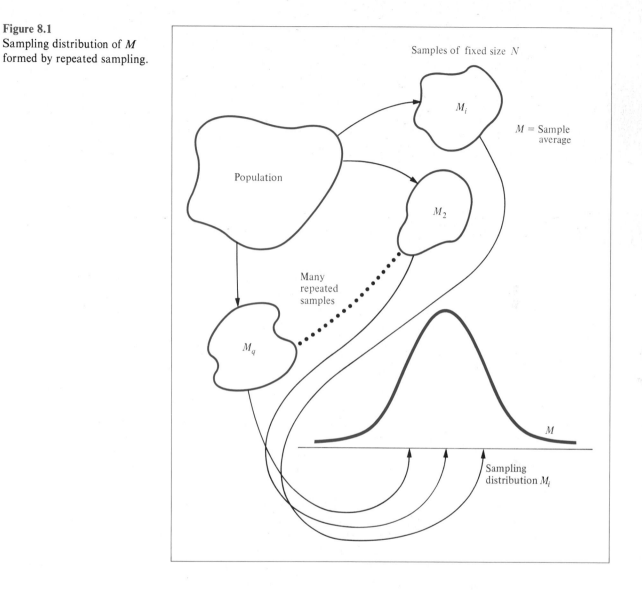

Samples of fixed size N

M_i

Population

$M = $ Sample average

M_2

Many
repeated
samples

M_q

M

Sampling
distribution M_i

(lowercase Greek letter mu) and a true standard deviation of σ (lowercase Greek letter sigma). Draw a sample of size N at random from this population and compute its mean to be M. Now draw a second sample of the same size, N, and find its mean, M_2. Repeat this process again and again to find M_3, M_4, $\cdots$, M_q where q is a very large number, always using the same sample size, N.

A frequency distribution of the M_i's form the sampling distribution of **sample means,** where the mean of the sample means is μ (Greek mu) and the standard deviation is defined with Formula 8.1a (fig. 8.1).

Suppose the population being treated here consists of all males in the United States and the variable being measured is height in inches. For simplicity, assume that $\mu = 68$ inches and $\sigma = 3$ inches. If this is true, about 68% of the males in the

sample mean   The mean, or average, of a sample.

Figure 8.2
Raw-score distribution and
sampling distribution.

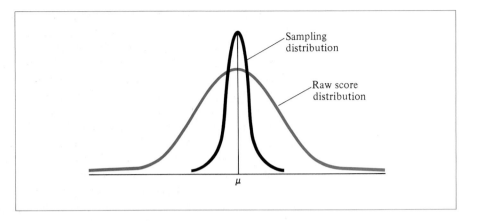

United States have heights between 65 inches and 71 inches, 95% have heights between 62 inches and 74 inches, and over 99% have heights between 59 inches and
77 inches, since height is approximately normally distributed in the population.
Scores of individuals clearly scatter over a very wide range.

The mean scores of samples of size N, however, do not scatter as widely as the
scores for individuals. This phenomenon is illustrated in figure 8.2. Both distributions center about the same point, μ, the population mean, but the variability of the
sampling distribution is considerably smaller than the variability of the raw-score
distribution. In fact, it can be proved mathematically that when the population size
is very large the variability of the sampling distribution of the mean is given by:

$$\sigma_M = \frac{\sigma}{\sqrt{N}} \tag{8.1[1]}$$

where σ_M = the true standard deviation of the sampling distribution of
means, the true **standard error of the mean,**

standard error of the mean
The standard deviation of the
sampling distribution of
sample means.

σ = the true population standard deviation, and

N = the sample size for sample means making up the sampling
distribution.

1. The true variability of the sampling distribution of the sample means is really represented by the
formula:

$$\sigma_M = \frac{\sigma}{\sqrt{N}} \sqrt{\frac{N_p - N}{N_p - 1}} \tag{8.1a}$$

where N_p = the size (number of observations) of the population, and
N = the sample size.

Formula 8.1 is used when the population size is unknown and/or assumed to be very large (unmeasurable). When the population is finite but uncountably large, the term $\sqrt{\dfrac{N_p - N}{N_p - 1}}$ approaches 1, leaving
the Formula 8.1. However, for finite countable populations where N is known and manageable, Formula
8.1a is used to obtain a more accurate measure of σ_M.

Thus Formula 8.1 shows that the true standard deviation of the sampling distribution, called the "true standard error of the mean," is always smaller than σ (except when $N = 1$) and gets progressively smaller by comparison as N increases in size. Since σ is divided by the square root of N, however, it is clear that the law of diminishing returns operates. That is, as N gets larger, it takes a bigger and bigger jump in N to get the same amount of decrease in the size of σ_M.

The smaller σ_M is, the more tightly sample means are clustered about μ, the population mean, which is at the center of both the raw-score distribution and the sampling distribution. This indicates that as the size of the sample increases, the mean of the sample is expected to be closer and closer to the true mean since *all* sample means tend to get closer and closer to the true mean as N increases.

The actual size of σ_M provides a way of determining how accurate the sample mean might be as an estimate of the population mean, μ. With a normal population raw-score distribution, the sampling distribution of means is normal in shape.

In the case of a non-normal population raw-score distribution, the sampling distribution of means approaches a normal distribution as the sample sizes increase. This particular sampling distribution has a mean of μ and a standard deviation of $\sigma/\sqrt{n}$. Although stated informally here, a formal mathematical derivation is available for this **central limit theorem.** The amazing aspect of this theorem is that no matter how extremely skewed or irregular the population distribution is, a nearly symmetrical normal distribution of the sample mean is generated for large sample sizes.

Knowing that the sampling distribution is normal or nearly normal, and knowing the value of σ_M permits a rather precise specification of the limits within which sample values scatter about μ, the true mean of the population. Thus 68% of sample means lie within the interval $\mu \pm 1.0\ \sigma_M$; 95% of the sample means lie within the interval $\mu \pm 1.96\ \sigma_M$; and 99% of the sample means lie within the interval $\mu \pm 2.58\ \sigma_M$. Of course these are theoretical values based on a theoretically derived sampling distribution. An empirically derived sampling distribution obtained in the manner just described might depart slightly from these precise figures. On the average, however, over many attempts the results should equal these theoretical values.

These facts permit some rather direct inferences concerning the accuracy of a sample mean as an estimate of the true mean. If the population is normal in form, then 95% of the sample means fall within 1.96 σ_M of the true mean and 99% of the sample means fall within 2.58 σ_M of the true mean. Suppose σ_M is known to be 0.3 inches for the variable, height, in a certain population. Then, multiplying 0.3 by 1.96 and 2.58 gives 0.588 and 0.774, respectively, indicating that 95% of the sample means of size N is within 0.588 inches of the true mean and 99% is within 0.774 inches of the true mean. Only 5% of the sample means differs from the true mean by more than 0.588 inches and only 1% of the sample means differs from the true mean by more than 0.774 inches.

The usual situation, of course, is that an investigator has only one sample and does not know the true mean. The investigator prefers to use the mean of the sample to get some information about the location of the true mean. If the information given in the example above is available, it is possible to assert that the sample mean is not

central limit theorem The mathematical theorem stating that for any population with finite variance, if the sample size is large enough, then the distribution of sample means will be approximately normal.

more than 0.588 inches or 0.774 inches away from the true mean, depending upon the investigator's willingness to take risks. If 0.588 is used, there is a 5% chance of being wrong; if 0.774 is used, there is a 1% chance of being wrong. In either case, however, a limit on the expected maximum discrepancy between the true mean and sample mean has been placed, and an associated probability value is determined to tell how likely it is that the value obtained is correct. This type of statement certainly provides useful information that has a direct bearing on the accuracy with which the sample mean is estimating the population mean.

Converting Sample Means to Z-Scores

As previously stated, the sampling distribution for the mean is normal in form provided that the samples are drawn randomly from a normally distributed population, that is, in such a way that each member of the population has an equally likely chance of being selected for the sample. In fact, the sampling distribution of means will approach normality as the size of the sample increases even if the population is not normally distributed. For normally distributed variables and all variables with large samples, therefore, the sampling distribution can be taken to be normal for all practical purposes. When sample means are converted to standard score (Z) values with respect to the sampling distribution, then, it is possible to use the normal curve tables to determine the percentage of sample means having Z-scores in various parts of the sampling distribution. Converting a sample mean to a Z-score relative to the sampling distribution requires subtracting μ, the population mean, from M, the sample mean, and dividing the difference by σ_M, the standard deviation of the sampling distribution. This is analogous to the conversion of a raw score to a standard score (Z) by the formula $Z = (X - M)/\text{S.D.}$ In this case, the mean of a raw-score distribution is subtracted from the raw score and the difference is divided by the standard deviation of the raw-score distribution. If the raw-score distribution is normal in form, the Z-scores derived from these raw scores are also normally distributed. Similarly, if the sample means are normally distributed, the Z-scores into which they are converted by the formula $Z = (M - \mu)/\sigma_M$ are also normally distributed.

When Z-scores are normally distributed, the normal curve tables are used to determine the percentage of Z-scores above or below a certain point in the distribution or between any two points. By taking a given sample mean and converting it into a Z-score with respect to its own sampling distribution using the formula $Z = (M - \mu)/\sigma_M$, it is possible to determine from the normal curve tables for that value of Z what proportion of sample means are larger or smaller. The percentage of sample means falling between the value of Z and the true mean, μ, can also be determined using the normal curve tables.

Converting the sample means to normally distributed Z-scores with respect to the sampling distribution of means is accomplished by using the formula $Z = (M - \mu)/\sigma_M$. Unfortunately, this conversion cannot be carried out in most cases because neither μ, the population mean, nor σ_M, the true standard error of the mean, is known. The value of σ_M is not known because by Formula 8.1, $\sigma_M = \sigma/\sqrt{N}$, it is a function of σ, the population standard deviation, which itself is unknown.

Not knowing μ, the true population mean, proves to be less bothersome, however, than not knowing σ, the population standard deviation. When a particular statistical hypothesis about means is to be tested, it ordinarily involves a specified hypothetical value for μ, the population mean and central value in the normal sampling distribution. The variability of the sampling distribution, however, is not so specified by a particular hypothesis. It is necessary, therefore, either to know what σ is or to estimate it so that a true or estimated value of σ_M can be obtained. Otherwise, it is not possible to convert the sample mean to a standard score (Z) relative to the sampling distribution.

ESTIMATING THE STANDARD ERROR OF THE MEAN

Formula 8.1 states that the true standard error of the mean, σ_M, is given by $\sigma/\sqrt{N}$, that is, the true standard deviation of the population for the variable in question divided by the square root of the sample size. If σ is known, then the value of σ_M becomes readily available since N is always known. Unfortunately σ is not known since, at least in theory, the population is infinite in size and it is impossible to compute the standard deviation for an infinitely large population.

Although σ is ordinarily unavailable making it impossible to compute σ_M, it is possible to obtain an estimated value of σ_M, (s_M called the "estimated standard error of the mean"), by substituting an estimate of σ in Formula 8.1. The estimated value of σ is derived from the sample that is drawn from the population, using the following formula:

$$s = \sqrt{\frac{\Sigma x^2}{N-1}} \tag{8.2}$$

where s = an estimate of the population standard deviation in variable X,

x = a deviation score in variable X, and

N = the sample size.

The only difference between Formula 8.2 and the formula for the standard deviation of a sample, S.D., is that the sum of the squared deviations is divided by $N-1$ instead of N. The effect of this is to make s slightly larger than S.D., especially for small samples when $N-1$ and N are sufficiently different to affect the outcome. With large samples S.D. and s become closer and closer to one another and either can be used as an estimate of σ.

The reason s must be used as the estimate of σ instead of the S.D. is that the S.D.^{2} is a **"biased"** estimate of σ^2. Bias here indicates that the expected value of the distribution estimates is not equal to the parameter estimated. If all possible samples of size N are drawn from the population and S.D.^{2} is calculated for each and then the mean of the S.D.^{2}'s is computed, this mean is not equal to σ^2. Figure 8.3 depicts the difference between biased and unbiased estimates. Theta (θ) is a general representation of a population parameter (μ and σ^2). Theta with a hat ($\hat{\theta}$) is a general representation of a sample statistic (M or S.D.). $E(\hat{\theta})$ is the expected value of the sample statistic (the mean of all the sample means).

biased estimator   An estimator that does not approximate the parameter.

Figure 8.3
Comparison of unbiased
estimator (a) and biased
estimator (b).

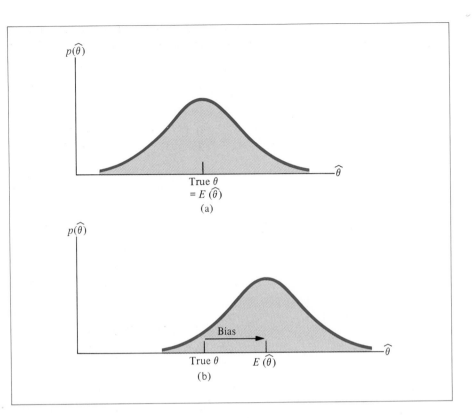

Mathematically the average value of S.D.^{2} taken over a very large number of random samples is slightly smaller than σ^2. Also, the slight correction involved in dividing Σx^2 by $N - 1$ instead of by N is just sufficient to adjust for this bias so that the average value of s^2 over a very large number of samples equals σ^2. Thus, s^2 is an **unbiased estimate** of σ^2. The sample mean, M, fortunately, is an "unbiased" estimate of μ, the population mean, so no correction is necessary in estimating μ.

Since the S.D. is sometimes known when s is needed, and vice versa, it is convenient to change one to the other using one of the following two formulas:

unbiased estimate An estimator whose expected value is equal to the value of the parameter to be estimated.

$$s = \sqrt{\frac{\Sigma x^2}{N - 1}} = \sqrt{\frac{\Sigma x^2}{N}} \sqrt{\frac{N}{N - 1}} = (\text{S.D.}) \cdot \sqrt{\frac{N}{N - 1}} \quad (8.3)$$

$$\text{S.D.} = \sqrt{\frac{\Sigma x^2}{N}} = \sqrt{\frac{\Sigma x^2}{N - 1}} \sqrt{\frac{N - 1}{N}} = s \cdot \sqrt{\frac{N - 1}{N}} \quad (8.4)$$

To convert s to S.D., multiply s by the square root of $N - 1$ over N. To convert S.D. to s, multiply S.D. by the square root of N over $N - 1$.

The standard error of the mean, an estimate of the true standard error of the mean, σ_M, then, can be obtained using the following formula:

$$s_M = \frac{s}{\sqrt{N}} \tag{8.5}$$

Symbols are defined as before. An alternate mathematically equivalent formula for the standard error of the mean is given by:

$$s_M = \frac{\text{S.D.}}{\sqrt{N-1}} \tag{8.6}$$

That these two are equivalent can be shown as follows:

$$s_M = \frac{s}{\sqrt{N}} = \frac{\sqrt{\dfrac{\Sigma x^2}{N-1}}}{\sqrt{N}} = \sqrt{\frac{\Sigma x^2}{N(N-1)}} = \frac{\sqrt{\dfrac{\Sigma x^2}{N}}}{\sqrt{N-1}} = \frac{\text{S.D.}}{\sqrt{N-1}}$$

The choice of Formula 8.5 over Formula 8.6 depends on whether s or S.D. is already known and sometimes on the convenience of taking a square root. If $N = 26$, obviously Formula 8.6 is easier to use for hand calculations.

Converting Means to *t*-Scores Using s_M   If μ_0 is taken to be a hypothetical true mean of a population distribution and hence the center of the normal sampling distribution of means as well, it is possible to convert mean scores to standard scores by the formula $t = (M - \mu_0)/s_M$. In this case, μ_0, the hypothetical true mean (generally the numerical value given in the null hypothesis), is substituted for the unknown μ, the true mean, and the estimated standard error of the mean, s_M, is substituted for σ_M, the unknown true standard error of the mean. These standard scores are designated as *t-scores,* instead of *Z*-scores, to distinguish them from the standard scores that are obtained by the formula $Z = (M - \mu)/\sigma_M$.

It has been stated that converting sample means to *Z*-scores by the formula $(M - \mu)/\sigma_M$ gives scores that are normally distributed in most situations. What about *t*-scores? Are they normally distributed? Unfortunately, they are not normally distributed unless $\mu_0 = \mu$ and $s_M = \sigma_M$. Again, testing a certain hypothesis about the true mean, μ, can be accomplished without knowing μ, using μ_0 for the purpose of calculating the standard score for a given sample mean. Distortions in the distribution of the *t*-scores resulting because μ_0 does not equal μ, do not interfere with the test of the hypothesis since if μ_0 does equal μ, the hypothesis being tested, there is no distortion from this source. Distortions in the distribution of *t*-scores because s_M does not equal σ_M are another matter. These distortions do affect the distribution of the *t*-scores so that they no longer follow a normal distribution. When s, the sample estimate of σ, is too small, s_M will be too small and t will therefore be too large. When s, the sample estimate of σ, is too large, s_M will be too large and t will be underestimated. Thus, when s is used as an estimate of σ in converting sample

t-distribution A distribution based on forming confidence intervals when we must estimate the population variance from the variance of a small sample.

degrees of freedom A technical term measuring the number of freely, or independently, varying data. Applies to the chi square distribution, the F distribution, and the t distribution.

means to standard scores (t-scores), some of the t values will be larger than they should be and some of them will be smaller than they should be as a consequence of errors in the values of s as estimates of σ. The shape of the distribution of t-scores, therefore, is not necessarily normal in form. As a consequence, the normal curve tables cannot be used to determine the percentage of t-scores above or below a certain t value.

Student's t Distribution   Means converted to standard scores using $t = (M - \mu_0)/s_M$ instead of $Z = (M - \mu)/\sigma_M$ yield scores that are not normal in form but instead follow the t distribution. The **t distribution** consists of a family of curves with a different curve for each value of df, the number of **degrees of freedom,** which is a function of sample size. These curves approach more and more closely to the normal curve as the number of degrees of freedom (df) increases. In computing a t-score for a sample mean relative to the sampling distribution of means, the number of degrees of freedom, df, equals $N - 1$ (the number of cases in the sample minus 1). Therefore, the particular t distribution that is appropriate for describing the distribution of t-scores being considered comes closer and closer in form to the normal distribution as the size of the sample increases, since $N - 1$ is very close to N. The normal distribution is a special case of the t distribution when df is infinite. In practice the particular t distribution appropriate for a given size sample becomes very close to the normal distribution long before the sample size approaches infinity. Even with samples of size 49 or 50 degrees of freedom, the t distribution is close to normal in shape. With a sample of size 100, or $df = 99$, there is very little difference between the t distribution and the normal distribution.

Figures 8.4 and 8.5 show a normal distribution and a t distribution that illustrate the difference between the two curves. The t distribution is taller and narrower in the middle and has higher tails, because some t values are overestimated and others are underestimated as a consequence of using the approximate s_M instead of using the true σ_M to compute the standard score. This indicates that the t distribution has more area in the tails than the normal distribution. Both the standard t distribution and the standard normal distribution have a total area under the curve equal to 1.00 but it is distributed differently in the two curves. In the two tails of the normal curve, above the point 1.96 and below -1.96, there is a total area of about 0.05, half in each tail. Beyond ± 2.58, there is a total area of about 0.01, 0.005 in each tail. The corresponding figures for the t distribution are larger, depending on the number of degrees of freedom. As the number of degrees of freedom increases, the areas in the tails of the t distribution get smaller and smaller, approaching those for the normal curve as a limit.

The difference between the t distribution and the normal curve is very apparent in the critical values associated with the commonly used levels of significance. The critical values of t are shown in the table of the t distribution (Appendix B, table C). At the bottom of the table, with $df = \infty$, the normal curve critical values are given. These are 1.64 for $p = 0.10$ (two-tailed test, or 0.05 for a one-tailed test), 1.96 for $p = 0.05$ (two-tailed test), 2.33 for $p = 0.02$ (two-tailed test, or 0.01 for

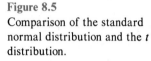

Figure 8.4
Normal curve and *t*
distribution.

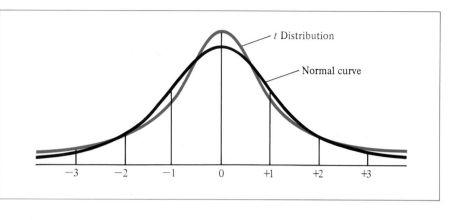

Figure 8.5
Comparison of the standard
normal distribution and the *t*
distribution.

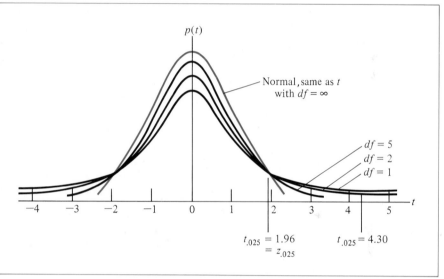

a one-tailed test), and 2.58 for $p = 0.01$ (two-tailed test). Notice that the critical values of *t* required to achieve significance get larger and larger as the number of degrees of freedom (and consequently the sample size) gets smaller and smaller. With only 10 degrees of freedom, for example, the 5% and 1% level two-tailed critical values are large, $t = 2.23$ and 3.17, respectively, compared to the critical *Z* values of 1.96 and 2.58 required for large samples.

Figure 8.5 shows the changes in the shape of the *t* distribution as the sample size changes. It also shows that as the sample size increases, the *t* distribution approaches the standard normal curve. For a sample size of 3, or degrees of freedom 2, the *t* value necessary to cut off 2 1/2% to the right of the value is 4.30. With increased degrees of freedom, the *t* value gets smaller and smaller until it reaches 1.96, the *t* value for the standard normal curve.

STATISTICS IN THE WORLD AROUND YOU

A child psychologist wonders whether children who have manic-depressive symptoms exhibit more or less creativity than other children. The psychologist wishes to compare a sample mean, M, of 108 with a known population mean, $\mu = 100$. Only 12 of the children's parents allow them to participate in the study.

With a $M = 108$, S.D. $= 1.42$, and $\alpha = 0.05$, what can the psychologist conclude? The first step is to state the hypotheses: h_0, that the sample mean of 108 is statistically equivalent to the population mean of 100; and alternatively, the sample mean of 108 is *not* equivalent to the population mean of 100.

All the conditions for a t-test are met. Since this is a two-tailed test, the psychologist wants an area $0.05/2 = 0.025$. Table C in Appendix B indicates that for $df = 11$ (there were 12 children in the sample) and an area of .025, the critical points are 2.20. Using the formula for the t-test, a test statistic of 18.6 is calculated. This test statistic is larger than the critical value 2.20, so the hypothesis that the sample mean of 108 is statistically equivalent to 100 is rejected. What does the child psychologist conclude about the creativity of manic-depressive children?

To summarize, sample means from normally distributed populations yield a normal distribution with a mean equal to the population mean, μ, and a standard deviation equal to the population standard deviation, σ, divided by the square root of N, the sample size. This standard deviation of the sampling distribution is called the "true standard error of the mean," and is denoted by the symbol σ_M. Sample mean values are converted to standard score values relative to this sampling distribution of means by the formula $Z = (M - \mu)/\sigma_M$. The normal curve tables are used to find the percentage of cases above and below any such Z values obtained from sample means. Since μ and σ are not ordinarily known, however, it is usually impossible to use this formula to convert sample means to standard scores relative to the sampling distribution of means. Instead, the standard error of the mean is estimated by Formulas 8.5 or 8.6, a hypothesized true mean, μ_0, is selected, and sample means are converted to standard scores by the formula $t = (M - \mu_0)/s_M$. Even if $\mu = \mu_0$, t-scores still will not be normally distributed due to distortions in the t values resulting because s_M is sometimes too large and sometimes too small. When $\mu = \mu_0$, however, the t-scores follow a known form, the t distribution, which varies with the number of degrees of freedom. As the number of degrees of freedom increases, the t distribution approaches the normal distribution more and more closely. Critical values of t required for significance are larger than those required for Z-scores and the normal distribution. These values may be found by consulting the table of the t distribution (Appendix B, table C).

PROBLEM 23

Test the Hypothesis That the Mean of the Population Is a Certain Value

There are occasions in social science research where it becomes important to test a hypothesis such as $\mu = C$, where μ is the population mean for a given variable and C is some specified constant. Then, μ_0, the hypothesized true mean, equals C. A theory might predict, for example, that the average number of trials to extinguish a particular response will be 50. To test the theory, the accuracy of this prediction must be tested. Thirty-seven subjects are trained until the specified response is established and then extinction trials are run to obliterate the response, counting the number of trials required for each subject.

With $N = 37$, the hypothesis is tested using the t distribution for $df = 36$, $N - 1$. The hypothesis to be tested is $h_0 : \mu = 50$. When $\mu = 50$, the sampling distribution of means is centered about 50 with a standard deviation that is estimated by $s_M = s/\sqrt{N}$ or $s_M = \text{S.D.}/\sqrt{N - 1}$. Suppose in this particular experiment, the average number of extinction trials required to obliterate the response is 58 and the standard deviation of the sample of values is 15. Then the standard error of the mean is:

$$s_M = \frac{\text{S.D.}}{\sqrt{N - 1}} = \frac{15}{\sqrt{37 - 1}} = \frac{15}{6} = 2.5$$

The obtained sample mean value of 58 must be converted to a standard score for evaluation relative to the t distribution with $df = 36$. This standard score is computed by the following formula:

$$t = \frac{M - \mu_0}{s_M} \tag{8.7}$$

The hypothesized true mean, 50, is substituted for μ_0. The computed standard error of the mean, 2.5, is substituted for s_M and the sample mean value, 58, is substituted for M to give:

$$t = \frac{58 - 50}{2.5} = 3.2$$

This obtained standard score relative to the t distribution must be compared with the critical values of t obtainable from the table of the t distribution (Appendix B, table C). If the 1% level, two-tailed test is chosen in advance as the appropriate test, the critical value is $t = 2.72$.

The obtained value of t, 3.2, is larger than 2.72, the critical value of $t_{.01}$ for 36 degrees of freedom, so the obtained result represents a significant departure from expectation under h_0. It is therefore appropriate to reject h_0 at the 1% level of confidence. This implies that $\mu \neq 50$ (is *not* equal to 50), although there is no statement of what actually is the correct value of μ.

The reasoning is as follows. If $\mu = 50$, then 99% of all sample means from this distribution yield standard-score values between $+2.72$ and -2.72. Only 1 time in 100 does a randomly drawn sample have a computed t-score larger than 2.72 in absolute value. In this one experiment, however, a t value of 3.2 is obtained. This is so unlikely (less than one chance in 100) under the assumption $\mu = 50$ that it must be concluded μ is *not* 50 but some other value, probably larger than 50, since the sample value is well above 50. Although h_0 is rejected there is some small chance that a t of 3.2 can occur even when h_0 is true.

In this particular experiment, rejection of h_0 represents a disconfirmation of the theory that predicted that the mean number of extinction trials would be 50. Since $h_0 : \mu = 50$ is rejected there is less than one chance in 100 that the theory is correct.

Two Types of Errors Revisited   In the previous example, only 1 time in 100 is a value of t larger in absolute value than 2.72 obtained. It does occur, however, 1 time in 100. If this does happen by chance in a given experiment, the investigator gets an obtained value of t greater than $+2.72$ or less than -2.72 and rejects h_0 thinking it is false on the basis of the fact that there is a large value of t ignoring the sign. Rejecting h_0, however, is an error, referred to as a Type I error. The Type I error was described in Chapter 7 and is briefly reviewed here. The hypothesis h_0 is rejected when it is actually true. For the 1% level of confidence this happens on the average of 1 time in every 100 experiments. For the 5% level of confidence, it happens 5 times in every 100 experiments.

A second type of error, called a Type II error, is made when h_0 is false, (different from the hypothesized μ_0 but the obtained t does not equal or exceed the critical value of t for that number of degrees of freedom). If t does not exceed the critical value, h_0 cannot be rejected even if h_0 is false. If h_0 is not rejected when it is false because t is too small, a Type II error is committed.

It is clear that there is a relationship between Type I and Type II errors. If the level of significance required is made stringent, for example the 0.01 level, to reduce the Type I errors, Type II errors increase. If the critical level is made easier to reduce Type II errors, Type I errors increase. As mentioned in Chapter 7, when selecting a significance level the investigators must decide for themselves which type of error they are more anxious to avoid. To make sure that a phenomenon of some importance is identified before reporting it, it is best to use a fairly stringent criterion of significance, such as the 0.01 level. On the other hand, if there is more concern to not miss something, a less stringent level such as the 0.05 level is best used.

Figure 8.6 shows the relationship between the two types of errors in a normal distribution. By adjusting the amount of α, the value of β (Type II error) is affected. As mentioned in Chapter 7, both α and β can be lowered by obtaining a larger sample size.

Figure 8.6
The inverse relationship between the probability of Type I and Type II errors.

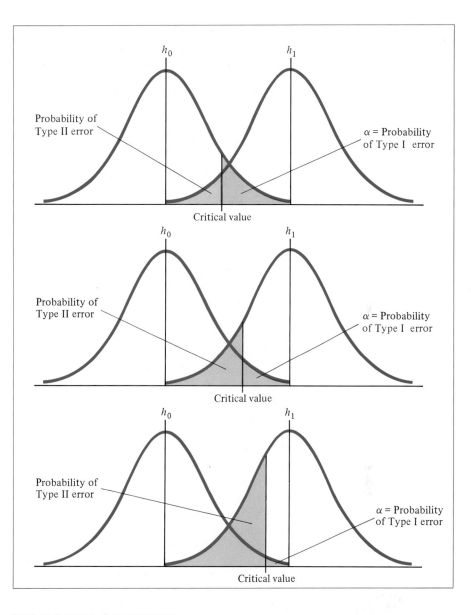

SUMMARY OF STEPS

Testing the Hypothesis That the Mean of the Population Is a Certain Value

Step 1   Compute the mean of the sample, M.

Step 2   Establish the value of μ_0, the hypothesized true value of the mean to be tested.

Step 3   Calculate the standard error of the mean, s_M, using Formula 8.5 or Formula 8.6.

Step 4   Convert the mean, M, to a standard t-score by subtracting M from μ, the hypothesized true mean, and dividing the resulting deviation score by s_M, the standard error of the mean.

Step 5   With $N - 1$ degrees of freedom, where N is the sample size, determine the critical values of t, $t_{.05}$, and $t_{.01}$, from the table of the t distribution (Appendix B, table C).

Step 6   If the obtained t-score from Step 4 is greater than $t_{.05}$ but less than $t_{.01}$, reject μ_0 as the population mean at the 5% level of significance. If the obtained t-score from Step 4 is greater than $t_{.01}$, reject μ_0 as the population mean at the 1% level of significance. If the obtained t-score is less than $t_{.05}$, do not reject the hypothesis that μ_0 is the population mean.

PROBLEM 24

Establish a Confidence Interval for the True Mean

In Problem 23 a hypothesized value for μ, the true mean, is available and the sample is used to investigate the tenability of this hypothesis. A more commonly encountered situation is when the mean of a sample is available but there is no hypothesis concerning where the true mean might be. In this situation the sample mean, M, is taken as an estimate of the population mean, μ, and the concern is with trying to establish a range of values that will probably include the population mean.

confidence interval   An interval of numbers that locates a parameter along the number line, together with a confidence level.

The problem is solved by setting up an interval about the sample mean that gives an idea of the limits within which the true mean can be expected to fall, together with an associated probability. The 95% **confidence interval** for the true mean, for example, is defined in such a way that if it is obtained for sample after sample, 95% of the time it will contain the true mean. The 99% confidence interval will contain the true mean 99 times out of 100 on the average if it is computed for a large number of samples. In practice, of course, the confidence interval, whether 95% or 99%, is computed only for a single sample and a statement is made that it contains the true mean. The statement may or may not be true in a given instance but if this is continuously repeated usually the statement will be correct and the true mean *will* be contained within the confidence interval.

Formulas for the 95% and 99% Confidence Intervals

The formulas for the 95% and 99% confidence intervals are as follows:

95% confidence interval: $M \pm t_{.05} \times s_M$ (8.8a)

99% confidence interval: $M \pm t_{.01} \times s_M$ (8.9a)

Thus, s_M is computed by Formula 8.5 or Formula 8.6, $s_M = \dfrac{s}{\sqrt{N}}$ or $\dfrac{\text{S.D.}}{\sqrt{N-1}}$ and multiplied by $t_{.05}$ or $t_{.01}$, for $N - 1$ degrees of freedom, taken from the table of the t distribution (Appendix B, table C). These intervals, $t_{.05} \times s_M$ and $t_{.01} \times s_M$

Figure 8.7
95% confidence intervals on μ.
95% of the intervals contain μ,
5% do not.

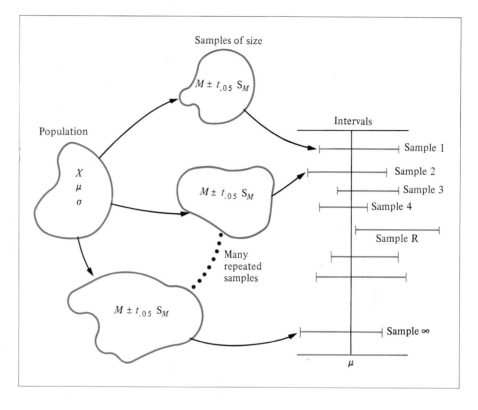

are added to and subtracted from the sample mean, *M,* to provide intervals with *M* at the middle, defined as the 95% and 99% confidence intervals, respectively.

Why do 95% and 99% of such intervals, respectively, contain the true mean on the average? It has already been established that sample means are distributed about the mean as follows:

95% of sample means on the average is contained within the interval $\mu \pm t_{.05} \times s_M$

99% of sample means on the average is contained within the interval $\mu \pm t_{.01} \times s_M$

Thus, the sample means are distributed about the true mean, μ, while each confidence interval has the sample mean at its center. The interval within which 95% of the samples is contained is twice as wide as $t_{.05} \times s_M$, half above μ, the true mean, and half below μ. A particular sample drawn at random from the population can have a mean located anywhere in the sampling distribution, but an average of 95% of them are contained in the region $\mu \pm t_{.05} \times s_M$. Suppose that successive samples are drawn at random and a 95% confidence interval computed for each, the location of these confidence intervals relative to the true mean can be schematized as in figure 8.7. Samples 1, 2, 3, 4, and ∞ have confidence intervals that contain the true mean, μ, although not at the center of the intervals. The sample means fall at the centers of the confidence intervals and, of course, the sample means do not all equal μ, the true mean, which lies at the center of the sampling distribution.

Figure 8.8
Typical results of constructing
20 interval estimates.

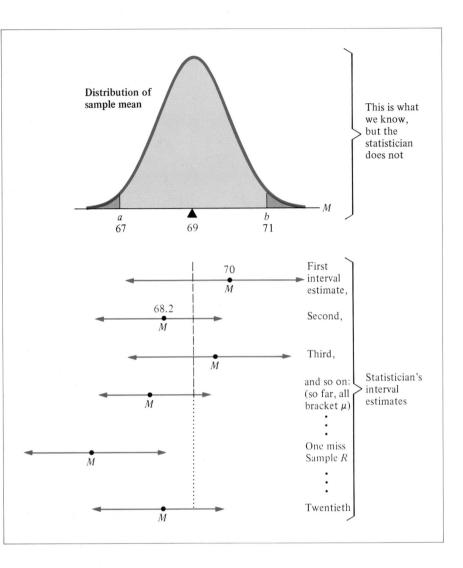

The total width of the 95% confidence interval is the same as the width of the
interval containing 95% of the sample means in the sampling distribution, namely
$t_{.05} \times s_M$. It is just that the confidence interval is centered about the sample mean
though the sample means are distributed about the true mean. If a sample mean,
however, is contained within the interval $\mu \pm t_{.05} \times s_M$ then the 95% confidence
interval for the true mean obtained from the sample will actually contain μ, the true
mean, as with samples 1, 2, 3, 4, and ∞ in figure 8.7.

This is because *no* sample within the interval $\mu \pm t_{.05} \times s_M$ is more than
$t_{.05} \times s_M$ points away from μ. Since the 95% confidence interval runs $t_{.05} \times s_M$
points on either side of the sample mean, the boundaries must extend beyond μ.
Only when the sample mean lies *outside* the interval $\mu \pm t_{.05} \times s_M$ will the 95%
confidence interval fail to extend far enough to encompass μ, as with sample R in
figure 8.7 and sample R in figure 8.8. Additionally, not only do the sample means

Figure 8.9
Normal distribution about lower and upper confidence limits.

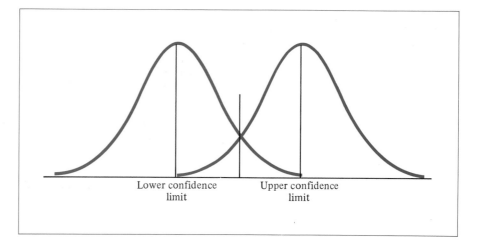

Lower confidence   Upper confidence
limit   limit

change from sample to sample, the sample standard deviations also change. Figure 8.9 shows that because of the variation from sample to sample, not all of the intervals have the same length. The lower confidence limit varies as well as the upper confidence limit.

Since 95% of the sample means lie in the interval $\mu \pm t_{.05} \times s_M$ on the average and all of the sample means in this region have confidence intervals large enough to encompass the true mean, it follows that only 5% of the time, on the average, will the true mean fail to be contained within the 95% confidence interval. The same line of reasoning dictates that only 1% of the time will the 99% confidence interval fail to encompass the true mean.

Computation of the confidence interval, then, provides a way of assessing the accuracy of the sample mean as an estimate of the true mean. It fixes an interval within which the true mean is likely to fall and provides a probability statement concerning the likelihood of being wrong in claiming that the true mean lies within that interval. In many cases, being able to tie down the location of μ to this degree is quite sufficient for the purposes at hand. If it is necessary to know only that μ is not less than 40, for example, and the 99% confidence interval runs from 43 to 47, there is little need to fix the location of μ any more precisely than it has been by the use of this confidence interval.

Note that the confidence interval can be used to solve Problem 23. After determining the 95% confidence interval, for example, h_0 is rejected at the 5% level of significance provided that the hypothesized true mean lies outside the 95% confidence interval. If the 95% confidence interval contains the hypothesized true mean within its boundaries, h_0 is not rejected.

Example   To illustrate the application of these principles, a 99% confidence interval for the true mean is computed given that $M = 20$, S.D. $= 5$, and $N = 16$. Then, using Formula 8.6:

$$s_M = \frac{\text{S.D.}}{\sqrt{N - 1}} = \frac{5}{\sqrt{16 - 1}} = \frac{5}{3.87} = 1.29$$

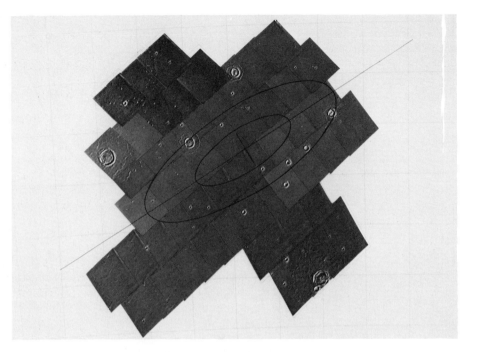

From the table of the t distribution (Appendix B, table C) with $N - 1 = 15$ degrees of freedom, $t_{.01} = 2.95$. The 99% confidence interval, then, using Formula 8.9 is:

$$99\% \text{ confidence interval} = M \pm t_{.01} \times s_M$$
$$= 20 \pm (2.95)(1.29)$$
$$= 20 \pm 3.80$$
$$99\% \text{ confidence interval} = 16.2 \text{ to } 23.8$$

The true mean should fall within the interval 16.2 to 23.8. When this process is applied, approximately 99 times out of 100, the 99% confidence interval contains the true mean if the variable in question is normally distributed or even approximately normal in form. To the extent that peculiarities in the parent population distribution result in distortions in the sampling distribution, some departures from this model can be anticipated.

Once the sample is drawn and the sample mean calculated, the interval estimate (such as the 95% confidence level) either contains the true population mean or it does not. The researcher does not know which confidence intervals contain the population mean and which do not. In fact, it is not known if the one that was just calculated contains the population mean. All that is known is that the interval estimate contains the population mean 95% of the time. The computed confidence interval can be one of the 95% or one of the 5%. A 95% confidence interval does *not* indicate that if 100 samples are drawn that 95 of them will contain the population mean, or that 95 of them are the same or have the same result. The 95% confidence interval statement says that there is a 95% chance that any single interval estimate computed using the appropriate information contains the true population mean. Figure 8.10 is an example of how confidence intervals are used in a real situation.

STATISTICS IN THE WORLD AROUND YOU

The head of your organization discovers that the salaries of the clerical workers are being criticized by the Board of Directors for being too high. The boss asks you to estimate the hourly wage of clerical employees in your city. You get busy on the telephone and determine from a random sample of $n = 100$ clerical employees that the mean salary of the clerks surveyed is $M = \$12.00$ per hour with a standard deviation of S.D. $= \$2.00$. It is important that you be reasonably sure about the wage, so you estimate the average hourly wage for *all* clerical employees using a 95% confidence interval.

By following the steps for finding a 95% confidence interval you discover that there is a 95% chance that the interval $11.60 to $12.40 contains the population hourly wage. Your organization's mean wage for clerical employees is $12.40 per hour with a standard deviation of $1.50. Does the Board of Directors have cause to be concerned?

SUMMARY OF STEPS

Establishing a Confidence Interval for the True Mean

Step 1   Compute the mean of the sample, M.

Step 2   Calculate the standard error of the mean, s_M, by Formula 8.5 or by Formula 8.6.

Step 3   With $N - 1$ degrees of freedom, determine the critical values of t, $t_{.05}$, and $t_{.01}$ using the table of the t distribution (Appendix B, table C).

Step 4   Multiply s_M by $t_{.05}$ and by $t_{.01}$.

Step 5   Add $t_{.05} \times s_M$ to M and subtract it from M to get the 95% confidence interval for the true mean, μ.

Step 6   Add $t_{.01} \times s_M$ to M and subtract it from M to get the 99% confidence interval for the true mean, μ.

PROBLEM 25

Test the Significance of a Correlation Coefficient from a Value of Zero

When the correlation between two variables in the population is zero, the computed correlation coefficient between those two variables in a randomly drawn sample from the population will not necessarily equal zero. In fact, the correlation values in the samples yield a sampling distribution of correlation coefficients just as sample means yield a sampling distribution of means. A given sample correlation, r, can be converted to a t-score value relative to the sampling distribution of correlations where the true r is zero by using the following formula:

$$t = \frac{r \sqrt{N - 2}}{\sqrt{1 - r^2}} \tag{8.10}$$

where N = the number of cases in the sample (the number of pairs of scores).

Formula 8.10 is appropriate for use in the situation where the investigator wishes to test whether the obtained sample correlation is significantly different from zero. In this case, h_0, the hypothesis to be tested, is that the true correlation is zero. The larger the sample correlation becomes, the bigger will be the t value computed by Formula 8.10. If the computed t value is larger than the critical value of t taken from the table of the t distribution (Appendix B, table C), h_0 is rejected and it is concluded that the true correlation is not equal to zero. Either a one-tailed or a two-tailed test can be used to test h_0 but a one-tailed test should be used only if it was perfectly clear in advance what sign the obtained correlation should have and only a correlation of that sign would be considered to be significant. If the intention is to reject h_0 with either a high positive or high negative correlation, a two-tailed test must be made.

Degrees of Freedom in Testing r   Chapter 6 demonstrated that the correlation coefficient is the slope of the regression line of best fit when standard scores are being plotted. Two constants must be specified to determine a straight line, the slope and the y-intercept. This means that 2 degrees of freedom are lost in fitting a straight line to a sample of score pairs, one for each of these constants. The number of degrees of freedom that must be used in connection with Formula 8.10, therefore, is $N - 2$ instead of $N - 1$, which is used with means. The t value at the 0.05 or 0.01 level for $N - 2$ df is taken from table C in Appendix B to obtain the critical value of t that must be equalled or exceeded to reject h_0. The df for the mean is $N - 1$ because only 1 degree of freedom is lost in specifying the value of the mean (any scores might be selected for $N - 1$ scores of the sample). The last score can always be chosen in such a way as to make the mean come out to be any specified value, no matter what the other scores are. In the cases of the correlation coefficient, two such values must be retained to make the correlation coefficient come out to be a particular value.

Example   Given a correlation between two variables of 0.24 in a randomly selected sample of 66 subjects, determine if the correlation is significantly different from zero. Using Formula 8.10 gives:

$$t = \frac{0.24 \sqrt{66 - 2}}{\sqrt{1 - (0.24)^2}} = \frac{(0.24)(8)}{\sqrt{0.942}} = \frac{1.92}{0.970} = 1.98$$

With $N - 2 = 64$ df, $t_{.05}$ for a two-tailed test is 2.00, so the t value of 1.98 is not significant and h_0 cannot be rejected.

A somewhat larger correlation coefficient in this sample results in a t value of 2.00 or larger which permits rejecting h_0 if the 5% level is selected in advance as the criterion. As it is, however, since h_0 cannot be rejected, there is no real evidence that the true correlation in the population is different from zero on the basis of the results from this random sample. This test also indirectly tests the hypothesis that the regression coefficient $r \dfrac{\text{S.D.}_y}{\text{S.D.}_x}$ is statistically different from zero. If the null hypothesis of the test on r is rejected, then the hypothesis that the regression coefficient is zero is also rejected.

SUMMARY OF STEPS

Testing the Significance of a Correlation Coefficient

Step 1   Compute the Pearson product-moment correlation coefficient, r.

Step 2   Using $N - 2$ degrees of freedom, determine the critical values of t, $t_{.05}$, and $t_{.01}$ from the table of the t distribution (Appendix B, table C).

Step 3   Substitute r and N in Formula 8.10 to obtain the t-score corresponding to this correlation coefficient.

Step 4   If the absolute value of the t-score obtained in Step 3 is greater than $t_{.05}$ but less than $t_{.01}$, reject at the 5% level of confidence the hypothesis that the true correlation is zero. If the t-score obtained in Step 3 is greater than $t_{.01}$, reject at the 1% level of confidence the hypothesis that the true correlation is zero. If the t-score is less than $t_{.05}$, do not reject the hypothesis that the true correlation is zero.

PROBLEM 26

Test the Significance of a Correlation Coefficient from a Value Other than Zero

The technique described in Problem 25 serves to test the hypothesis that the sample correlation coefficient, r, is significantly different from zero. It cannot be used to compare the sample correlation value with a correlation value other than zero. The problem lies in the sampling distribution of r. This distribution is not normal when the test value differs from zero. Additionally, the distribution takes on a different shape for different values. R. A. Fisher developed a method using the logarithmic function of the sample correlation coefficient. He called this transformation z'. The sampling distribution of z' is approximately normal irrespective of the true value of the correlation coefficient. The standard error of z' is independent of the true value. Using these properties of z', inferences about the correlation coefficients can be made. Also, the properties of the normal curve can be used in making the evaluation.

The z' Formula

The formula for translating r to Fisher's z' is

$$z' = 1/2 \, (\log_e (1 + r) - \log_e (1 - r)) \qquad (8.11)$$

Fisher's z' transformation A logarithmic transformation of correlation coefficients, r to z', where z' is approximately normally distributed with a standard error independent of rho, the population correlation.

The value of z' is entirely dependent on r. The **Fisher's z' transformation** is *not* a standard score (Z-score) and it is not related to the Z-score. Since the above formula can be a problem for those who are not mathematically sophisticated or do not have a natural logarithm key on their calculator, table F in Appendix B is provided for converting r to z' and vice versa. If the r value is negative, the z' value is also negative.

For comparing a sample correlation coefficient r to a true population correlation value other than zero, use the following formula:

$$Z = \frac{z' - z'_0}{\sqrt{1/(N - 3)}} \qquad (8.12)$$

The Z value is a standard score and its evaluation is made exactly like the Z value obtained in Problem 21. The z' value is the transformed r (sample) and z'_0 is the transformed comparison (population) correlation. In Problem 25, z'_0 is the transformed value for zero.

Example From previous research, the correlation between salt intake and blood pressure was determined to be 0.30. In a current research project, the correlation is 0.39 for a sample of 84 participants. It is desired to establish whether or not the sample correlation coefficient is statistically different from the value found in previous research (h_0 is $\rho = 0.30$, h_1 is $\rho \neq 0.30$). The pertinent values in this example include: $r = 0.39$ (sample); population correlation $= 0.30$; z' for 0.39 is 0.412; z' for 0.30 is 0.310; and the z' of 0.30 is z'_0. Substituting these values in Formula 8.12 gives:

$$Z = \frac{0.412 - 0.310}{\sqrt{1/(84 - 3)}} = \frac{0.102}{0.111} = 0.918$$

Using table B, Appendix B, the probability for this Z value is 0.358 (two-tailed), which is greater than 0.05. Therefore, it cannot be concluded that there is a significant difference between the sample correlation coefficient in the current study and the correlation coefficient established by previous research.

The probability for the Z value, 0.918, is obtained by first rounding 0.918 to 0.92. Entering table B, the normal curve table, with this Z value the area from $-\infty$ to Z is found to be 0.8212 or 0.179 to the right of $Z = 0.92$. Doubling this figure gives 0.358. This corresponds to the area in the right tail plus the area in the left tail.

An alternative procedure is to decide on the size of the Type I error and then find the appropriate Z value that divides the normal distribution into a "rejection region" and a "do not reject region" for the null hypothesis. If the computed Z value falls into the rejection region, the null hypothesis is rejected.

Example   Let's say we decide on a level of significance equal to 0.05, two-tail. The rejection and nonrejection regions are:

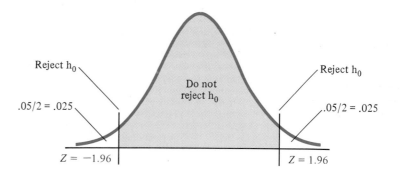

If the calculated Z value is ≥ 1.96 or ≤ -1.96, the null hypothesis is rejected. If the Z value is less than 1.96 and greater than -1.96, the null hypothesis is not rejected. The use of this alternative method avoids the computation of probabilities for the calculated Z value.

SUMMARY OF STEPS

Testing the Significance of a Correlation Coefficient from a Value Other than Zero

Step 1   Using table F, Appendix B convert the sample correlation to z'.

Step 2   Using table F convert the hypothesized true correlation value to z'_0.

Step 3   Set the current sample size to N and find $N - 3$.

Step 4   Form the difference between the values obtained in Steps 1 and 2.

Step 5   Find the test statistic Z by dividing the square root of the reciprocal of the value found in Step 3 into the value found in Step 4.

Step 6   Enter the normal curve table (Appendix B, table B) with the Z-score obtained in Step 5 and find the proportion of area under the normal curve beyond this value.

Step 7   If a one-tailed test is being conducted, let p equal the probability obtained in Step 6. If a two-tailed test is being conducted, let p be twice the probability obtained in Step 6.

Step 8   If p is less than 0.05, reject the hypothesis that the actual true correlation value equals the hypothesized correlation value, at the 5% level of significance. Or, reject h_0 if Z equals or exceeds the chosen critical value of Z.

To test the difference between two sample correlation coefficients, the following formula is used to obtain the test statistic Z. Evaluating the Z value is the same as for the sample case described on page 226.

The formula for comparing two sample correlations is:

$$Z = \frac{z'_1 - z'_2}{\sqrt{(1/(N_1 - 3)) + (1/(N_2 - 3))}} \qquad (8.13)$$

where z'_1 = the Fisher transformation for the sample correlation for the first group,

z'_2 = the Fisher transformation for the second correlation,

N_1 = the sample size for the first group, and

N_2 = the sample size for the second group.

Example   The correlation between creativity and academic achievement for a sample of 154 freshmen is 0.43. The corresponding correlation for a sample of 103 seniors is 0.31. Do these correlations differ significantly? Are these two samples random samples from a common population? Use $\alpha = 0.10$. The null hypothesis is $h_0 : \rho_1 = \rho_2$. The first step is to convert each r using the Fisher transformation:
    For freshmen:

$$r = 0.43 \quad z'_1 = 0.460 \quad N_1 = 154$$

For seniors:

$$r = 0.31 \quad z'_2 = 0.321 \quad N_2 = 103$$

$$Z = \frac{0.460 - 0.321}{\sqrt{\dfrac{1}{154 - 3} + \dfrac{1}{103 - 3}}} = \frac{0.139}{\sqrt{\dfrac{1}{151} + \dfrac{1}{100}}}$$

$$= \frac{0.139}{\sqrt{0.0166}} = \frac{0.139}{0.129} = 1.0775$$

The critical region is found using table B of Appendix B. $Z = 1.0775$ is in the nonrejection region, therefore, there is insufficient evidence to claim a difference in correlations between freshmen and seniors.

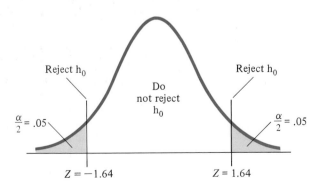

PROBLEM 27

Test the Significance of the Difference between Two Variance Estimates with the *F*-Test

random sample   A subset of the population where each element had an equal probability of being chosen.

Given two populations with equal variances so that $\sigma_1^2 = \sigma_2^2$, **random samples** taken from these respective populations can be expected to have variance estimates that differ from one another, $s_1^2 \neq s_2^2$. A statistic, F, can be defined in this context as the ratio of two variance estimates as follows:

$$F = \frac{s_x^2}{s_y^2} \tag{8.14}$$

F ratio   A statistic used to test whether several populations all have the same mean. *See* analysis of variance.

where $F =$ the **F ratio,**

$s_x^2 =$ the variance estimate for the sample that has the larger value of s, and

$s_y^2 =$ the variance estimate for the sample that has the smaller value of s.

Thus, for the two samples, s^2 is computed as $\Sigma x^2/(N - 1)$ and the larger one is divided by the smaller one to get F, which in this situation is always greater than or equal to 1.00.

If $\sigma_1^2 = \sigma_2^2$ and the estimates of these parameters, s_1^2 and s_2^2, are accurate, F will equal 1.00. In practice, however, random samples from the population produce variance estimates that differ from the true value in the population. The probability that a pair of samples will produce a particular F value becomes smaller and smaller as the size of F increases, other things being equal, when $\sigma_1^2 = \sigma_2^2$. If a large value of F is obtained, for example, one must conclude either that a very rare thing has happened even though $\sigma_1^2 = \sigma_2^2$, or that $\sigma_1^2 \neq \sigma_2^2$. A test of significance of the difference between two variance estimates with the **F-test** involves calculating F by Formula 8.14 and then evaluating that F to see if it is too large to continue entertaining the hypothesis, h_0, that $\sigma_1^2 = \sigma_2^2$.

F-test   Same as the F ratio. The test of the ratio of two variances.

How large F can become with random sampling from populations where $\sigma_1 = \sigma_2$ depends on the sample size, just as the standard error of the mean depends on the sample size. More precisely, the distribution of F can be shown mathematically to be a function of the number of degrees of freedom in the two samples, $(N_x - 1)$ and $(N_y - 1)$. Each combination of values df_x (for the larger variance estimate) and df_y (for the smaller variance estimate) results in its own distribution of F so that there is a specified probability of getting that F or a larger one when sampling at random from two populations with $\sigma_1^2 = \sigma_2^2$.

Two values from these distributions are given in table E, Appendix B, namely the 5% level value of F (lightface type) and the 1% level value of F (boldface type). Table E is entered with the two degrees of freedom values, df_x and df_y. The df for the larger variance estimate is located across the top of table E and the df for the smaller variance estimate is located down the left-hand column. At the intersection of this column and row are located $F_{.05}$ and $F_{.01}$ for that particular combination of degrees of freedom in the two samples. If $\sigma_1^2 = \sigma_2^2$, only 5 times in 100 on the average the samples yield a value of F as large or larger than $F_{.05}$. Only 1 time in 100 on the average the samples yield a value of F as large or larger than $F_{.01}$. If the obtained value of F exceeds $F_{.05}$ for this particular combination of df values for the two samples, the hypothesis, $h_0 : \sigma_1^2 = \sigma_2^2$, is rejected at the 5% level of significance. If the obtained F exceeds $F_{.01}$, h_0 is rejected at the 1% level of significance. If the obtained value of F is less than $F_{.05}$, then h_0 is not rejected. As with the mean, h_0 is never "accepted," it is merely "not rejected" without further information. This F-test is very useful in many situations. It can be used to determine whether two variances are similar enough to be considered equal. In Problem 29, the variances of two samples are pooled to form the standard error. This can be legitimately done if the variances do not differ significantly. The F-test is further explained and expanded upon in Chapters 10 and 11.

Example   Two methods of instruction with randomly selected groups yield mean achievement scores of $M_1 = 95$ and $M_2 = 99$, respectively. Other sample statistics are S.D.$_1$ = 10, S.D.$_2$ = 7, $N_1 = 50$, and $N_2 = 65$. Before making a further statistical comparison that assumes $\sigma_1^2 = \sigma_2^2$, it is necessary to test the hypothesis, $h_0 : \sigma_1^2 = \sigma_2^2$, to see if these other comparisons can be made legitimately in this situation. That is, it is necessary to test the validity of the assumption, $\sigma_1^2 = \sigma_2^2$, upon which the other statistical comparison is based.

Since in this case the S.D. is given for each sample rather than s, the S.D. values are converted to s values by Formula 8.3:

$$s = \text{S.D.} \sqrt{\frac{N}{N-1}}$$

$$s^2 = (\text{S.D.})^2 \left(\frac{N}{N-1}\right) \tag{8.3}$$

For the two samples:

$$s_1^2 = (10)^2 \cdot \left(\frac{50}{49}\right) = \frac{5000}{49} = 102.04$$

$$s_2^2 = (7)^2 \cdot \left(\frac{65}{64}\right) = \frac{49 \cdot 65}{64} = \frac{3185}{64} = 49.77$$

Since s_1^2 is larger than s_2^2, s_1^2 becomes s_x^2, s_2^2 becomes s_y^2 and Formula 8.14 gives:

$$F = \frac{s_x^2}{s_y^2} = \frac{102.04}{49.77} = 2.05$$

with $df_x = N_x - 1$, and $df_y = N_y - 1$,

df_x is equal to $N_1 - 1$ $(50 - 1 = 49)$

and df_y is $N_2 - 1$ $(65 - 1 = 64)$

Entering Appendix B table E with 49 degrees of freedom for the larger variance estimate (across top of table) and 64 degrees of freedom for the smaller variance estimate (down the left side), $F_{.05} = 1.54$ and $F_{.01} = 1.84$. Where necessary, interpolation is used to obtain values of $F_{.05}$ and $F_{.01}$ for nontabled intermediate degrees of freedom. Since the obtained value of F, 2.05, is larger than 1.84, the $F_{.01}$ level value of F, if the 1% level of confidence is chosen as the criterion, it is necessary to reject h_0 at that level and to conclude that $\sigma_1^2 \neq \sigma_2^2$. This renders technically inappropriate the application of further statistical comparisons with these data that rest upon the assumption that $\sigma_1^2 = \sigma_2^2$. In practice, however, such tests have been shown to be relatively insensitive to violations of this particular assumption.

SUMMARY OF STEPS

Testing the Significance of the Difference between Two Variance Estimates by Means of the F-Test

Step 1   Compute s^2, the estimated population variance, for each of two random samples using the formula $s^2 = \Sigma x^2/(N - 1)$.

Step 2   Let the larger of these two variance estimates be designated as s_x^2 and the smaller variance estimate as s_y^2.

Step 3   Substitute s_x^2 and s_y^2 into Formula 8.14 to obtain the value of F.

Step 4   Using $N_x - 1$ degrees of freedom for the larger variance estimate and $N_y - 1$ degrees of freedom for the smaller variance estimate enter the table of the F distribution (Appendix B, table F) to find $F_{.05}$ and $F_{.01}$, the critical values of F. N_x is the sample size of the larger variance.

Step 5   If the value of F obtained in Step 3 is greater than $F_{.05}$ but less than $F_{.01}$, reject at the 5% level of confidence the hypothesis that these two samples were drawn at random from populations with equal variances. If the value of F obtained in Step 3 is greater than $F_{.01}$, reject the same hypothesis at the 1% level of confidence. If the obtained value of F is less than $F_{.05}$, do not reject the hypothesis that these two samples were drawn at random from populations with equal variances.

PROBLEM **28**

Test the Hypothesis That the Variance of a Population Is a Certain Value

Even though a test between a sample mean and the population mean might not be statistically significant, scores from the sample might not have come from the population of interest. A significant difference might have come from a marked difference between the sample variance and the population variance. Problem 28 describes a test so that it can be statistically determined whether or not the sample variance obtained from the data is the same as the variance from the population. In other words, is the difference observed between the sample and population due to chance? Hypothesis tests concerning a single sample variance (or standard deviation) versus a population variance (or standard deviation) are based on the chi square (χ^2) distribution. The chi square test statistic is given as:

$$\chi^2 = \frac{(N)\,\text{S.D.}^2}{\sigma_0^2} \tag{8.15}$$

where $N =$ the number of elements in a random sample from the population,

$\text{S.D.}^2 =$ the sample variance computed using formula 4.4, and

$\sigma_0^2 =$ the value of the population variance under test.

Since this test involves variances, the χ^2 value is never negative. The assumptions underlying the use of this statistic are (1) the population distribution of measurements, X, is normal and (2) the X's are a random sample from the population. If the X's are not normally distributed, but approximately so, then this test can be used only for large samples ($N \geq 100$). The **chi square distribution,** which will be used in a different application in Chapter 9, is a family of distributions like the t distribution, with a shape dependent on the degrees of freedom. The number of degrees of freedom, *df,* is equal to $N - 1$. However, unlike the t or Z distributions, the chi square distribution is positively skewed for low degrees of freedom. The left tail is anchored at zero, while the right tail is asymptotic, or not quite touching, to the horizontal axis (fig. 8.11).

chi square distribution   The distribution of the chi square statistic.

Figure 8.11
Chi square distribution.

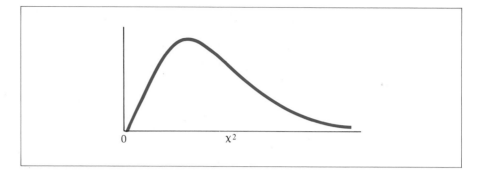

Example   A manufacturer of a coffee packaging machine claims that the machine can load cans at a given weight with a variance of no more than 0.01 of an ounce. The mean and variance of a sample of 30 three pound cans are found to equal 3.05 and 0.018 respectively. Test the hypothesis that the variance of the population weight is really 0.01. Using Formula 8.15 with $N = 30$, $S.D.^2 = 0.018$ and $\sigma_0^2 = 0.01$ gives:

$$\chi^2 = \frac{(30)\ 0.018}{0.01} = 54$$

$$df = N - 1 = 30 - 1 = 29$$

To evaluate this test statistic, look in Appendix B, table D for the value of χ^2 for 0.05 and 0.01 levels of significance for 29 degrees of freedom (df). Then determine if the computed chi square is bigger than the critical value obtained from the table, χ^2 at the 0.05 level or χ^2 at the 0.01 level. If the obtained chi square is greater than or equal to χ^2 at 0.05 for example, the hypothesis of equality between sample variance and population variance is rejected at the 5% level of significance. If not, the hypothesis of equality is not rejected. This hypothesis of equality is described in Problem 29 as the null hypothesis.

From the upper-tailed portion of table D, for 29 degrees of freedom, χ^2 at the 0.05 level $= 42.6$ and χ^2 at the 0.01 level $= 49.6$. Since the computed chi square is 54, and it is greater than both critical values 42.6 and 49.6, the null hypothesis stating that the sample variance is statistically equivalent to the population variance is rejected. This indicates that the evidence does not support the manufacturer's claim.

SUMMARY OF STEPS

Testing the Hypothesis That the Variance of a Population Is a Certain Value

Step 1   Determine the sample size N.

Step 2   Set the sample variance equal to S.D.^{2}.

Step 3   Set the population variance equal to σ_0^2.

Step 4   Compute the chi square test statistic by multiplying N by S.D.^{2} and dividing by σ_0^2.

Step 5   Using $N - 1$ degrees of freedom, determine the critical value of the chi square: χ^2 at 0.05 and χ^2 at 0.01 from table D, Appendix B.

Step 6   Test the value of χ^2 obtained in Step 4 against the critical value of chi square found in Step 5 to determine whether the null hypothesis is to be rejected or not.

PROBLEM 29

Test the Significance of the Difference between Two Independent Random Sample Means Using the *t*-Test

One of the most common applications of statistics in the social sciences is to compare the means of two random samples, such as experimental and control groups, to determine if a difference obtained is significant or not. In a typical experiment, individuals are drawn from a common pool and are randomly assigned to an experimental group (group E) and a control group (group C). Then, the experimental group (E) is given one type of treatment while the control group (C) is given another type of treatment. At the end of the experiment both groups are tested on a common variable, presumably related to the treatments, and mean values are computed. Then M_E, the mean of the experimental group, is compared with M_C, the mean of the control group, to see if the difference between the two groups is large enough to conclude that the two kinds of treatments had different effects on the experimental variable being studied.

Sampling Distribution of Differences between Means

Given two populations that are normally distributed with means μ_1 and μ_2 and standard deviations of σ_1 and σ_2, draw two random samples of sizes N_1 and N_2 from the two populations and compute their means, M_1 and M_2. Let $D_1 = M_1 - M_2$, the difference between these two sample means. Draw two more samples of the same sizes, compute the sample means, and obtain D_2 as the difference between them. If this process is repeated over and over again to get values $D_1, D_2, D_3, \ldots, D_\infty$, the D values can be used to form the sampling distribution of differences between means (fig. 8.12). This is a hypothetical distribution, of course, since it is not possible to draw an infinite number of sample pairs to get an infinite number of D values. Mathematically it can be proved that this distribution is a normal curve if the two populations from which the samples are drawn are normal curves. Even if the populations

Figure 8.12
A sampling distribution of the difference between means.

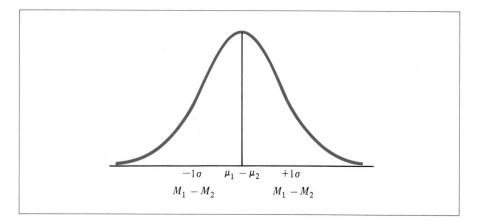

are not normally distributed, the sampling distribution of differences between means comes closer and closer to normality in shape as the sample sizes increase. As the mean of a single sample is converted to a standard score relative to the sampling distribution of means, so can the difference between two sample means be converted to a standard score relative to the sampling distribution of differences between means. Unless the true difference between means ($\mu_1 - \mu_2$) and the true standard deviations of the two distributions, σ_1 and σ_2, are known, the differences between means cannot be converted to normally distributed standard scores. Using a hypothesized true difference between means, however, and an estimated standard deviation of the sampling distribution of differences between means, s_{D_M}, it is possible to obtain standard scores, t values, that follow the t distribution.

The particular t distribution that is appropriate for converting a given difference between means to a standard score depends on the number of degrees of freedom, as was the case with one sample and the sampling distribution of means. With one single sample and the sampling distribution of means, the number of degrees of freedom is $N - 1$, with two independent random samples, the number of degrees of freedom for the sampling distribution of differences between means is $N_1 + N_2 - 2$. This is just the sum of the degrees of freedom for the two separate samples, $(N_1 - 1) + (N_2 - 1) = N_1 + N_2 - 2$.

The Null Hypothesis   The center of the sampling distribution of differences between means is $\mu_1 - \mu_2$, the difference between population means. For most applications, it is assumed that $\mu_1 = \mu_2$ or $\mu_1 - \mu_2 = 0$, there is *no* difference between the population means from which these samples were drawn at random. This is the so-called null hypothesis, $h_0 : \mu_1 = \mu_2$; or $h_0 : \mu_1 - \mu_2 = 0$. When two sample means are being compared, such as the experimental group mean, M_E, and the control group mean, M_C, it is customary to test the null hypothesis, $h_0 : \mu_1 - \mu_2 = 0$. This hypothesis assumes that there is no difference in the effect of the experimental and control treatments on the variable in question. The investigator might wish and expect to find h_0 to be false, but assumes it as something to be tested rather than as something believed to be true.

The Standard Error of the Difference between Means

The standard deviation of the sampling distribution of the differences between means is estimated by the following formula:

$$s_{D_M} = \sqrt{\frac{\Sigma x_1^2 + \Sigma x_2^2}{N_1 + N_2 - 2} \left(\frac{N_1 + N_2}{N_1 \cdot N_2} \right)} \qquad (8.16)$$

standard error of the difference between means The standard deviation of a sampling distribution of the difference between two sample means.

where

s_{D_M} = the **standard error of the difference between means,**

Σx_1^2 = the sum of the squared deviation scores for sample 1,

Σx_2^2 = the sum of the squared deviation scores for sample 2,

N_1 = the number of cases in sample 1, and

N_2 = the number of cases in sample 2.

Remember that Σx^2 is difficult to compute directly because it involves fractional deviation scores. It is expressed in terms of raw scores as:

$$\Sigma x^2 = \Sigma X^2 - \frac{(\Sigma X)^2}{N} = \text{S.D.}^2 (N) \qquad (8.17)$$

The sum of the squared deviations is computed for each sample separately, using Formula 8.17 and the totals substituted into Formula 8.16 to find s_{D_M}.

In the typical situation, then, the sampling distribution of differences between means is presumed to center about zero, the hypothesized true difference between population means, and has a standard deviation (the standard error), s_{D_M}, estimated by Formula 8.16. The t-scores, computed by converting differences between sample means to standard scores, follow the t distribution with $N_1 + N_2 - 2$ degrees of freedom.

Testing the Null Hypothesis   In order to test whether the null hypothesis, $h_o : \mu_1 - \mu_2 = 0$, is tenable or not when comparing the means of two samples, the difference between two means is converted to a standard score as follows:

$$t = \frac{(M_1 - M_2) - 0}{s_{D_M}} \qquad (8.18)$$

The difference $(M_1 - M_2)$ is a score in the sampling distribution of differences between means. The hypothesized mean of this distribution, zero, is subtracted from $(M_1 - M_2)$, to get a deviation score, and this deviation score is divided by s_{D_M}, the standard deviation of the sampling distribution of differences between means to convert the difference score to a standard score in the t distribution. If the difference between means is small, other things being equal, t in Formula 8.18 will be small, and if the difference between means is large, other things being equal, t will be large.

If the absolute value of t computed using Formula 8.18 is larger than the critical values of t given in the table of the t distribution (Appendix B, table C), the null hypothesis, h_0, is rejected. If the absolute value of t computed by Formula 8.18 is smaller than the selected critical value of t, then the null hypothesis is not rejected. The null hypothesis is *never* "accepted," only "not rejected." The critical value of t depends on the number of degrees of freedom, $N_1 + N_2 - 2$, and the type of test selected, two-tailed 1% level, one-tailed 1% level, and so on. Testing the null hypothesis in this way is referred to as "making a *t*-test of the difference between means."

t-test A hypothesis test based on a small sample, leading to a *t*-score and examination of a critical value of the *t* distribution.

In making a *t*-test, it sometimes happens that a negative value of t is obtained for example, when a larger mean is subtracted from a smaller mean. The t distribution is symmetric, however, just as the normal curve is, so the area above a positive t and the area below minus t are the same. When a t comes out negative, therefore, ignore the sign (absolute value) and treat it as though it is positive.

Example Subjects are drawn at random from a common pool to form an experimental group (E) and a control group (C). Group E receives visual imagery training for a pursuit rotor task while group C receives a control type of training. In the pursuit rotor task, the subject tries to hold a stylus in electrical contact with a small metal dot that moves on a rotating turntable. The score is the number of seconds in contact. Both groups are then tested on a pursuit rotor task with the following results:

Group C:

$$\Sigma X = 1433, \Sigma X^2 = 77435, N = 31$$

Group E:

$$\Sigma X = 1033, \Sigma X^2 = 55525, N = 21$$

The means are $M_E = 1033/21 = 49.2$ and $M_C = 1433/31 = 46.2$. Thus, there is a difference, $M_E - M_C$, of 3 seconds on the average in favor of the experimental group, suggesting that the visual imagery training is effective. It remains to be tested, however, whether this is a statistically significant difference or whether it is one that can be attributed to random sampling fluctuations in the means.

Before the *t*-test is applied to test the null hypothesis in this case, it must be shown that the variance estimates, s_E^2 and s_C^2, are not significantly different from each other since the *t*-test is based on the assumption that $\sigma_E^2 = \sigma_C^2$.

Formula 8.2 for s requires a knowledge of Σx^2, the sum of squares of the deviation scores. This value is obtained for each sample using Formula 8.17.

For group E:

$$\Sigma x^2 = \Sigma X^2 - \frac{(\Sigma X)^2}{N} = 55525 - \frac{(1033)^2}{21} = 4711$$

For group C:

$$\Sigma x^2 = \Sigma X^2 - \frac{(\Sigma X)^2}{N} = 77435 - \frac{(1433)^2}{31} = 11193$$

Using Formula 8.2, and squaring both sides,

$$s_E^2 = \frac{\Sigma x^2}{N-1} = \frac{4711}{20} = 235.5$$

$$s_C^2 = \frac{\Sigma x^2}{N-1} = \frac{11193}{30} = 373.1$$

Computing the F ratio by means of Formula 8.14 gives:

$$F = \frac{s_x^2}{s_y^2} = \frac{373.1}{235.5} = 1.58$$

Entering table E Appendix B with 30 degrees of freedom for the larger variance estimate and 20 degrees of freedom for the smaller variance estimate yields $F_{.05} = 2.04$ as the value of F required for significance at the 5% level. Since the obtained F of 1.58 is well below the 2.04 required to reject h_0 at the 5% level of confidence, it is not unreasonable to entertain the idea that $\sigma_E^2 = \sigma_C^2$ so the t-test can proceed.

Having obtained a nonsignificant F ratio for comparing variance estimates, the t-test proceeds by first calculating the standard error of the difference between means, s_{D_M}, the standard deviation of the sampling distribution of the differences between means, using Formula 8.16.

$$s_{D_M} = \sqrt{\frac{\Sigma x_1^2 + \Sigma x_2^2}{N_1 + N_2 - 2}\left(\frac{N_1 + N_2}{N_1 \cdot N_2}\right)} \tag{8.16}$$

where $\Sigma x_1^2 = (S.D._1)^2\, N_1$

$\qquad\qquad = s_1^2\,(N_1 - 1)$ and

$\quad\Sigma x_2^2 = (S.D._2)^2\, N_2$

$\qquad\qquad = s_2^2\,(N_2 - 1).$

Since Σx^2 was already computed for both samples in the F ratio test, those values and the sample sizes are substituted in Formula 8.16 to give:

$$s_{D_M} = \sqrt{\frac{4711 + 11193}{21 + 31 - 2}\left[\frac{21 + 31}{(21)(31)}\right]}$$

$$s_{D_M} = \sqrt{\frac{15904}{50} \cdot \frac{52}{651}} = \sqrt{(318.08)(0.0799)}$$

$$s_{D_M} = \sqrt{25.41} = 5.04$$

Calculating a standard score, t, for the difference between means in the sampling distribution of differences between means, using Formula 8.18 gives:

$$t = \frac{(M_E - M_C) - 0}{s_{D_M}}$$

$$t = \frac{(49.2 - 46.2) - 0}{5.04} = \frac{3}{5.04} = 0.6$$

With $N_E + N_C - 2$, or 50, degrees of freedom, the table of the t distribution (table C, Appendix B) shows $t_{.05} = 2.01$ for a two-tailed test. Since the experimental group (E) was expected to do better, a one-tailed test might have been justified. This would give $t_{.05} = 1.68$ for a one-tailed test. The obtained value of $t = 0.6$ is far below even the lower of these two critical values of t so it is clearly not possible to reject the null hypothesis on the basis of this experiment. The visual imagery training, therefore, did not result in significantly superior performance for the experimental group.

Example   Although the difference in the preceding example was not significant, it was in the predicted direction so the experimenter decides to repeat the experiment with larger samples to see if the trend holds and will reach significance with more cases. The data for the second study give the following results:

$$M_E = 50.1, \quad \text{S.D.}_E = 16, \quad N_E = 50$$
$$M_C = 45.0, \quad \text{S.D.}_C = 20, \quad N_C = 60$$

Applying Formula 8.3 to get variance estimates from the standard deviations gives:

$$s_E^2 = (\text{S.D.}_E)^2 \cdot \left(\frac{N}{N-1}\right) = (16)^2 \cdot \left(\frac{50}{49}\right) = (256)(1.020) = 261.1$$

$$s_C^2 = (20)^2 \cdot \left(\frac{60}{59}\right) = (400)(1.017) = 406.8$$

The F ratio is $406.8/261.1 = 1.56$. Since $F_{.05}$ is 1.58, this result is still not significant although it is approaching the 5% level value rather closely. Many investigators would still consider the t-test applicable as long as the obtained F fails to reach the 1% level of significance.

Since $(\text{S.D.})^2 = \Sigma x^2/N$, the value of Σx^2 for each sample is obtained by multiplying S.D.^{2} by N as follows:

$$(\Sigma x^2)_E = (\text{S.D.})_E^2 \cdot N_E = (16)^2(50) = (256)(50) = 12800$$

$$(\Sigma x^2)_C = (20)^2(60) = (400)(60) = 24000$$

Substituting these values in Formula 8.16 for s_{D_M} gives:

$$s_{D_M} = \sqrt{\frac{\Sigma x_E{}^2 + \Sigma x_C{}^2}{N_E + N_C - 2}\left(\frac{N_E + N_C}{N_E \cdot N_C}\right)}$$

$$s_{D_M} = \sqrt{\frac{12800 + 24000}{50 + 60 - 2}\left[\frac{50 + 60}{(50)(60)}\right]}$$

$$s_{D_M} = \sqrt{\frac{36800}{108} \cdot \frac{110}{3000}} = \sqrt{(340.74)(0.0367)}$$

$$s_{D_M} = \sqrt{12.505}$$

$$s_{D_M} = 3.54$$

Computing the t ratio by Formula 8.18 gives:

$$t = \frac{(M_E - M_C) - 0}{s_{D_M}} = \frac{50.1 - 45.0}{3.54} = \frac{5.1}{3.54} = 1.44$$

Since this value is still below even the large sample value of t for a 5% level one-tailed test, $t_{.05} = 1.65$, the obtained value of t is not significant and the null hypothesis cannot be rejected.

The obtained value of t is pushing closer to significance, however, and if the trend holds, increasing the size of the samples will eventually result in a significant value of t. When samples as large as these fail to result in a significant difference between means, however, it must be concluded that the effect, if it exists at all, is certainly not very pronounced. Effects of experimental treatments that will make an important difference in practice are ordinarily readily detected by samples as large as 50 in each group. On the other hand, even if there is only a tiny difference in true effect of diverse treatments, it is detected as being statistically significant if the samples are made large enough. A statistically significant difference with very large samples, therefore, is of relatively little practical import.

SUMMARY OF STEPS

Testing the Significance of the Difference between Two Independent Random Sample Means Using the t-Test

Step 1   Compute the means of the two independent random samples, M_1 and M_2.

Step 2   Find the sum of the squared deviation scores, Σx^2, for each of the two samples using Formula 8.17.

Step 3   Find the standard error of the difference between means, s_{D_M}, by substituting the sample sizes, N_1 and N_2, and the sums of squared deviation scores into Formula 8.16.

Step 4 Obtain the t-score for this difference between means by dividing $(M_1 - M_2)$ by the standard error of the difference between means as called for in Formula 8.18. Ignore the sign of the t-value.

Step 5 Using $N_1 + N_2 - 2$ degrees of freedom, enter the table of the t distribution (Appendix B, table C) to obtain the critical values of t, $t_{.05}$, and $t_{.01}$.

Step 6 If the t-score obtained in Step 4 is greater than $t_{.05}$ but less than $t_{.01}$, reject the null hypothesis at the 5% level of significance. If the t-score obtained in Step 4 is greater than $t_{.01}$, reject the null hypothesis at the 1% level of significance. If the obtained t-score is less than $t_{.05}$, do not reject the null hypothesis.

PROBLEM 30

Establish a Confidence Interval for the True Mean Difference between Two Independent Random Sample Means

Problem 24 demonstrated the construction of a confidence interval for a single mean. In this problem a confidence interval is constructed for the true difference between group means. Just as Problem 24 represented an extension of the information and method of Problem 23, Problem 30 builds on the information and method used in Problem 29.

The 95% and 99% confidence intervals are given by the following formulas:

95% confidence interval: $(M_1 - M_2) \pm t_{.05} \times s_{D_M}$ $\qquad$ (8.8b)

99% confidence interval: $(M_1 - M_2) \pm t_{.01} \times s_{D_M}$ $\qquad$ (8.9b)

where M_1 and M_2 = the appropriate sample means,

$t_{.05}$ and $t_{.01}$ = the t values obtained from table C of Appendix B for $N_1 + N_2 - 2$ degrees of freedom,

N_1 and N_2 = the sample sizes, and

s_{D_M} = the standard error of the difference between means defined in Formula 8.16.

Just as in Problem 24, the confidence interval provides a way of assessing the accuracy of the difference between sample means as an estimate of the true difference between population means.

The confidence interval can be used in solving Problem 29. After constructing the 95% confidence interval, the null hypothesis is rejected at the 5% level of significance if the value of zero lies outside the confidence interval. If the 95% confidence interval contains the value of zero within its boundaries, the null hypothesis, h_0, is not rejected. The value of zero is derived from hypothesizing no difference between the two group means. If there is no difference, then the separation between the two means is statistically zero.

Example   The closing prices of two stocks on the New York Stock Exchange were recorded for a 21 day period. The mean prices and variances were computed and are listed below:

$$M_1 = 40.33 \qquad\qquad M_2 = 42.54$$

$$\text{S.D.}_1{}^2 = 1.58 \qquad\qquad \text{S.D.}_2{}^2 = 2.95$$

Construct a 99% confidence interval for the true mean difference. The appropriate formula to use is:

$$(M_1 - M_2) \pm t_{.01} \times s_{D_M}$$

M_1 and M_2 are available without further computation. The sample sizes, N_1 and N_2 are equal to 21 (21 days). Hence, the degrees of freedom are $21 + 21 - 2 = 40$. Use table C, Appendix B and 40 degrees of freedom, to find the value for $t_{.01}$. Remember, the confidence interval is like a two-tailed test, so the correct column in table C is the one labeled 0.01. This value is 2.71.

The value s_{D_M} is not given, but can be computed from the information given. Use Formula 8.16 to find s_{D_M}.

$$s_{D_M} = \sqrt{\frac{\Sigma x_1{}^2 + \Sigma x_2{}^2}{N_1 + N_2 - 2}\left(\frac{N_1 + N_2}{N_1 \cdot N_2}\right)} \qquad (8.16)$$

where $\Sigma x_1{}^2 = (\text{S.D.}_1{}^2) \cdot N_1 = 1.58 \times 21 = 33.18$

$\Sigma x_2{}^2 = (\text{S.D.}_2{}^2) \cdot N_2 = 2.95 \times 21 = 61.95$

$$s_{D_M} = \sqrt{\frac{33.18 + 61.95}{21 + 21 - 2}\left(\frac{21 + 21}{21 \times 21}\right)} = \sqrt{\frac{95.13}{40}\left(\frac{42}{441}\right)}$$

$$= \sqrt{0.2265} = 0.4759$$

99% confidence interval

$$(40.33 - 42.54) \pm 2.71 \times 0.4759 = -2.21 \pm 1.29$$

$$\text{Upper limit} = -2.21 + 1.29 = -0.92$$

$$\text{Lower limit} = -2.21 - 1.29 = -3.5$$

This interval states that there is a probability of 0.99 that the true mean difference lies between these two limits.

To test the hypothesis that the means are equal, look to see if the value zero is contained within the limits of the confidence interval. If it is, do not reject the hypothesis of equal means. If zero is not contained within the limits of the interval, reject the hypothesis. For this example, zero is not within the limits so reject the hypothesis. There is sufficient information to claim a reliable difference between stocks.

SUMMARY OF STEPS

Establishing a Confidence Interval for the True Mean Difference between Two Independent Random Sample Means

Step 1   Compute the means for each sample M_1 and M_2.

Step 2   Calculate the standard error of the difference between means.

Step 3   With $N_1 + N_2 - 2$ degrees of freedom, determine the confidence coefficient of t, $t_{.05}$, and $t_{.01}$ from table C.

Step 4   Multiply s_{D_M} by $t_{.05}$ and by $t_{.01}$.

Step 5   Add $t_{.05} \times s_{D_M}$ to $(M_1 - M_2)$ and subtract it from $(M_1 - M_2)$ to get the 95% confidence interval for the true difference between group means $(M_1 - M_2)$.

Step 6   Add $t_{.01} \times s_{D_M}$ to $(M_1 - M_2)$ and subtract it from $(M_1 - M_2)$ to get the 99% confidence interval for the true difference between group means $(M_1 - M_2)$.

PROBLEM 31

Test the Difference between Two Correlated Means Using Difference Scores

matched samples   Data from two groups that are paired, where a correlation is assumed to exist between the pair. Also see within-subjects design.

independent groups (samples)   Two or more groups of subjects chosen from the population where no logical or meaningful relationship is assumed to exist between the members of each group.

difference scores   The resulting scores formed by subtracting X from Y, or Y from X, where X and Y are paired scores. Also known as change scores.

In Problem 29, the significance of the difference between the means of two independent random samples was tested. In certain experimental situations it is better to use **matched samples** rather than **independent samples** or to use the same subjects in two different situations. For example, if an experiment is to prove very costly per subject, it is important to obtain the necessary test of the null hypothesis with as few subjects as possible. One way to reduce the number of subjects necessary is to reduce the standard deviation of the sampling distribution. This can be done by using matched samples or the same subject in both conditions.

One type of matched sample design is to start with a pool of identical twin pairs. Members of the twin pairs are assigned at random to the experimental (E) and control (C) groups, respectively. It is intuitively evident that the difference between means of two such matched samples on the average will be less than the difference between the means of two independent random samples. When each person in one group has an identical twin in the other group, scores will be fairly comparable on any variable for matched individuals in the two groups, resulting in similar means. The D values, **difference scores,** making up the sampling distribution for matched samples will be less scattered about zero than for independent, random samples, giving a smaller standard error. With a smaller standard error, a given size difference between means becomes more significant. Hence matching experimental and control groups is a way of achieving the same level of significance with fewer cases, other things being equal. It increases the "precision" of the experiment. The actual difference due to the treatment stands out better because the background "noise" is reduced by the matching process.

With independent random samples, the standard error of the sampling distribution must be reduced by increasing sample sizes. In many cases it is easier to do this, and less expensive, than it is to use matched groups. Getting a sample of twins is not easy and other types of matching might be less effective.

When groups of identical twins cannot be obtained but matched groups are desirable to reduce sampling error, it is often possible to obtain good results by matching subjects on one or more particularly relevant variables. If an experiment with two teaching methods is to be carried out that involves expensive equipment used on a concurrent basis, one way of proceeding to match groups is to choose university student volunteers as subjects. Pairs of students of the same sex, approximate grade point average, major, and age are chosen and assigned at random to the experimental (E) and control (C) groups, one member from each pair to group E and one to group C. This type of matching is usually not as good as having identical twins but it probably functions effectively to reduce the variability of the sampling distribution of differences between means in most cases.

Using a subject in both experimental and control groups is called a **within-subject design.** The subject serves as its own control. By measuring a subject twice, the two resulting scores can be correlated. Use the statistical method presented in Problem 31 to determine whether the change measured on the same set of subjects under two different conditions is significant.

within-subjects design An experimental design using the same subjects in both experimental and control groups. The subject serves as its own control and is measured at least twice.

Distribution of Differences between Matched Pairs

When samples are matched, rather than independent, Formula 8.16 does not give a suitable estimate of the variability of the sampling distribution of differences between means because the matching process has reduced (hopefully) the average size of the differences between sample means. It is necessary, therefore, to approach the problem of testing the null hypothesis in a different manner when samples are matched.

An effective and simple way to test the difference between means of matched samples is to first compute the difference scores for each pair of matched subjects. That is, for every person in the experimental group, there is a person in the normal group who is their twin or the person who is matched with them on one or more matching variables. The difference between the scores of these pairs of subjects on the experimental variable is obtained leading to a series of difference scores d_1, d_2, . . . , d_N, where N is the number of pairs of subjects, also the number of cases in each group. In a matched group design, N_E must, of course, equal N_C. These d_i scores are sometimes positive, sometimes negative. The signs are maintained and utilized in all calculations.

Assuming that the experimental variable being studied is normally distributed in the population from which the subjects are selected, the difference scores (d_1, d_2, d_3, . . . , d_N) taken from matched samples should also be normally distributed. Thus the distribution of difference scores is treated as a normally distributed population of raw scores from which a sample (d_1, d_2, . . . , d_N) was selected. This sample of difference scores, selected at random from the population, has a mean value, M_d, the mean difference score. These difference scores also have a standard deviation, S.D.$_d$, since some are larger than others.

Note also that the mean difference score, M_d, is equal to the difference between the means of the two matched samples, which is shown as follows:

$$M_E = \frac{1}{N}\,(X_{E_1} + X_{E_2} + \cdots + X_{E_N})$$

$$M_C = \frac{1}{N}\,(X_{C_1} + X_{C_2} + \cdots + X_{C_N})$$

$$M_E - M_C = \frac{1}{N}\,(X_{E_1} - X_{C_1}) + (X_{E_2} - X_{C_2}) + \cdots + (X_{E_N} - X_{C_N})$$

$$M_E - M_C = \frac{1}{N}\,(d_1 + d_2 + \cdots + d_N) = M_d$$

Standard Error of the Mean Difference   Since the difference scores (d_1, d_2, . . . , d_N) function essentially as raw scores of a certain kind, they have a mean and standard deviation. Accordingly, they also have a standard error of the mean. Formulas 8.5 and 8.6 give two methods of estimating the standard error of the mean for ordinary raw scores:

$$s_M = \frac{s}{\sqrt{N}} = \frac{\text{S.D.}}{\sqrt{N-1}}$$

Comparable formulas for use with difference scores derived from matched samples are given by:

$$s_{d_M} = \frac{\text{S.D.}_d}{\sqrt{N-1}} \tag{8.19}$$

$$s_{d_M} = \frac{s_d}{\sqrt{N}} \tag{8.20}$$

where s_{d_M} = the standard error of the mean for difference scores in matched samples,

s_d = the estimate of the population standard deviation of difference scores,

S.D.$_d$ = the standard deviation of a sample of difference scores, and

N = the number of pairs.

Formulas 8.19 and 8.20 merely apply Formulas 8.5 and 8.6 to a sample of measurements that are difference scores instead of the usual kind of raw measurements.

Testing the Null Hypothesis with Matched Samples   If the means of the two populations from which the matched pairs are drawn are equal, $\mu_1 = \mu_2$, then the differences between pairs will average to zero if an infinitely large number of difference scores is used. The means of the samples of difference scores, therefore, are expected to distribute normally about zero when $\mu_1 = \mu_2$. Standard scores computed for these means of difference scores are distributed according to the t distribution when the standard deviation of the sampling distribution is estimated, as in the case of the means of samples of raw scores. Since one difference score becomes one measurement, the number of degrees of freedom will be $N - 1$, where N is the number of differences. If the samples are not matched, but independent of one another, the number of degrees of freedom will be twice as great.

To test the null hypothesis with matched samples first take the difference scores between matched pairs of subjects, always subtracting in the same direction for all pairs, E − C or C − E, and keeping algebraic signs on the d values. Next, compute the average of the difference scores (d values) as M_d. Obtain the standard error of the mean of the difference scores using either Formula 8.19 or Formula 8.20. This is just the same as applying Formula 8.5 or Formula 8.6 on difference scores as raw scores. Then compute a standard score, t, for this mean difference score in the sampling distribution of means of difference scores as follows:

$$t = \frac{M_d - 0}{s_{d_M}} \tag{8.21}$$

where M_d = the mean of the difference scores, and

s_{d_M} = the standard error of the mean for the difference scores.

Relationship of Problem 23 to Problem 31   In Problem 23, the mean of a sample of raw scores was evaluated to test whether it could have come from a population with a specified mean value, μ. Problem 31 reduces to an example of Problem 23 with difference scores being used as raw scores and the population mean value, μ, being zero. In Problem 31, then, the mean of the sample of difference scores is being evaluated to see if it could have come from a population with a mean of zero. The formulas for Problem 31 are identical to those for Problem 23 except that a subscript, d, has been added to make it clear that difference scores are being used instead of raw scores.

Example   Drug A is now used routinely in a certain large psychiatric in-patient facility to control symptoms in paranoid schizophrenic patients. A new drug, Drug B, which the manufacturer claims to be superior to Drug A in controlling symptoms has recently become available. Hospital officials decide to run a double-blind study to compare the efficacy of the two drugs. A pool of matched pairs of patients is formed with individuals in the pairs matched for sex, body weight, severity of symptoms, and previous history of hospitalization. Members of the pairs are randomly assigned, by coin toss, to groups E and C. Patients will be maintained on the drugs for a two week period under intensive observation with extensive record keeping relative to their symptomatic behavior.

Testing the significance of the difference between means of matched samples

Pair number	Drug A	Drug B	$d = A - B$	d^2
1	6	4	2	4
2	7	6	1	1
3	9	7	2	4
4	6	7	-1	1
5	3	4	-1	1
6	4	2	2	4
7	7	5	2	4
8	2	1	1	1
9	1	1	0	0
10	8	5	3	9
Σ	53	42	11	29

$$M_d = \frac{\Sigma d}{N} = \frac{11}{10} = 1.1$$

$$\text{S.D.}_d = \frac{1}{N}\sqrt{N\Sigma X^2 - (\Sigma X)^2} = \frac{1}{N}\sqrt{N\Sigma d^2 - (\Sigma d)^2}$$

$$\text{S.D.}_d = \frac{1}{10}\sqrt{10(29) - (11)^2} = \frac{1}{10}\sqrt{290 - 121} = \frac{\sqrt{169}}{10} = \frac{13}{10}$$

$$\text{S.D.}_d = 1.3$$

$$s_{d_M} = \frac{\text{S.D.}_d}{\sqrt{N-1}} = \frac{1.3}{\sqrt{10-1}} = \frac{1.3}{\sqrt{9}} = \frac{1.3}{3} = 0.433$$

$$t = \frac{M_d - 0}{s_{d_M}} = \frac{1.1 - 0}{0.433} = 2.54^*$$

$t_{.05}$ with 9 $df = 2.26$

The drugs are administered by personnel who do not know who is receiving which drug. The personnel evaluating the patients do not know which patients are receiving which drug. The observations over the entire two weeks are summarized into overall ratings of severity of symptoms displayed. The ratings are converted to a scale from 1 to 9.

There were 10 pairs of patients. The overall symptom severity ratings for those receiving Drug A and Drug B are shown in the second two columns of table 8.1. The first column gives the patient pair number. The difference scores, d, appear in the next to last column and the squares of these difference scores appear in the last column. The various column sums appear at the bottom of the table.

At the bottom of table 8.1 the computations for testing the null hypothesis are shown. The mean difference score, M_d, is computed as the sum of the difference scores divided by N and equals 1.1. The standard deviation of the difference scores, S.D.$_d$, is computed substituting d for X in the usual raw-score formula for the standard deviation. It also is appropriate to calculate s_d by the formula:

$$s_d = \sqrt{\frac{\Sigma(d - M_d)^2}{N - 1}} = \sqrt{\frac{\Sigma x_d^2}{N - 1}} \tag{8.22}$$

TABLE 8.2		
Blood pressure levels of patients		
Patient	Before drug	After drug
1	155	146
2	170	150
3	152	140
4	162	137
5	155	141
6	168	159

Estimate the true mean decrease in blood pressure resulting from the drug using a 95% confidence interval. For each patient, compute a difference score: Before drug—After drug as follows:

$$d = 155 - 146 = 9,$$
$$170 - 150 = 20,$$
$$152 - 140 = 12,$$
$$162 - 137 = 25,$$
$$155 - 141 = 14,$$
$$168 - 159 = 9$$

The mean of the difference scores d is:

$$M_d = \frac{\Sigma d}{N} = \frac{9 + 20 + 12 + 25 + 14 + 9}{6} = \frac{89}{6} = 14.833$$

Where $N = 6$, the number of pairs or number of patients. The estimated true standard deviation of the difference scores is:

$$
\begin{aligned}
s_d &= \frac{\Sigma d^2 - (\Sigma d)^2/N}{N-1} \\
&= \sqrt{\frac{[(9)^2 + (20)^2 + (12)^2 + (25)^2 + (14)^2 + (9)^2] - [(89)^2/6]}{6-1}} \\
&= \sqrt{\frac{1527 - 7921/6}{5}} = \sqrt{\frac{1527 - 1320.167}{5}} = \sqrt{\frac{206.833}{5}} \\
&= \sqrt{41.367} = 6.432
\end{aligned}
$$

The standard error of the mean for the difference scores is:

$$s_{d_M} = \frac{s_d}{\sqrt{N}} = \frac{6.432}{\sqrt{6}} = \frac{6.432}{2.4495} = 2.626$$

For $6 - 1 = 5$ degrees of freedom, $t_{.05}$ (for 95% confidence interval) from Appendix B, table C is 2.57 (look it up). The confidence interval is:

$$14.833 \pm 2.57 \times 2.626, \text{ or } = 14.833 \pm 6.75$$

$$\text{Upper limit} = 14.833 + 6.75 = 21.583$$

$$\text{Lower limit} = 14.833 - 6.75 = 8.082$$

There is a 0.95 probability that this interval contains the true mean decrease in blood pressure level.

SUMMARY OF STEPS

Constructing a Confidence Interval for the True Mean Difference between Two Correlated Means

Step 1   Pair the matched scores and take the difference between each pair of matched scores.

Step 2   Compute the mean of these difference scores, M_d.

Step 3   Compute the estimated population standard deviation for these difference scores, s_d, using Formula 8.22.

Step 4   Compute the standard error of the mean for these difference scores, s_{d_M}, using Formula 8.20.

Step 5   With $N - 1$ degrees of freedom, where N is the number of pairs, determine the critical values $t_{.05}$ and $t_{.01}$ from table C in Appendix B.

Step 6   Multiply s_{d_M} by $t_{.05}$ and $t_{.01}$.

Step 7   Add $t_{.05} \times s_{d_M}$ to M_d and subtract it from M_d to get the 95% confidence interval for the true mean difference.

Step 8   Repeat Step 7 for $t_{.01}$ to get the 99% confidence interval.

PROBLEM 33

Test the Significance of the Difference in Mean Change for Two Independent Random Samples

In Problem 29, the difference between the means of two independent random samples is evaluated to determine if it can be reasonably attributed to random sampling variations when $\mu_1 = \mu_2$. It is sometimes possible to improve the precision of an experiment of this type by taking before and after measures in both the experimental and control groups.

Consider an experiment in which two methods of teaching high jumping are being considered: the traditional straight-up scissors jump and the back flip method.

Using the design in Problem 29, persons with no previous high jumping experience are assigned at random to the two groups. Both groups are trained, but by different methods. At the end of the experiment a measure of each person's jumping skill is obtained and the means of the two groups are compared on this measure of jumping skill.

If the mean initial jumping ability happens by chance to be lower in one group than the other, as it will be in most cases, it puts the training method applied to that group at a disadvantage. The training has to overcome the deficit of ability in the group in addition to achieving an increment in real performance greater than that in the other group.

One way of partially controlling for the difference in ability in the two randomly chosen groups is to compare how much one group *gains* in performance with how much the other group gains. In this design, subjects are measured on their jumping performance *both before and after* training. The net improvement in performance is the raw score for each person. The mean net change in performance is computed for each group and this difference in means is evaluated in the same way that was described in Problem 29. The raw score is, then, a change in score from pre- to postexperiment measurement instead of just a postexperiment measurement.

This approach does not eliminate all effects of random differences in ability in the two groups, but it does control such effects to some extent. The same increase in precision can be achieved by using larger samples. In any given situation, the experimenter has to evaluate the advantages and disadvantages of the various options open for increasing the probability of rejecting a false null hypothesis. Using matched samples, using before and after measurement, and increasing sample size are all ways of achieving this objective.

Example   Thirty-six male subjects are selected from a defined population and are randomly assigned to an experimental group (E) and a control group (C). Two of the experimental group subjects fail to complete the experiment, leaving 16 subjects in group E and 18 subjects in group C. Each subject is given three high jump trials at each level of the bar using any method he prefers and his highest successful jump is recorded as his initial score. These values are given in the E_B column of table 8.3 for the experimental group (*Before* scores) and in column C_B of table 8.3 for the control group. None of the subjects was ever trained in high jumping. The experimental group is given a two week training course in high jumping using the "back flip" method and the control group is given a two week training course in high jumping using the "scissors" method. At the end of the training period, both groups of subjects are given three test jumps at each level and their highest successful jump is recorded. Those values are shown in columns E_A and C_A, the *After* measures for the experimental and control groups, respectively. All measures are given in inches.

The differences between the *Before* and *After* measures are shown in columns D_E and D_C of table 8.3 for the experimental and control groups, respectively. Once the D values are obtained, comparing their means becomes an example of Problem 29, testing the difference between independent sample means using the *t*-test. The D score becomes a raw score for Problem 29. The calculations are shown at the bottom of table 8.3. The null hypothesis is that there is no difference between the two methods.

TABLE 8.3

Testing the significance of the difference between independent, random sample mean change scores

Subject	E_B	E_A	D_E	$D_E{}^2$	Subject	C_B	C_A	D_C	$D_C{}^2$
1	38	45	7	49	1	48	44	−4	16
2	60	68	8	64	2	50	55	5	25
3	42	44	2	4	3	46	53	7	49
4	48	50	2	4	4	45	55	10	100
5	52	59	7	49	5	57	62	5	25
6	58	66	8	64	6	52	58	6	36
7	46	56	10	100	7	38	45	7	49
8	55	62	7	49	8	42	40	−2	4
9	51	52	1	1	9	47	50	3	9
10	48	60	12	144	10	36	40	4	16
11	36	35	−1	1	11	39	43	4	16
12	47	54	7	49	12	54	60	6	36
13	49	61	12	144	13	50	60	10	100
14	51	57	6	36	14	61	66	5	25
15	44	50	6	36	15	43	50	7	49
16	47	54	7	49	16	50	52	2	4
					17	46	50	4	16
					18	48	51	3	9
Σ			101	843	Σ			82	584

$$M_{D_E} = \frac{\Sigma D_E}{N} = \frac{101}{16} = 6.31$$

$$M_{D_C} = \frac{\Sigma D_C}{N} = \frac{82}{18} = 4.56$$

Σx^2 for group E:

$$\Sigma x^2 = \Sigma X^2 - \frac{(\Sigma X)^2}{N} = \Sigma D_E{}^2 - \frac{(\Sigma D_E)^2}{N} = 843 - \frac{(101)^2}{16} = 843 - \frac{10201}{16}$$

$$= 843 - 637.56 = 205.44$$

Σx^2 for group C:

$$\Sigma x^2 = \Sigma X^2 - \frac{(\Sigma X)^2}{N} = \Sigma D_C{}^2 - \frac{(\Sigma D_C)^2}{N} = 584 - \frac{6724}{18} = 584 - 373.56 = 210.44$$

$$s_{D_M} = \sqrt{\frac{\Sigma x_1{}^2 + \Sigma x_2{}^2}{N_1 + N_2 - 2}\left(\frac{N_1 + N_2}{N_1 N_2}\right)}$$

$$s_{D_M} = \sqrt{\left(\frac{205.44 + 210.44}{16 + 18 - 2}\right)\left(\frac{16 + 18}{16 \cdot 18}\right)} = \sqrt{\left(\frac{415.88}{32}\right)\left(\frac{34}{288}\right)}$$

$$s_{D_M} = \sqrt{(12.996)(0.118)} = \sqrt{1.534}$$

$$s_{D_M} = 1.24$$

$$t = \frac{M_1 - M_2 - 0}{s_{D_M}} = \frac{M_{D_E} - M_{D_C} - 0}{s_{D_M}}$$

$$t = \frac{6.31 - 4.56 - 0}{1.24} = \frac{1.75}{1.24}$$

$$t = 1.41$$

With 32 *df*, $t_{.05} = 2.04$ (two-tailed test)

change scores   *See* difference scores.

First obtain the means of the **change scores,** D_E and D_C, as $M_{D_E} = 6.31$ and $M_{D_C} = 4.56$. Next compute the sum of the squared deviation scores, using Formula 8.17, for each group. Use the sum of the squared deviation scores to compute the standard error of the difference between means. These sums of squared deviations, 205.44 for group E and 210.44 for group C, are entered into Formula 8.16 to obtain the standard error of the difference between means, $s_{D_M} = 1.24$. Finally, t is computed by Formula 8.18, dividing the difference between means by the standard error, giving $t = 1.41$. With $N_1 + N_2 - 2$, or 32, degrees of freedom, $t_{.05}$ from the table of the t distribution (Appendix B, table C) is 2.04 for a two-tailed test. Since the obtained value of t is less than 2.04, the null hypothesis is not rejected. This indicates that the experiment failed to establish a definite superiority of one training method over the other. The "back flip" method gave a larger average improvement but it was not statistically significant.

PROBLEM 34

Test the Significance of the Difference in Mean Change for Two Matched Samples

Just as it is possible to compare mean change scores with independent, random samples, so is it possible to compare mean change scores with matched groups. In either case, the net effect, as a rule, is to obtain a slight increase in the precision of the experiment where this technique is applicable. With independent, random samples, once the before and after measures are used to obtain change scores, means of the change scores are treated like means of raw scores and the significance of the difference between such means is evaluated by the methods of Problem 29.

When a before and after measures design is applied to matched samples, again the process calls for using the before and after measures to obtain differences or change scores. Once these change scores are obtained, they are treated like raw scores and the difference between their means is evaluated by the methods of Problem 31. Problem 31 treated the significance of the difference between raw-score means of matched samples using the t-test. Problem 34, therefore, becomes an example of Problem 31 once before and after measures are converted to change scores. It is customary to arrange the data to subtract the generally smaller scores from the generally larger scores so that the difference scores are predominantly positive. The same direction of taking the difference, however, must be used over all subjects in both groups.

Example   Table 8.4 shows a completely calculated example of Problem 34. A new memory drug is proposed as a method of accelerating the rate at which students can master college curricular material. It is suggested that under the influence of the drug, students can learn in three years what otherwise takes four years to learn. If all students in state institutions finished in three years instead of four, this would save the taxpayers a good deal of money. Psychologist X is skeptical, however, and decides to run an experiment to verify these claims.

TABLE 8.4

Testing the significance of the difference between means of matched sample change scores

Subject	E_B	E_A	C_B	C_A	D_E	D_C	d	d^2
1	21	22	23	20	−1	3	4	16
2	35	30	30	25	5	5	0	0
3	30	28	28	24	2	4	2	4
4	26	27	29	26	−1	3	4	16
5	23	20	26	22	3	4	1	1
6	27	30	24	20	−3	4	7	49
7	26	24	27	21	2	6	4	16
8	22	24	24	24	−2	0	2	4
9	15	20	18	19	−5	−1	4	16
10	24	22	21	16	2	5	3	9
11	27	25	25	21	2	4	2	4
12	25	21	25	20	4	5	1	1
13	28	25	29	27	3	2	−1	1
Σ							33	137

$$M_d = \frac{\Sigma d}{N} = \frac{33}{13} = 2.54$$

$$\Sigma x^2 = \Sigma X^2 - \frac{(\Sigma X)^2}{N} = \Sigma d^2 - \frac{(\Sigma d)^2}{N} = 137 - \frac{(33)^2}{13}$$

$$\Sigma(d - M_d)^2 = \Sigma x_d^2 = 137 - 83.77 = 53.23$$

$$s_d = \sqrt{\frac{\Sigma x^2}{N-1}} = \sqrt{\frac{\Sigma(d-M_d)^2}{N-1}} = \sqrt{\frac{53.23}{12}} = \sqrt{4.436} = 2.11$$

$$s_{d_M} = \frac{s_d}{\sqrt{N}} = \frac{2.11}{\sqrt{13}} = \frac{2.11}{3.61} = 0.58$$

$$t = \frac{M_d - 0}{s_{d_M}} = \frac{2.54}{0.58} = 4.38**$$

With 12 degrees of freedom, $t_{.01} = 3.06$, two-tailed test

Thirteen pairs of identical twins are located for this experiment and a twin of each pair is assigned by coin toss to either the experimental group and or the control group. The experimental group receives injections of the drug and the control group receives placebo injections. A double-blind procedure is employed to control for experimenter bias.

Subjects in both groups learn a list of nonsense syllables. The number of trials to obtain one perfect recitation of the list is recorded for each subject. These values are given in table 8.4 in column E_B (*Before* measures) for the experimental group and in column C_B (*Before* measures) for the control group. Following this, the injections are administered and after a two hour interval both groups are tested on a new but comparable list of nonsense syllables. Again the number of trials to reach one perfect recitation of the entire list is recorded for each subject. These values are given in column E_A (*After* measures) for the experimental group and in column C_A (*After* measures) for the control group.

The change scores for the two groups are more often positive than negative if the *after* scores are subtracted from the *before* scores. Column D_E in table 8.4 records these differences for the experimental group and column D_C records these differences for the control group. In both groups, these difference scores are more often positive than negative, indicating that fewer trials are required to learn the list of nonsense syllables after the injections. The change scores are more on the positive side for the control group than they are for the experimental group, however, contrary to what is expected on the basis of claims made for the drug.

From this point on, the calculations are the same as in Problem 31. The d scores in the d column of table 8.4 are obtained by subtracting the D_E scores from the D_C scores. Generally, if the D_E scores had been larger, the subtraction would have been done the other way to make the d scores predominately positive. The last column gives the squared d values. The sums of the last two columns give Σd and Σd^2, values needed for further calculations.

The mean of the difference scores, M_d, is the average of the difference scores, 2.54, as shown at the bottom of table 8.4. The estimate of the population standard deviation, s_d, is calculated using Formula 8.2 but substituting the sum of squared deviation difference scores as Σx_d^2. These calculations are shown at the bottom of table 8.4 with 53.23 as Σx_d^2 for the difference scores and 2.11 as s_d, the estimated population standard deviation of the difference scores. The standard error of the mean for the difference scores is computed next, using Formula 8.20, yielding a value of 0.58. The t ratio is then computed by dividing M_d, the mean difference score by s_{d_M}, the standard error of the mean of the difference scores, yielding a t of 4.38. This value is marked by two asterisks (**) to indicate that it is significant at the 1% level of confidence, two-tailed test, since $t_{.01}$ for a two-tailed test with 12 degrees of freedom (number of pairs, minus one) from Appendix B, table C, is only 3.06.

If a one-tailed test had been planned in advance here, under the assumption that the drug condition at worst is equivalent to the no drug condition, it would not be theoretically correct to attribute **statistical significance** to these unexpected findings. If the investigator clearly intended in advance to make a two-tailed test, then it is appropriate to claim significance for these findings. Taken at face value the obtained outcome suggests that the "memory" drug tended to depress performance although both groups did better on the second test, perhaps due to practice effect. In any event, the experiment does not lend scientific support for a social action program involving use of the drug to speed learning.

statistical significance   The likelihood that an event has occurred by chance is less than some experimenter-established limit. Rejection of h_0 at some α level.

References

Brunk, H. D. (1965). *Mathematical Statistics* (2d ed.). New York: Blaisdell.

Hoel, P. G. (1971). *Introduction to mathematical statistics* (4th ed.). New York: Wiley.

Exercises

8–1.   An investigator who has been told by a friend, "Don't listen to stock brokers, they give bad advice," decides to test this hypothesis. The investigator picks 12 brokers at random and buys 100 shares of recommended stock through each broker. One year later all the stocks are sold. The number of points (rounded to nearest point) gained or lost on the different stocks are as follows: -2, -10, $+5$, -7, -6, -8, -3, $+1$, $+2$, -12, -8, -4. State h_0, carry out an appropriate test, and draw conclusions.

    a. State the null and alternative hypotheses.
    b. Determine n, the sample size.
    c. Find the mean and standard deviation for the 12 values.
    d. Find the standard error.
    e. Find the t value.
    f. Determine the degrees of freedom.
    g. Find the critical t value using $\alpha = 0.01$.
    h. Determine if h_0 should be rejected.

8–2.   A consulting psychologist advises a firm that industrial output per worker can be increased by more than 10% in one month using a special training program. The present output is 30 units a day per worker. Fifteen workers are randomly selected to take the training program and are followed for one month. Their production rates at the end of one month are as follows: 25, 32, 35, 45, 40, 50, 47, 38, 36, 30, 37, 32, 35, 37, 35. State h_0, select an appropriate test, make the test, and state conclusions.

    a. State the null and alternative hypotheses.
    b. Find the mean and standard deviation.
    c. Find the standard error.
    d. Find the t value, $t = \dfrac{M - \mu_0}{s_M}$.
    e. Determine the degrees of freedom.
    f. Find the critical t value using $\alpha = 0.01$.
    g. Determine if the computed t value is in rejection region.

8–3.   Goaded by taunts of rival UCLA students, USC students decide to establish once and for all that they are not "dumb, rich kids." They draw a random sample of the USC student body and have the Wechsler Adult Intelligence Scale administered to each student. They wish to establish a 95% confidence interval for the true mean. With a mean I.Q. score of 117 and a standard deviation of 8 what is the interval with $N = 50$?

    a. Determine the mean and standard deviation for population.
    b. Determine s_M, $s_M = \dfrac{\text{S.D.}}{\sqrt{N - 1}}$.
    c. Substitute the value into Formula 8.8: $M \pm t_{.05} \times s_M$.

8–4.   Obtain the 99% confidence interval for μ when data from the sample are as follows: $\Sigma X = 512$; $\Sigma X^2 = 19451$; $N = 17$.

    a. Find the mean using $M = \Sigma X/N$.
    b. Find the standard deviation using Formula 4.2.
    c. Find the standard error using $s_M = \text{S.D.}/\sqrt{N - 1}$.
    d. Form the confidence interval using Formula 8.9.

8–5.   Given $N = 100$ and $r_{12} = 0.30$, test to see if r is significant at the 0.01 level.

a. Insert the appropriate values into Formula 8.10 $t = \dfrac{r\sqrt{N-2}}{\sqrt{1-r^2}}$.

b. Determine the critical t value.
   1. $df = N - 2$
   2. $\alpha = 0.01$
   3. $t_{.01} =$
c. Reject h_0 if the computed t value is in rejection region.

8–6.   With $N = 81$ and $r = 0.20$, test to see if the correlation coefficient is significantly different from zero at the 5% level.

a. Insert the appropriate values into Formula 8.10 $t = \dfrac{r\sqrt{N-2}}{\sqrt{1-r^2}}$.

b. Determine the critical t value.
   1. $df = N - 2$
   2. $\alpha = 0.05$
   3. $t_{.05} =$
c. Reject h_0 if the computed t value is in rejection region.

8–7.   Given the following data on two independent random samples: $\Sigma X_1 = 226$, $\Sigma X_1^2 = 7128$, $N_1 = 20$, $\Sigma X_2 = 275$, $\Sigma X_2^2 = 8321$, $N_2 = 18$, test to see if these samples could have been drawn from populations with $\sigma_1^2 = \sigma_2^2$. Is it appropriate to proceed with a t-test of the difference between means?
a. State the null and alternative hypotheses.
b. Find the variance for samples 1 and 2.
   1. S.D.$_1 = 1/N_1 \sqrt{N_1 \Sigma X_1^2 - (\Sigma X_1)^2} =$
   2. $s_1 = $ S.D.$_1 \sqrt{N_1/N_1 - 1} =$
   3. S.D.$_2 = 1/N_2 \sqrt{N_2 \Sigma X_2^2 - (\Sigma X_2)^2} =$
   4. $s_2 = $ S.D.$_2 \sqrt{N_2/N_2 - 1} =$
c. Form the F ratio using Formula 8.14.
d. Find the critical F value, $\alpha = 0.05$.
   1. $df_1 =$
   2. $df_2 =$
   3. $\alpha = 0.05$
   4. $F_{.05} =$
e. Reject h_0 if the calculated F ratio is in rejection region.

8–8.   Two independent, random samples, an experimental and control group, gave the following results: $M_E = 42$, S.D.$_E = 6.1$, $N_E = 26$; $M_C = 36$, S.D.$_C = 10.2$, $N_C = 17$. Make an F-test to see if these samples could have come from populations with equal variances. Is it appropriate to make a t-test of the difference between means?
a. State the null and alternative hypotheses.
b. Find the variance for the experimental group and the control group.
   1. $S_E = $ S.D.$_E \sqrt{N_E/N_E - 1}$
   2. $s_E^2 =$
   3. $S_C = $ S.D.$_C \sqrt{N_C/N_C - 1}$
   4. $s_C^2 =$
c. Form the F ratio using Formula 8.14.

    d. Find the critical F value.
      1. $df_x =$
      2. $df_y =$
      3. $F_{.05} =$
      4. $F_{.01} =$
    e. Reject h_0 if the calculated F ratio is in the rejection region.

8–9.   Test the difference between means for the data in question 8–7. State h_0, select an appropriate test, make the test, and draw conclusions.
    a. State the null and alternative hypotheses.
    b. Determine the standard deviation for each sample.
    c. Compute the standard error using Formula 8.16.
    d. Compute the sample means.
    e. Compute the t values using Formula 8.1.
    f. Find the critical t values, $\alpha = 0.05$.
    g. Reject h_0 if the calculated t value is in the rejection region.

8–10.   Whether or not F was significant in question 8–8, test the difference between means. In light of the F-test result in question 8–8 and the t-test result, what conclusion do you draw?
    a. State the null and alternative hypotheses.
    b. Determine the standard deviation for each sample.
    c. Compute the standard error using Formula 8.16.
    d. Compute the t value using Formula 8.1.
    e. Find the critical t value.
      1. $df =$
      2. $t_{.05} =$
    f. Reject h_0 if the calculated t value is in the rejection region.

8–11.   Eight pairs of twins are recruited and assigned at random to an experimental (E) and a control group (C). Group E has a tape recorder that recites a poem over and over to them during times that their rapid eye movements indicate they are asleep. Group C has music of the same sound level instead of the poem. The next day, both groups memorize the poem given on tape to the experimental group to one perfect recitation. The numbers of trials to reach one perfect recitation for the two groups are given below:

Pair	Group E	Group C
1	25	28
2	18	23
3	23	23
4	30	35
5	28	30
6	26	28
7	25	29
8	24	27

Make an appropriate statistical test of the difference between means and state conclusions.
    a. State the null and alternative hypotheses.
    b. Form the difference scores per pair.
    c. Find the mean and standard deviation for the difference scores.
    d. Find the standard error and the degrees of freedom.

e. Form the t value.

f. Find the critical value, $t_{.05}$.

g. Reject h_0 if the calculated t value is in the rejection region.

8–12.   Pairs of subjects that are matched on I.Q., age, sex, and education level are chosen from a given population. Group E, the experimental group, is subjected to a "memory course" while Group C, the control group, is given intellectual tasks to perform that are equivalent in time and attention to the experimental tasks. Afterward, both groups are given a list of 20 names to remember on one presentation. The number of names retained are as follows:

Pair	Group E	Group C
1	18	15
2	16	17
3	10	9
4	10	9
5	9	10
6	13	8
7	12	9
8	13	10
9	7	5
10	11	8
11	12	13
12	13	13

State h_0, make an appropriate test, and draw conclusions.

a. State the null and alternative hypotheses.

b. Form the differences between each pair.

c. Find the mean and standard deviation for the difference scores.

d. Find the standard error and the degrees of freedom.

e. Form the t value.

f. Find the critical value, $t_{.05}$.

g. Reject h_0 if the calculated t value is in the rejection region.

8–13.   The following data represents before and after measurements for experimental and control groups, using two independent, random samples. Make a t-test and state the outcome.

Pair	E_B	E_A	C_B	C_A
1	120	130	100	112
2	115	125	97	95
3	80	95	110	105
4	95	95	118	116
5	100	110	·75	80
6	110	105	83	86
7	102	108	94	95
8	98	104	96	97
9	103	110	104	110
10	85	94	106	103
11	90	97	98	100
12	99	106		

a. State the null and alternative hypotheses.
b. Form the difference scores (gain) for the experimental and control group respectively.
c. Find the mean and ΣX^2 for D_E and D_C scores.
d. Find s_{D_M}.
e. Compute the t value.
f. Find the critical t value and the degrees of freedom.
g. Reject h_0 if the computed t value is in the rejection region.

8–14. Given the following data:

Pair	E_B	E_A	C_B	C_A
1	22	32	24	23
2	15	30	16	17
3	35	40	32	31
4	23	28	20	21
5	25	24	28	30
6	31	39	26	22
7	26	26	30	34
8	23	28	23	24
9	16	30	17	19
10	19	32	18	16
11	20	28	22	20

These are before and after measures in an experiment with matched groups. Find M_d, s_{d_M}, t, and test the significance of the difference between means.

a. State the null and alternative hypotheses.
b. Form the difference between After and Before scores for groups E and C respectively.
c. Form the difference between D_C and D_E.
d. Find M_d.
e. Find ΣX^2, s_d, s_{d_M}.
f. Compute the t value and the degrees of freedom.
g. Find the critical t value.
h. Reject h_0 if the computed t value is in the rejection region.

8–15. Identical twin calves are assigned at random to group E and group C, one twin pair member in each group. Calves in the experimental group are fed raw milk and twins in the control group are fed pasteurized milk. Weights are recorded before and after the experiment period as shown below. Test the significance of the mean change scores, state the results of the test, and draw conclusions.

Pair	E_B	E_A	C_B	C_A
1	40	80	42	64
2	35	71	38	70
3	38	78	35	60
4	45	92	43	72
5	39	80	36	64
6	42	78	40	59
7	37	78	35	56
8	39	82	39	70

a. State the null and alternative hypotheses.
b. Form the difference between Before and After scores for groups E and C respectively.
c. Form the difference between D_E and D_C.
d. Find M_d.
e. Find ΣX^2, s_d, s_{d_M}.
f. Compute the t value and determine the degrees of freedom.
g. Find the critical t value.
h. Reject h_0 if the computed t value is in the rejection region.

8-16. Compute the 95% confidence interval for the difference between two sample means using the information in question 8–7.

a. Compute s_{D_M}. $s_{D_M} = \sqrt{\dfrac{\Sigma x_1^2 + \Sigma x_2^2}{N_1 + N_2 - 2} \left(\dfrac{N_1 + N_2}{N_1 \cdot N_2} \right)}$
b. Compute the sample means.
c. Find appropriate t value for 95% confidence interval at $N + N - 2$ degrees of freedom.
d. Form the 95% confidence interval using Formula 8.8b.

8-17. Given $N_1 = 20$, $N_2 = 105$, $r_1 = 0.36$ and $r_2 = 0.42$, conduct the appropriate test to determine if the two correlations are statistically different from one another. Use α (the level of significance) $= 0.05$.

a. State the null and alternative hypotheses.
b. Convert the correlations r_1 and r_2 to z'_1 and z'_2 using either Formula 8.11 or table F of Appendix B.
c. Substitute the values into Formula 8.13.
d. Find the probability of observing a Z value smaller than $Z = -0.27$
e. Reject h_0 if the probability found in d is smaller than 0.05.

8-18. Refer to question 8–11 and find the 99% confidence interval for the true difference.

a. Form the difference scores.
b. Find the mean and standard deviation for the difference scores.
c. Find the standard error and the degrees of freedom.
d. Find the t value for the 99% confidence interval.
e. Substitute the values into Formula 8.9c, $M \pm t_{.01} \times s_{d_M}$.

8-19. In previous research studies, the established relationship between height and weight was a correlation of 0.60. In a recent research study, the correlation is found to be 0.75. This latter research used a sample of 52 people. Develop the appropriate test to see if the sample correlation of 0.75 is statistically different from 0.60.

a. State the null and alternative hypotheses.
b. Convert the established correlation and the sample correlation to z'_0 and z', respectively.
c. Substitute appropriate values into Formula 8.12, $Z = \dfrac{z' - z'_0}{\sqrt{1/(N - 3)}}$.
d. Find the probability of observing a Z greater than the Z value found in (c).
e. Reject h_0 if the probability is less than 0.05.

8–20. Referring to question 8–12, find the 95% confidence interval for the true difference.
  - a. Form differences between each pair.
  - b. Find mean and standard deviation for the differences.
  - c. Find the standard error and degrees of freedom.
  - d. Find the t value for the 95% confidence interval.
  - e. Substitute the values into Formula 8.9c, $M_d \pm t_{.05} \times s_{d_M}$.

8–21. A written claim included with each stereo made by a certain electronics appliance manufacturer states that each stereo unit is guaranteed to last 5 years without any defects. Six individuals purchased units, and all six units fail before 5 years. The failure times in years are 3.9, 4.6, 3.8, 4.7, 4.8, and 4.1, respectively. Develop the appropriate hypothesis test to determine whether or not there is evidence to contradict the manufacturer's claim.
  - a. State the null and alternative hypotheses.
  - b. Compute the mean and standard deviation.
  - c. Compute the standard error.
  - d. Compute the t value using Formula 8.7, $t = M - \mu_0/s_M$.
  - e. Find the degrees of freedom and the critical t value.
  - f. Reject h_0 if the calculated t value is in the rejection region.

8–22. A very precise medical instrument is guaranteed to be accurate within 2.5 units. A sample of five instrument readings on the same item give measurements of 162, 160, 158, 161, and 160. Test the hypothesis that $\sigma = 0.89$ using the 0.01 level of significance.
  - a. State the null and alternative hypotheses.
  - b. Compute the standard deviation of the five measurements.
  - c. Compute the chi square value using Formula 8.15

    $$\chi^2 = \frac{(N)\ S.D.^2}{\sigma_0^2}.$$

  - d. Determine the degrees of freedom and the critical value.
  - e. Reject h_0 if computed χ^2 is greater than the critical value.

8–23. Using the data in question 8–15 (a) conduct the appropriate test to see if the obtained correlation coefficient is statistically different from zero, and (b) conduct the appropriate test to see if the correlation coefficient is statistically different from 0.23.
  - a. State the null and alternative hypotheses.
  - b. Compute the correlation coefficient using Formula 6.14.
  - c. Find the t value using Formula 8.10 and the Z value using Formula 8.12.
  - d. Find the degrees of freedom and critical t value and the probability of observing a Z less than the Z value found in (c).

Chapter 9

Comparing Frequencies Using Chi Square

Many situations arise in social science research where it becomes important to compare two sets of frequencies for the purpose of ascertaining if they are significantly different from each other. One of these situations involves comparing obtained frequencies in a frequency distribution with theoretically determined frequency values to see if the discrepancies between the two are statistically significant. For example, the theoretical distribution of native-born American Caucasians into various blood types might be compared with the numbers of native-born American Blacks in these same categories to test some hypothesis about inheritance. Wherever it is necessary to compare two sets of frequencies, there is a good chance that the **chi square statistic** is appropriate.

chi square statistic The statistic that measures the discrepancy between the observed values and the expected values in a contingency table.

In some psychological experiments, the researcher determines whether the data collected on participants fit a particular mathematical model. The research of Prokasy (1972) involves the fitting of data to a mathematical learning model and the goodness of that fit. Prokasy's mathematical model creates expected values for certain learning trials. An experiment is conducted to collect empirical data. A chi square statistic is computed to determine if the model is valid for real learning situations.

In situations where frequencies in only two categories are collected as data, the methods used in Chapter 7 on the binomial distribution can be used instead of the chi square test. The tests used on the binomial distribution are from a class of tests more specialized than the multi-category (multinomial) χ^2 test.

The following is the formula for chi square:

$$\chi^2 = \Sigma \frac{(f_o - f_t)^2}{f_t} \tag{9.1}$$

where χ^2 = chi square,

f_o = an obtained frequency, and

f_t = a theoretical frequency.

Where obtained frequencies are compared with theoretically determined frequencies, the f_o values used are the actual obtained frequencies in the distribution and the f_t values are the corresponding theoretical or expected frequencies. Some statistics books and articles in the literature use the symbol f_e for the theoretical frequencies. In applying Formula 9.1 to a problem of this kind, the difference between each obtained and theoretical frequency is squared and divided by the theoretical frequency f_t. These values are determined for each category and summed to give the overall chi square value for the entire frequency distribution.

THE DISTRIBUTION OF CHI SQUARE

Examination of Formula 9.1 readily reveals that χ^2 is zero if the obtained frequencies for a given distribution exactly equal the theoretical frequencies, since all the $(f_o - f_t)$ values are zero. If the obtained frequencies depart to some extent from the theoretical frequencies, as they do in almost every case even if the sample is drawn at random from a normally distributed population, then chi square is some positive number greater than zero. The squaring of $(f_o - f_t)$ removes the negative sign from any negative $(f_o - f_t)$ value. It is evident that chi square increases in size as the departure of the f_o and the f_t values become more pronounced.

A distribution of chi square values for a particular number of class intervals is empirically developed as follows. Draw a sample of 200 cases from a very large, normally distributed population and determine the number of cases falling into each of a specified set of class intervals. All cases falling above or below the end intervals are placed in the end intervals. Then, determine the theoretical normal curve frequencies by the methods outlined in Chapter 5 (see page 106) and compute the chi square value using Formula 9.1. Repeat this experiment using another 200 cases, and compute another chi square. This process must be repeated a maximum number of times. Set up a frequency distribution that shows how many times each chi square value was obtained. Convert this to a smoothed curve, which is to determine the proportion of chi square values that fall above, below, or between any two specified chi square values.

This process is analogous to an empirical determination of the sampling distribution of means, as described in Chapter 8 (see page 204). In practice, of course, the sampling distribution of means is not determined empirically in such a manner. In the case of chi square, it is also unnecessary to determine the distribution of chi square empirically in the manner specified. It is possible to prove mathematically that under proper conditions the distribution takes a particular mathematical form, which is called the chi square distribution.

The normal distribution is a family of curves, requiring specification of mean (M) and standard deviation (S.D.) to designate a particular member of the family. The t distribution is also a family of curves requiring specification of mean (M), standard deviation (S.D.), and the degrees of freedom (df) to designate a particular t distribution. Of course, it is the standard normal curve, the one with $M = 0$ and S.D. $= 1$, that is tabled in the back of the book. In the table of the t distribution (Appendix B, table C), selected values of t are given from the standard t distributions for different values of df, the degrees of freedom.

The chi square distribution also is a family of curves, not just one curve. There is a different curve for each value of df, the number of degrees of freedom. In the experiment just described as a method for empirically determining the chi square distribution, the number of degrees of freedom (df) is held constant for each determination of χ^2 and hence the total obtained distribution is appropriate for that particular value of df associated with that situation. The method of determining the number of **degrees of freedom** for an application of chi square varies from situation to situation. In comparing obtained and normal curve frequencies, the number of degrees of freedom (df) depends on the number of class intervals in the distribution.

A different theoretical chi square distribution can be determined mathematically for each value of df. Table D in Appendix B gives the critical values of chi square for the various numbers of degrees of freedom. A specified percentage of obtained chi square values are larger than a critical value by chance. For example, with 10 degrees of freedom, the value of chi square in table D for $p = 0.05$ is 18.3

degrees of freedom A technical term measuring the number of freely, or independently, varying data. Applies to the chi square distribution, the F distribution, and the t distribution.

STATISTICS IN THE WORLD AROUND YOU

Congress has just appropriated $5 billion annually for a new job training act. The act places most of its emphasis on providing jobs for members of ethnic and language minority groups. Demographic data from 359 individuals who applied for assistance from a jobs program were analyzed in an effort to determine if race was a factor in individuals being unemployed. The data collected are shown below.

Employment status

Race	Unemployed	Employed	Total
A	108	10	118
B	27	2	29
C	34	14	48
D	135	14	149
E	10	1	11
F	3	1	4
Total	317	42	359

The researcher considers employment status as a frequency and uses chi square to determine if there are differences among the several races (A–F) in frequency of unemployment. A chi square of 17.53 with 5 degrees of freedom is calculated. What can the researcher conclude from this sample of data?

---

using upper-tail probabilities. In testing normal curve frequencies against obtained sample frequencies where the df value is 10, only 5% of the time is the obtained chi square value larger than 18.3 if the samples are drawn at random from a normally distributed population.

Testing the Null Hypothesis

The null hypothesis in tests using chi square can take on different forms. Problem 28 in Chapter 8 is one type of application of chi square. Problems 35, 36, and 37 are other examples. To apply chi square, determine the number of degrees of freedom in the situation, consult Appendix B, table D for the value of chi square for the 0.05 or 0.01 level, or whatever standard is being used, and then determine if the computed chi square is bigger than the critical value, $\chi^2_{.05}$ or $\chi^2_{.01}$. If the obtained χ^2 is greater than or equal to $\chi^2_{.05}$, for example, for that number of degrees of freedom, the null hypothesis is rejected at the 0.05 level using uppertail probabilities. If the obtained chi square is less than $\chi^2_{.05}$, the null hypothesis is not rejected. Of course, the 1% level value of χ^2, $\chi^2_{.01}$, can be used as the critical value, or any other tabled value of χ^2. The most commonly used critical values are $\chi^2_{.05}$ and $\chi^2_{.01}$. Since the differences $(f_o - f_t)$ are squared, obscuring whether f_o was greater than f_t, or vice versa, χ^2 is always positive and the test is always a two-tailed test.

Calculating χ^2 in the proper fashion and determining the correct number of degrees of freedom involve certain complications depending on the particular application of chi square. Some of the most commonly encountered contingencies are described in connection with the individual problems presented in this chapter.

PROBLEM 35

Test the Difference between Obtained and Theoretical Frequency Distribution Values (Goodness-of-Fit)

goodness-of-fit test A test of the hypothesis that data observed from a sample belongs to a stated population distribution.

When observations are being taken at random from a specified population, they can be expected to fall into certain categories with frequencies based on those in the population. Or, if observations are being determined according to a certain model, the distribution of those observations into categories should follow a pattern inferred from the model. The chi square test can be used to determine whether or not the obtained distribution of observations into the available categories conforms to expectations, based on the theoretical distribution or model, that is, **goodness-of-fit.** As the discrepancies between obtained and expected frequencies increases, the computed chi square increases in size. When the computed chi square exceeds the critical chi square value for the particular situation being investigated, then it must be concluded that the obtained observations are not distributing themselves in accordance with the theoretical distribution or model being tested.

Example   A college instructor has been using multiple-choice questions with five possible answers for the examinations in a large introductory course. The intent in writing the questions is to avoid any systematic preference for a, b, c, d, or e as the correct answer position. The instructor decides to conduct an investigation to see if she has avoided systematic bias in the choice of correct and incorrect answer positions. There are 435 items in the question file with the following distribution of correct answers: a = 89, b = 72, c = 95, d = 105, e = 74. If the response alternative positions are equally preferred, there will be 435/5, or 87, questions with correct answers in each of the five response positions. Clearly, the numbers of questions in the various categories depart from the theoretical expected value of 87 in each category. The question to be answered by the chi square test is whether these departures are large enough to represent a statistically significant bias (lack of fit).

Applying Formula 9.1 to this problem gives:

$$\chi^2 = \frac{(89 - 87)^2}{87} + \frac{(72 - 87)^2}{87} + \frac{(95 - 87)^2}{87} + \frac{(105 - 87)^2}{87} + \frac{(74 - 87)^2}{87}$$

$$\chi^2 = \frac{4}{87} + \frac{225}{87} + \frac{64}{87} + \frac{324}{87} + \frac{169}{87}$$

$$\chi^2 = \frac{786}{87} = 9.03 = 9.0$$

Degrees of Freedom

cells   The categories in a contingency table, or the categories in a two-way analysis of variance; the items in a matrix.

In this situation there are five categories of response into which the observations are distributed. The only constraint on these frequencies is that they must add up to 435, the total number of items. No matter what frequencies occur in four of the **cells,** the total is kept equal to 435 by making the last cell frequency 435 minus the sum of the other cell frequencies. Thus, once four cell frequencies are determined, the last cell frequency is fixed by the requirement that the cell frequencies must add up to 435. Since four cell frequencies can vary freely, there are 4 degrees of freedom for this chi square test. Entering table D, Appendix B, the critical values (upper tail) of χ^2 with 4 *df* are found to be:

$$\chi^2_{.05} = 9.49 \text{ and } \chi^2_{.01} = 13.3.$$

Using the Chi Square Table

Notice that table D is a complete chi square table in two parts: for lower and upper tail probabilities. The first part contains the chi square values for lower tail probabilities. The second part contains the chi square values for upper tail probabilities; the ones used in the problems found in this text. To use the table, first determine if the upper tail probabilities are being used. Then look down the left side to find the number of degrees of freedom. Look across the table to the predetermined level of significance (indicated at the top of the table). Determine the chi square value at the level and compare it to the one computed from formula 9.1.

Interpretation of Results

Since the obtained χ^2 value of 9.0 was less than $\chi^2_{.05} = 9.49$, the 5% level critical value of chi square for 4 degrees of freedom, it is not possible to reject the null hypothesis (observed frequencies fit expected frequencies) in this situation. It cannot be concluded, therefore, that the instructor is departing from random choice of correct answer position in writing the test items. The closeness of the obtained chi square value to the 5% level critical value of chi square tends to suggest, however, that there is a strong possibility that there might be bias in the choices. In particular, the frequency for the d category is elevated, so the instructor should make a conscious effort to use it less frequently while using b and e more often.

SUMMARY OF STEPS

Testing the Difference between Obtained and Theoretical Frequency Distribution Values (Goodness-of-Fit)

Step 1   Determine the theoretical frequency for each category in the frequency distribution, making certain that the sum of these values equals the sum of the actual obtained frequencies.

Step 2   If any of the theoretical frequencies is less than 10, if possible, combine adjacent categories in both theoretical and obtained frequency distributions to eliminate categories with small theoretical frequencies. Do not compute chi square by this method if categories cannot be combined to eliminate all theoretical frequencies less than 5.

SALLY FORTH By Greg Howard

Step 3   Take the difference between the theoretical and the obtained frequency for each category in the frequency distribution.

Step 4   For each category, square the difference obtained in Step 3 and divide it by the theoretical frequency value.

Step 5   Add up the values computed in Step 4 to get chi square, χ^2.

Step 6   Using the appropriate number of degrees of freedom, usually one less than the number of categories in the frequency distribution, find the critical values of chi square, $\chi^2_{.05}$ and $\chi^2_{.01}$.

Step 7   If the value of χ^2 obtained in Step 5 is greater than $\chi^2_{.05}$ but less than $\chi^2_{.01}$, reject the null hypothesis at the 5% level of confidence. If the obtained value of χ^2 is greater than $\chi^2_{.01}$, reject the null hypothesis at the 1% level of confidence. If the obtained value of χ^2 is less than $\chi^2_{.05}$, do not reject the null hypothesis.

PROBLEM 36

Test the Association between Two Variables Using Frequencies in a Fourfold Table (Contingency Tables)

contingency table   The rectangular array resulting from the partitioning of two variables, such as political preference and age, into classes.

When variables are measured on a continuous, normally distributed scale, the best way to determine whether or not they are related to each other is to compute the product-moment correlation coefficient between them, assuming that the regression line of best fit is a straight line. Many variables of interest, however, are not continuously measured on a normally distributed scale. In fact many such variables are dichotomous (in only two categories, binomial). Examples of two-category variables are sex membership, high school graduate versus nongraduate, Republican versus Democrat, serviceman versus civilian, and employed versus unemployed. Many social science research projects involve the study of such dichotomous variables in relation to each other and to questionnaire data, themselves often in dichotomous form. Chi square is an appropriate statistic for determining whether or not there is a significant association between such variables.

TABLE 9.1			
Computing chi square in a fourfold table			

		Obtained frequencies (f_o)		
		Sex		
		Males	Females	Total
Opinion	Yes	32	62	94
	No	59	25	84
	Total	91	87	178

		Expected frequencies (f_t)		
		Sex		
		Males	Females	Total
Opinion	Yes	48.1	45.9	94.0
	No	42.9	41.1	84.0
	Total	91.0	87.0	178.0

$$\chi^2 = \Sigma \frac{(f_o - f_t)^2}{f_t}$$

$$\chi^2 = \frac{(32.0 - 48.1)^2}{48.1} + \frac{(59.0 - 42.9)^2}{42.9} + \frac{(62.0 - 45.9)^2}{45.9} + \frac{(25.0 - 41.1)^2}{41.1}$$

$$\chi^2 = \frac{(-16.1)^2}{48.1} + \frac{(16.1)^2}{42.9} + \frac{(16.1)^2}{45.9} + \frac{(-16.1)^2}{41.1}$$

$$\chi^2 = 5.39 + 6.04 + 5.65 + 6.31$$

$$\chi^2 = 23.39$$

$$\chi^2 = 23.4**$$

With 1 df, $\chi^2_{.05} = 3.84$, $\chi^2_{.01} = 6.63$

fourfold table   A square table consisting of two rows and two columns (four cells).

Example   Table 9.1 shows a worked out example of the application of χ^2 to a **fourfold table.** The data represent frequencies of *yes* and *no* replies to a questionnaire item by male and female respondents in a randomly selected sample of voters. It is clear that men answered this question very differently from women, but the question remains whether the difference is significant. The null hypothesis, h_0, in this situation is that there is no association between sex of respondent and opinion on this issue as manifested by replies to a questionnaire item. Chi square is used to determine if the proportion of men and women giving *yes* and *no* answers is very different.

Determining the Theoretical Frequencies

If there is no difference between men and women in their replies, the split between *yes* and *no* replies should be the same in both groups. The f_t values are determined in such a way as to make this occur. This is done in various ways. The simplest procedure is to multiply the column total by the row total for the cell in question and then divide by the overall N. In table 9.1, therefore, the f_t (theoretical frequencies) for the upper left cell is $(91 \times 94)/178 = 48.1$; the upper right cell f_t is $(87 \times 94)/178 = 45.9$; the lower left cell f_t is $(91 \times 84)/178 = 42.9$; and the lower right cell f_t is $(87 \times 84)/178 = 41.1$.

TABLE 9.2

Fourfold table layout

f_{o11}	f_{o12}	R_1
f_{o21}	f_{o22}	R_2
C_1	C_2	Total

$R_1 = f_{o11} + f_{o12}$ $C_1 = f_{o11} + f_{o21}$

$R_2 = f_{o21} + f_{o22}$ $C_2 = f_{o12} + f_{o22}$

This method of obtaining the f_t values is just a matter of making the *yes–no* splits equal proportionally in the males and females columns, although the frequencies themselves are not equated since the number of males overall is different from the number of females. The ratios 48.1/42.9, 45.9/41.1, and 94/84 all equal the same value, 1.12. Also, 48.1/91.0 = 45.9/87.0 = 94.0/178.0 and 42.9/91 = 41.1/87.0 = 84/178.0. Corresponding ratio relationships among the f_t values in table 9.1 hold for the rows. That is, 48.1/45.9 = 42.9/41.1, 48.1/94.0 = 42.9/84.0 = 91.0/178.0, and 45.9/94.0 = 41.1/84.0 = 87.0/178.0. These proportionality relationships provide the basis for obtaining the correct f for any cell by multiplying the row and column totals for a cell and dividing the product by the overall N.

Given a fourfold table as in table 9.2 where $f_{o_{ij}}$ are the observed frequencies, the theoretical frequencies, $f_{t_{ij}}$ are found by using the formula:

$$f_{t_{ij}} = \frac{R_i C_j}{N}$$

where: i = the row number,

j = the column number,

R_i = the total for row i,

C_j = the total for column j, and

N = the total.

For example, the expected (theoretical) frequency for males answering "yes" in table 9.2 are computed as:

$$f_{t_{11}} = \frac{R_1 C_1}{N} = \frac{(94)(91)}{178} = 48.1$$

Once the table of f_t values is determined, as shown in table 9.1, the computation of chi square proceeds by taking the difference between the f_o and f_t values for each cell. These differences are squared, divided by the f_t values, and summed to obtain χ^2, as shown at the bottom of table 9.1. The value of χ^2 is ordinarily rounded to one decimal place unless it is very close to a critical value. In this example, $\chi^2 = 23.4$, a substantial value for a fourfold table.

Determining the Number of Degrees of Freedom

Before $\chi^2_{.05}$ and $\chi^2_{.01}$ can be determined from the table of the chi square distribution (table D, Appendix B), it is necessary to determine the number of degrees of freedom for the particular situation being studied. In the case of a fourfold table, it is always true that $df = 1$. By considering the table of frequencies in table 9.1, it can be seen that this is true. With the row and column totals fixed, the specification of any one of the four cell frequencies automatically determines what the others must be since the other cell frequencies are then determinable as the difference between the row or column total and the already fixed cell frequency. Since only one cell frequency can be specified without fixing the others, there is only 1 degree of freedom in the fourfold table situation. In the example shown in table 9.1, χ^2 is significant beyond the 1% level of confidence since with 1 degree of freedom, $\chi^2_{.01}$ is 6.63 and the obtained χ^2 is 23.4. It is marked with two asterisks (**) in table 9.1 to indicate significance at the 0.01 level.

When Theoretical Frequencies Are Small

As with other comparisons of obtained and theoretical frequencies, the use of chi square in the fourfold table situation requires theoretical frequencies of adequate size. As a guide, a good minimum to use is 10. In planning a research investigation, make every effort to ensure that no f_t will be less than 10 if chi square is to be used to evaluate the results.

There is a gradual deterioration in the fit of the chi square distribution to the data as the size of the smallest f_t value decreases. This deterioration does not take place suddenly at 10 as a sharp dividing point. It begins slowly long before 10 and continues to increase in severity as the expected frequency for a cell drops below 10. Some investigators feel that it is still appropriate to apply chi square as long as no theoretical frequency is less than 5.0. In the case where one or more f values are between 5.0 and 10.0, it is common to apply the Yates' (1934) correction, which involves subtracting 0.5 from the absolute value (ignoring the sign) of each $(f_o - f_t)$ value. This has the effect of reducing the overall size of the obtained chi square, making it more difficult to reject h_0. This correction is always applied in 2 cell by 2 cell tables. In table 9.2, if it were necessary to apply **Yates' correction,** all the $(f_o - f_t)$ values would be 15.6 (after removing the sign of the difference) instead of 16.1. Of course, no f_t value in table 9.2 is less than 10 so Yates' correction is unnecessary.

Yates' correction factor A correction factor used to more accurately approximate the continuity of the chi square distribution when N is relatively small and the expected value of any cell in a contingency table is less than 10. The factor reduces the absolute difference between observed and expected frequency before squaring.

When the f_t value for any cell drops below 5.0, chi square definitely should *not* be computed by the methods described here. It is possible, however, to carry out an "exact" chi square test when one (or more) of the f_t values in a fourfold table is less than 5.0. A more advanced text should be consulted if such a test is needed (see Ferguson, 1971, pp. 340–42). Note that these restrictions on the use of the chi square test apply to the theoretical frequencies, not to the obtained frequencies. Some f_o values can be small without invalidating the chi square test as long as the f_t values are large enough.

SUMMARY OF STEPS

Testing the Association between Two Variables Using Frequencies in a Fourfold Table (Contingency Tables)

Step 1   Arrange the obtained frequencies in a fourfold table for the four possible outcomes: high on both variables, high on one but low on two, low on one but high on two, and low on both variables.

Step 2   Determine the theoretical frequency for each cell by multiplying the column sum by the row sum in the fourfold table and dividing this product by the sum of the frequencies in all four cells, N.

Step 3   For each cell, take the difference between the obtained and theoretical frequencies, square the difference, and divide by the theoretical frequency.

Step 4   Sum the four values obtained in Step 3 to obtain the value of χ^2.

Step 5   If any theoretical frequency obtained in Step 2 is between 5.0 and 10.0, reduce the absolute size of the difference between all four obtained and theoretical frequencies in Step 3 by 0.5 before squaring (Yates' correction). If any theoretical frequency is less than 5.0, do not compute χ^2 by this method.

Step 6   Using 1 degree of freedom, determine the critical values of chi square, $\chi^2_{.05}$ and $\chi^2_{.01}$.

Step 7   Test the value of χ^2 obtained in Step 4 against the critical values of χ^2 to determine whether the null hypothesis is to be rejected or not.

PROBLEM 37

Test the Association between Two Variables Using Frequencies in Larger Contingency Tables

The fourfold table in Problem 36 has only two categories for each variable, *yes–no* responses for *males* and *females*. Many tables occur in which one or both variables have more than two categories. Responses to questions, for example, can fall into *yes, no,* or *?,* instead of just *yes* or *no*. Political affiliation can be classified as *Democrat, Republican,* or *Independent*. For these multiple category investigations of association between noncontinuous variables, chi square is often the only practical statistic to use.

Example   In a large organization with members representing a wide range of educational backgrounds, an anonymous questionnaire is circulated asking for political affiliation and amount of education completed, among other things. Out of 250 questionnaires mailed out, 190 are completed and returned. The investigator wishes to study the relationship between educational level and political affiliation.

TABLE 9.3					
Computing chi square in a multiple contingency table					

	Obtained frequencies (f_o)				
	Elementary school	High school	College degree	Advanced degree	Total
Democrat	16	36	24	15	91
Republican	15	19	13	9	56
Other	4	10	13	16	43
Total	35	65	50	40	190

	Theoretical frequencies (f_t)				
	Elementary school	High school	College degree	Advanced degree	Total
Democrat	16.8	31.1	23.9	19.2	91
Republican	10.3	19.2	14.8	11.7†	56
Other	7.9	14.7	11.3	9.1	43
Total	35.0	65.0	50.0	40.0	190

Cell	$(f_o - f_t)$	$(f_o - f_t)^2$	$(f_o - f_t)^2/f_t$
1	−0.8	0.64	0.04
2	4.7	22.09	2.14
3	−3.9	15.21	1.93
4	4.9	24.01	0.77
5	−0.2	0.04	0.00
6	−4.7	22.09	1.50
7	0.1	0.01	0.00
8	−1.8	3.24	0.22
9	1.7	2.89	0.26
10	−4.2	17.64	0.92
11	−2.7	7.29	0.62
12	6.9	47.61	5.23
			$\chi^2 = 13.63^*$

$df = 3 \times 2 = 6$; $\chi^2_{.05} = 12.6$; $\chi^2_{.01} = 16.81$

†This f_t had to be lowered from 11.8 to 11.7 to make $\Sigma f_t = \Sigma f$. This discrepancy was due to rounding error.

Educational level is broken down into four categories: (1) those who have only an elementary school education; (2) those who have completed high school; (3) those who have a bachelor's degree or equivalent; and (4) those with an advanced degree of some kind. Political affiliation is classified into the two major parties with all others going into the *other* category. Each of the 190 respondents is placed into one and only one of the 12 possible cells created by this joint classification of educational level and political affiliation. The frequencies for the 190 cases in these 12 cells are shown in table 9.3 under *Obtained Frequencies*.

The theoretical frequencies, f_t, are shown below the obtained frequencies, f_o, in table 9.3. These are calculated in the same way they are calculated in Problem 36. For a given cell, the cell's row total is multiplied by its column total and the product is divided by the overall N. In table 9.3, for the upper left cell f_t is $(35 \times 91)/190 = 16.8$; for the cell below it f_t is $(35 \times 56)/190 = 10.3$; for the lower right cell $f_t = (40 \times 43)/190 = 9.1$. The f_t values for the other cells are found in the same manner.

The calculations for chi square are shown at the bottom of table 9.3. Each discrepancy between corresponding f_o and f_t cell frequencies is squared and divided by its own f_t value. These figures are summed for the 12 cells to obtain the overall chi square value or $\chi^2 = 13.63$.

Degrees of Freedom in Larger Contingency Tables

In the fourfold table, the degrees of freedom is always 1, since only one cell frequency can be specified without fixing all the other cell frequencies, holding constant the row and column totals. In contingency tables larger than 2 by 2, the degrees of freedom may be calculated by the rule, $df = (r - 1) \cdot (c - 1)$, where r is the number of rows and c is the number of columns in the contingency table. Only the rows and columns of the table representing variable categories count. Thus, in table 9.3, there are three row categories and four column categories, so $df = (3 - 1) \cdot (4 - 1) = 6$ for this contingency table. In a fourfold table, this rule gives $df = (2 - 1) \cdot (2 - 1) = 1$, which is, of course, consistent with what has already been stated about degrees of freedom in that situation.

That the rule $df = (r - 1) \cdot (c - 1)$ holds can be grasped intuitively by a consideration of table 9.3. If two of the cell frequencies in the first column are fixed arbitrarily, the third one is automatically determined since the first column total must be 35. This gives 2 df for the first column. The same is true for each of the next two columns, adding 4 more df. When the frequencies are determined for the first three columns, with a total of 6 df accumulated, no more degrees of freedom are available because the frequencies in the last column must make the row totals come out right. Analysis of contingency tables of various sizes will convince the reader that the $df = (r - 1) \cdot (c - 1)$ always gives the correct number of degrees of freedom.

Interpreting the Results

With 6 degrees of freedom, the values of chi square for significance at the 0.05 and 0.01 levels of significance, respectively, are (see table D, Appendix B) $\chi^2_{.05} = 12.6$ and $\chi^2_{.01} = 16.8$. When the obtained value of χ^2 for the data in table 9.3, $\chi^2 = 13.63$, is compared with these critical values, it is found to be greater than $\chi^2_{.05}$ and less than $\chi^2_{.01}$. It is, therefore, significant at the 0.05 level. This fact is indicated in table 9.3 by one asterisk (*) beside the obtained chi square value.

Notice that two of the f_t values in table 9.3 are less than 10, one is 7.9 and the other is 9.1. This is not an ideal situation. It is better if the *other* category has more cases so the f_t values are all 10 or higher. The overall N is a good size, however, and there are 10 other cells with f_t values greater than 10 so it is probably not unreasonable to apply chi square to this table, especially since the two low f_t values are above 7.0. The obtained chi square is about halfway between the 0.05 and 0.01 critical values, so to consider the result significant at the 0.05 level is not unreasonable even if the rule of "no f_t below 10" is violated.

The significant chi square obtained suggests that there is a relationship between political preference and educational background in the population sampled, although the effect on the results of those who did not respond to the questionnaire remains in doubt. The significance of the chi square value merely indicates the presence of some type of relationship without telling what kind, linear or curvilinear.

Inspection of the contributions to chi square from the 12 different cells shows only one to be large, the lower right cell. The number of individuals in this organization with advanced degrees who do not prefer either of the major political parties is greater than expected. Elsewhere in the table there appears to be no noteworthy discrepancy between obtained and expected frequencies. Since the one large discrepancy in this table is for one of the two cells with a low f_t value, a further note of caution must be injected into any statement of conclusions drawn from this study.

SUMMARY OF STEPS

Testing the Association between Two Variables Using Frequencies in Larger Contingency Tables

Step 1   Prepare an r by c table of obtained cell frequencies where r is the number of categories in the first variable and c is the number of categories in the second variable.

Step 2   Compute the theoretical frequency for each cell of the contingency table obtained in Step 1 by multiplying the row sum by the column sum and dividing by N, the sum of frequencies over the entire table.

Step 3   If any theoretical frequency is less than 10.0, combine categories to eliminate the small theoretical frequency if possible. Do not compute χ^2 if categories cannot be combined to eliminate cells with theoretical frequencies less than 5.0.

Step 4   Take the difference between the obtained and theoretical frequencies for each cell in the contingency table, square the difference, and divide by the theoretical frequency.

Step 5   Sum the values computed in Step 4 to obtain χ^2.

Step 6   Using $(r - 1) \cdot (c - 1)$ degrees of freedom, find the critical values of chi square, $\chi^2_{.05}$ and $\chi^2_{.01}$.

Step 7   Test the value of χ^2 obtained in Step 5 against the critical values of χ^2 to determine whether the null hypothesis is to be rejected or not.

COMMON ERRORS IN THE USE OF CHI SQUARE

Chi square is widely and often incorrectly used. Some of the more common errors to be avoided are the following:

1. *Small theoretical frequencies.* As previously emphasized, it is preferable that theoretical frequencies not be less than 10.0 in any cell and certainly under no circumstances less than 5.0. Even if one of the f_t values is as low as 10.0, the others should be considerably larger. It is best to plan the study so that all f_t values will be well above 10.0.

2. *Failure to equate theoretical and obtained frequencies.* Attempts have been made to compare two sets of frequencies by means of chi square where the sums of f_o and f_t are different. This is unacceptable since it promotes larger discrepancies than normally occur by chance, thereby invalidating the test.

3. *Improper categorization.* Small categories often must be combined with other categories to increase the f_t values over the minimum. Also, variables on a more or less continuous scale are frequently broken down into a limited number of categories to make a chi square test. The decisions about categorization in such cases must not be made with an eye to increasing the size of the chi square value. Some uninformed individuals might try different categorizations to find the one that gives the highest chi square value before "deciding" how to categorize the data. Such decisions should be made in advance of data collection if possible, or at least on a rational basis that can be justified. Generally, it is not ethical to make such decisions to enhance the likelihood of supporting one's favored hypothesis.

4. *Lack of independence.* A common and often subtle error in the application of chi square involves the situation where each subject frequency is not independent of each other data observation. Each of the N trials must be free of dependence on any of the other trials for chi square to be appropriate. This requirement can be violated in many ways. In evaluating sex in relation to opinion on a given political question, for example, if subjects include husbands and wives, there is apt to be a linkage between the opinions of married individuals, invalidating the chi square test. Some investigators have applied chi square where several observations are taken from each of a number of individuals. This type of situation is very apt to introduce a lack of independence based on linkages between the trials for the same individual. For chi square to be appropriate, knowledge of the results of one trial should not enable the investigator to say what is apt to happen on some other trial because of any connection between those particular trials, for example, they are for the same or related people. Where individuals are the source of data, it is best to have only one frequency from each subject and not have the subjects related to each other in any way that would determine the location of their frequency tallies in the contingency table.

MATHEMATICAL
NOTES

The test statistic used for the chi square test on frequency data is Formula 9.1. This formula is not necessarily the easiest formula to be used in all situations involving frequency data or contingency tables. An alternative formula that involves only one subtraction is

$$\chi^2 = \left(\Sigma \frac{f_o^2}{f_t} \right) - N \tag{9.2}$$

Some researchers may find this formula easier to use. Formulas 9.1 and 9.2 are equivalent. Given below is a set of algebraic steps showing the equivalence:

$$\chi^2 = \Sigma \frac{(f_o - f_t)^2}{f_o} = \left(\Sigma \frac{f_o^2}{f_t} \right) - N$$

The derivation starts with Formula 9.1

$$\Sigma \frac{(f_o - f_t)^2}{f_t} =$$

$$= \Sigma \frac{(f_o^2 - 2f_o f_t + f_t^2)}{f_t} = \Sigma \frac{f_o^2}{f_t} + \Sigma \frac{(-2f_o f_t + f_t^2)}{f_t}$$

$$= \Sigma \frac{f_o^2}{f_t} + \Sigma(-2f_o + f_t)$$

$$= \Sigma \frac{f_o^2}{f_t} + \Sigma - 2f_o + \Sigma f_t$$

$$= \Sigma \frac{f_o^2}{f_t} + (-2)\Sigma f_o + \Sigma f_t$$

$$= \Sigma \frac{f_o^2}{f_t} - 2\Sigma f_o + \Sigma f_t$$

Since

$$\Sigma f_o = \Sigma f_t = N$$

then

$$\chi^2 = \Sigma \frac{f_o^2}{f_t} - 2N + N \qquad (9.2)$$

$$= \Sigma \frac{f_o^2}{f_t} - N$$

For a fourfold contingency table (Problem 36), there is an alternative formula that does not make use of the theoretical or expected frequencies. That is, the χ^2 can be computed directly from the observed frequencies. This is possible because the differences between the observed frequencies and expected (theoretical) frequency in each cell are equal (ignoring the sign). That is, $|f_o - f_t|$ (absolute value of the difference) are all equal. Knowing this,

$$\chi^2 = \Sigma \frac{(f_o - f_t)^2}{f_t} = (f_o - f_t)^2 \Sigma \frac{1}{f_t}$$

If the fourfold table layout of table 9.2 is rewritten as

a	b	$a + b$
c	d	$c + d$
$a + c$	$b + d$	N

then

$$\chi^2 = \frac{N(ad - bc)^2}{(a + b)(c + d)(a + c)(b + d)} \tag{9.3}$$

To show that Formula 9.1 is equivalent to Formula 9.3 for a fourfold contingency table, we must show

$$|f_o - f_t| = |ad - bc|/N \quad \text{and} \tag{9.4}$$

$$\Sigma \frac{1}{f_t} = N^3/(a + b)(c + d)(a + c)(b + d) \tag{9.5}$$

To show that Formula 9.4 is algebraically true, we start with

$$f_o = a, f_t = \frac{(a + b)(a + c)}{N}, N = a + b + c + d$$

$$|f_o - f_t| = \left| a - \left(\frac{a^2 + bc + ab + ac}{N} \right) \right|$$

Note:

$$\frac{N}{N} = 1 = \frac{a + b + c + d}{N}$$

Multiplying a by $\dfrac{N}{N}$ gives $\dfrac{aN}{N}$ or $\dfrac{a^2 + ab + ac + ad}{N}$ so

$$|f_o - f_t| = \left| \frac{a^2 + ab + ac + ad}{N} - \left(\frac{a^2 + bc + ab + ac}{N} \right) \right|$$

$$= \left| \frac{a^2 + ab + ac + ad}{N} + \left(\frac{-a^2 - bc - ab - ac}{N} \right) \right|$$

$$= \left| \frac{a^2 + ab + ac + ad - a^2 - bc - ab - ac}{N} \right|$$

$$= \left| \frac{ad - bc}{N} \right| = \frac{|ad - bc|}{N}$$

The same steps can be repeated for each f_o (i.e., $f_o = b, c, d$) resulting in the same answer. Hence, Formula 9.4 is true.

To show that Formula 9.5 is algebraicly correct, we begin with

$$\Sigma \frac{1}{f_t} = N^3/[(a + b)(c + d)(a + c)(b + d)]$$

The f_t values are

$$\frac{(a + b)(a + c)}{N}, \frac{(a + b)(b + d)}{N}, \frac{(a + c)(c + d)}{N}, \frac{(b + d)(c + d)}{N}$$

$$\Sigma \frac{1}{f_t} = \frac{1}{\dfrac{(a + b)(a + c)}{N}} + \frac{1}{\dfrac{(a + b)(b + d)}{N}} + \frac{1}{\dfrac{(a + c)(c + d)}{N}} + \frac{1}{\dfrac{(b + d)(c + d)}{N}}$$

$$= \frac{N}{(a + b)(a + c)} + \frac{N}{(a + b)(b + d)} + \frac{N}{(a + c)(c + d)} + \frac{N}{(b + d)(c + d)}$$

Multiply each numerator by the appropriate terms to get $(a + b)(a + c)(b + d)$
$(c + d)$ in denominator, gives:

$$\Sigma \frac{1}{f_t} = \frac{N(b + d)(c + d) + N(a + c)(c + d) + N(a + b)(b + d) + N(a + b)(a + c)}{(a + b)(a + c)(b + d)(c + d)}$$

$$= \frac{N[(b + d)(c + d) + (a + c)(c + d) + (a + b)(b + d) + (a + b)(a + c)]}{(a + b)(a + c)(b + d)(c + d)}$$

$$= \frac{N[a^2 + b^2 + c^2 + d^2 + 2ab + 2ac + 2ad + 2bc + 2bd + 2cd]}{(a + b)(a + c)(b + d)(c + d)}$$

Since

$$N = a + b + c + d$$

and

$$N^2 = a^2 + b^2 + c^2 + d^2 + 2ab + 2ac + 2ad + 2bc + 2bd + 2cd,$$

$$\Sigma \frac{1}{f_t} = \frac{N[N^2]}{(a + b)(a + c)(b + d)(c + d)} \tag{9.5}$$

$$= \frac{N^3}{(a + b)(a + c)(b + d)(c + d)}$$

This shows Formula 9.5 is true.
Since

$$\chi^2 = \Sigma \frac{(f_o - f_t)^2}{f_t} = (f_o - f_t)^2 \Sigma \frac{1}{f_t}$$

is true for a fourfold contingency table, combining Formulas 9.4 and 9.5 gives:

$$\chi^2 = (f_o - f_t)^2 \, \Sigma \, \frac{1}{f_t} = \left(\frac{(ad - bc)^2}{N^2}\right)\left(\frac{N^3}{(a + b)(a + c)(b + d)(c + d)}\right)$$

$$= \frac{(ad - bc)^2 N}{(a + b)(a + c)(b + d)(c + d)}$$

Rearranging terms,

$$\chi^2 = \frac{N(ad - bc)^2}{(a + b)(c + d)(a + c)(b + d)} \tag{9.3}$$

With Yates' correction for continuity, Formula 9.3 is modified to read

$$\chi^2 = \frac{N\left(|ad - bc| - \dfrac{N}{2}\right)^2}{(a + b)(c + d)(a + c)(b + d)} \tag{9.6}$$

PROBLEM D

Test the Differences between Obtained and Normal Curve Frequency Distribution Values

In this problem, differences between obtained and normal curve frequencies are tested to ascertain whether or not it is reasonable to suppose that the obtained distribution was drawn by random sampling from a normally distributed population. If the computed chi square value proves to be larger than the selected critical value, $\chi^2_{.05}$ or $\chi^2_{.01}$, the null hypothesis is rejected and the obtained distribution cannot be considered to vary only negligibly from normality in shape. On the other hand, if the obtained chi square value is less than the selected critical value of χ^2, then the null hypothesis is not rejected and the discrepancies between the obtained and the normal distribution are treated potentially or conceivably as being due to chance. Of course, a nonsignificant χ^2 does not *prove* the null hypothesis any more than a nonsignificant difference between means *proves* the null hypothesis in that situation.

Example   Application of the chi square to this type of situation is illustrated by means of a worked out example. Methods for computing normal curve frequencies corresponding to obtained frequencies are described in Problem 15, Chapter 5. Table 5.3 gives an example of a score distribution for which normal curve frequencies are computed both by the ordinate method and by the area method.

TABLE 9.4					

Computation of chi square for the difference between obtained and normal curve frequencies

X	f_o	f_t	$(f_o - f_t)$	$(f_o - f_t)^2$	$(f_o - f_t)^2/f_t$
35–54	17	16.4	0.6	0.36	0.02
30–34	10	9.5	0.5	0.25	0.03
25–29	8	8.8	−0.8	0.64	0.07
5–24	10	10.3	−0.3	0.09	0.01
Σ	45	45.0	0.0		$\chi^2 = 0.13$ N.S.

With $(k - 3) = 4 - 3 = 1$ df, $\chi^2_{.05} = 3.84$, $\chi^2_{.01} = 6.64$

Table 9.4 reproduces the actual and theoretical frequencies for the distribution shown in table 5.3 using the actual frequencies to obtain the f_o values and the area method normal curve frequencies to obtain the f_t values. Note two important changes between table 5.3 and table 9.4. First, it is necessary to modify the area method frequencies so that they add up to the same number total as the obtained frequency total, N. In this case, the area method frequencies total is 44.9, only 0.1 off from the actual frequency total. This discrepancy is due to rounding error. It will be eliminated here by arbitrarily adding 0.1 to the top frequency of 1.1, making it 1.2. The 0.1 can be added to any other area method frequency, but preferably not to reduce the obtained $(f_o - f_t)$ value.

The second, and more critical difference between the values in tables 5.3 and 9.4 is that some of the intervals at both the top and bottom of the distribution are combined in table 9.4. Thus, the intervals 50–54, 45–49, 40–44, and 35–39 are combined, summing their obtained frequencies of 1, 2, 5, and 9, respectively, to obtain the f_o value of 17 for the interval 35–54 in table 9.4. The corresponding area method normal curve frequencies for the intervals 1.2, 2.3, 5.0, and 7.9 are summed to give the f_t value of 16.4 for the interval 35–54 in table 9.4. Correspondingly, the bottom intervals 5–9, 10–14, 15–19, and 20–24 are combined to give f_o and f_t values of 10 and 10.3, respectively.

These intervals at the top and bottom of the distribution are combined because the f_t values are too small for these end intervals.

To maintain all theoretical frequencies at 10.0 or greater requires, in most cases, the combining of at least some end intervals in the distribution, especially when the overall N is not very large. In the case of table 9.4, a strict adherence to the rule of 10.0 for f_t values also requires the combination of the two intervals 25–29 and 30–34. In this example, the overall N is too small, making it impossible to obtain large enough f_t values for an ideal use of the chi square test. The f_t values of 9.5 and 8.8 for the two middle intervals in table 9.4 are large enough, however, so that the χ^2 value is usable unless it falls near the critical value.

After the intervals have been combined, as shown in table 9.4, the next step is to take the differences between the obtained and theoretical frequencies $(f_o - f_t)$, square them, and divide by their theoretical frequencies. These computations are shown in table 9.4 with the sum of the last column being the chi square value. In this case, the χ^2 is 0.13, a very small value since f_o and f_t values were very close to each other.

Determining the Number of Degrees of Freedom

It is necessary to determine the number of degrees of freedom in this situation before the Table of the Chi Square Distribution can be used to find $\chi^2_{.05}$ and $\chi^2_{.01}$. These critical values of χ^2, when obtained, provide the standard of comparison against which the obtained $\chi^2 = 0.13$ is compared. If the obtained χ^2 is less than $\chi^2_{.05}$, as it is expected to be in this case, then h_o cannot be rejected and there is no significant difference between the obtained and the corresponding normal curve frequencies.

On the other hand, if the discrepancies between the f_o and f_t values are large, such as to give an obtained chi square as large or larger than $\chi^2_{.05}$ or $\chi^2_{.01}$, then it is appropriate to reject h_0. This implies that the obtained distribution did not arise as a result of random sampling from a normal distribution with the specified μ and σ.

To determine the number of degrees of freedom in this situation it is necessary to see how many data values can be specified at liberty before the other values are of necessity fixed by the restraints created by the column total and the demand that the frequencies assigned result in the mean and standard deviation actually found in the sample. One degree of freedom is lost by the requirement that the assigned frequencies add up to N since, if all but one of the frequencies had been fixed, the last one would be predetermined to make the frequency total add up to N. This reduces the df to $k - 1$ where k is the number of class intervals in this case. The requirement that the frequencies lead to a specified mean and standard deviation takes away 2 more degrees of freedom, leaving $k - 3$ degrees of freedom for comparing obtained and normal curve frequencies. In the problem shown in table 9.4, this gives $4 - 3$ or 1 df, which leads to $\chi^2_{.05} = 3.84$ and $\chi^2_{.01} = 6.63$ (see table D, Appendix B). Since the obtained chi square of 0.13 is well below the $\chi^2_{.05} = 3.84$, the obtained chi square is not significant and therefore h_0 is not rejected in this situation.

References

Ferguson, G. A. (1971). *Statistical analysis in psychology and education* (3d ed.). New York: McGraw-Hill.

Prokasy, W. F. (1972). Developments with the two phase model applied to human eyelid conditioning, in Black, A. H. & Prokasy, W. F. (eds.) *Classical Conditioning II*. New York: Appleton-Century-Crofts, pp. 119–47.

Yates, F. (1934). Contingency tables involving small numbers and the χ^2 test. *Supplemental Journal of the Royal Statistical Society, 1,* 217–35.

Exercises

9–1.   Given the following obtained and theoretical normal curve frequencies, combine class intervals so that no f_t is less than 10.0 and make a chi square test of the fit.

X	f_o	f_t
110–119	3	2.3
100–109	8	5.9
90–99	12	13.8
80–89	20	24.5
70–79	36	32.6
60–69	33	33.0
50–59	24	25.3
40–49	16	14.6
30–39	7	6.4
20–29	2	2.6

 a. Starting on either end of the distribution, combine class intervals until f_t is $> = 10$.
 b. Combine the f_o for the corresponding class intervals used in (a).
 c. Rewrite the frequency distribution.
 d. Add the columns $f_o - f_t$, $(f_o - f_t)^2$, and $(f_o - f_t)^2/f_t$.
 e. Sum the values in the $(f_o - f_t)^2/f_t$ column and set the sum equal to χ^2.
 f. Determine the degrees of freedom and the critical χ^2 value.
 g. If the computed χ^2 value exceeds the critical χ^2 value, then the observed frequencies do *not* fit the theoretical frequencies.

9–2.   Given the following obtained and theoretical normal curve frequencies, combine class intervals so that no f_t is less than 10.0 and make a chi square test of the fit.

X	f_o	f_t
50–54	3	2.6
45–49	11	7.2
40–44	22	18.3
35–39	31	36.0
30–34	42	54.3
25–29	56	64.1
20–24	70	55.8
15–19	48	37.9
10–14	18	19.8
5–9	6	11.0

    a. Starting at the top end, combine class intervals until f_t is $> = 10$.

    b. Combine the f_o for the corresponding class intervals used in (a).

    c. Rewrite the frequency distribution.

    d. Add the columns $f_o - f_t$, $(f_o - f_t)^2$, and $(f_o - f_t)^2/f_t$.

    e. Sum the values in the $(f_o - f_t)^2/f_t$ column and set the sum equal to χ^2.

    f. Determine the degrees of freedom and the critical χ^2 value.

    g. If the computed χ^2 value exceeds the critical χ^2 value, then the observed frequencies do not fit the theoretical frequencies.

9–3.   In a small coeducational college a controversy developed between student factions over whether students who do not belong to fraternities and sororities are doing their part to support student functions. The student council decides to carry out a survey. At random, 76 students are selected and asked whether they attended the recent college-wide dance and whether they belong to a fraternity or sorority. The results show that of the 76 students surveyed, 15 organizationals and 11 nonorganizationals went to the dance; 11 organizationals and 39 nonorganizationals did not go to the dance. Use the χ^2 test to analyze the data and reach a conclusion. Would you use Yates' correction?

    a. State the null and alternative hypotheses.

    b. Set up a 2-by-2 table with a column and row labeled "total." At the top of table label one column "org" and the other "nonorg." On the left side, label one row "go" and the other "no-go."

    c. Place the appropriate frequency counts into each cell. These are obtained or observed frequencies.

    d. Find the theoretical or expected frequencies (see page 273).

    e. Examine the theoretical frequencies for any value below 10. If one or more exists, Yates' correction is necessary.

    f. Compute the χ^2 value using Formula 9.1.

    g. Find the degrees of freedom and the critical χ^2 value.

    h. If the computed χ^2 is greater than the critical χ^2 value, reject h_0.

9–4.   A sociologist is interested in studying the relationship between age at time of marriage and durability of the marriage. Her hypothesis is that early marriages for males tend to be unstable. She takes a random sample of males from the general population in her city and determines that of 200 males married over 10 years ago, 96 were over 21 at the time of their marriage and were still married 10 years later to the same person; 24 were over 21 when married, but did not stay married for 10 years; 44 of the 200 were under 21 at time of marriage and were still married 10 years later; 36 were under 21 at the time of marriage and did not stay married 10 years. Test the hypothesis using χ^2 and interpret the results of the study.

    a. State the null and alternative hypotheses.

    b. Set up a 2-by-2 table with top labels of "not separated," "separated," and "total." Create left margin labels of "over 21," "under or at 21," and "total."

    c. Place the appropriate frequency counts into each cell. These are the obtained or observed frequencies.

    d. Find the theoretical or expected frequencies (see page 273).

    e. Examine the theoretical frequencies for any value below 10. If one or more exists, Yates' correction is necessary.

f. Compute the χ^2 value using Formula 9.1.

g. Determine the degrees of freedom and the critical χ^2 value.

h. If the computed χ^2 is greater than the critical χ^2 value, reject h_0.

9-5.   A personality test developer wishes to exclude from the test any items on which males and females differ significantly in their responses. The following table for one of the three-choice items is obtained. Use χ^2 to determine if the item should be eliminated:

	No	?	Yes	Total
Males	40	20	40	100
Females	30	10	60	100
Total	70	30	100	200

a. State the null and alternative hypotheses.

b. Find the theoretical frequencies (see page 273).

c. Examine the theoretical frequencies for any value below 10. If one or more exists, Yates' correction is necessary.

d. Compute the χ^2 value using Formula 9.1.

e. Determine the degrees of freedom and the critical χ^2 value.

f. If the computed χ^2 is greater than the critical χ^2 value, reject h_0.

9-6.   A movie producer is anxious to learn what group should be the target for the advertising relative to a new release. At a preview, the data shown below are obtained. Do the data suggest anything about the advertising plans?

	under 20	20–29	30–39	40–49	over 50	Total
Liked	33	31	20	10	11	105
Indifferent	10	15	22	8	6	61
Disliked	12	19	16	25	22	94
Total	55	65	58	43	39	260

a. State the null and alternative hypotheses.

b. Find the theoretical frequencies.

c. Examine the theoretical frequencies for any value below 10. If N is small and one or more exists, Yates' correction is necessary.

d. Compute the χ^2 value using Formula 9.1.

e. Determine the degrees of freedom and the critical χ^2 value.

f. If the computed χ^2 is greater than the critical χ^2 value, reject h_0.

9-7.   The track coaches suspect that the track at their school is not uniform and decide to check the records. They locate records for 180 runnings of the 100-yard dash and find that the winners distributed themselves by lanes as follows: lane 1, 20; lane 2, 22; lane 3, 40; lane 4, 45; lane 5, 32; and lane 6, 21. Determine if there is any evidence of nonrandomness in the pattern of wins by lane. Use the 0.01 level of confidence for χ^2.

a. State the null and alternative hypotheses.

b. Determine the theoretical value for each lane.

c. Compute the χ^2 using Formula 9.1.

d. Determine the degrees of freedom and the critical χ^2 value.

e. If the computed χ^2 is greater than the critical χ^2 value, reject h_0.

9-8.   A die is suspected of being loaded. To test this hypothesis, the die is thrown 120 times and a record kept of the number of times each face shows up. The frequencies are as follows: 1, 35; 2, 22; 3, 16; 4, 15; 5, 17; and 6, 15. Use the chi square test to determine if the obtained frequencies depart significantly from the theoretical frequencies. Use the 0.05 level of confidence.

a. State the null and alternative hypotheses.

b. Determine the theoretical frequencies for each face.

c. Compute the χ^2 value using Formula 9.1.

d. Determine the degrees of freedom and the critical χ^2 value.

e. If the computed χ^2 is greater than the critical χ^2 value, reject h_0.

9-9.   A manufacturer of soft drinks conducted a consumer-preference study to determine whether Brand X was preferred over those made by four competitors. A sample of 100 consumers taste-tested each soft drink and indicated which one they preferred. The results of the survey are:

Soft drink	X	Y	Z	W	T
Frequency	27	15	22	17	19

Do these data present sufficient evidence to indicate a preference of one soft drink over the others? If the hypothesis of "no preference" for this study is rejected, does this mean that soft drink X is superior to the others?

a. State the null and alternative hypotheses.

b. Determine the theoretical frequencies for each soft drink.

c. Compute the χ^2 value using Formula 9.1.

d. Determine the degrees of freedom and the critical χ^2 value.

e. If the computed χ^2 is greater than the critical χ^2 value, reject h_0.

f. If the hypothesis of no preference is rejected, does this mean that soft drink X is superior to the others?

9-10.   A survey is conducted to determine student, faculty, and administration attitudes toward a new track stadium. The distribution of those favoring or opposed to the stadium is given below:

	Student	Faculty	Administration
Favor	252	107	40
Oppose	139	81	43

Do the data provide sufficient evidence to indicate that attitudes regarding the stadium are independent of student, faculty, or administration status?

a. State the null and alternative hypotheses.

b. Determine the theoretical frequencies of each cell. Find row totals and column totals.

c. Compute the χ^2 value using Formula 9.1.

d. Determine the degrees of freedom and the critical χ^2 value.

e. If the computed χ^2 is greater than the critical χ^2 value, reject h_0.

9–11. An analysis of cereal sales is made to determine the distribution of high sales for packages of three sizes. The data for 346 cereal packages sold are as follows:

	Size of package		
	Small	Medium	Large
High	128	63	46
Low	67	26	16

Do the data indicate that the frequency of high sales is dependent on the size of package? Use $\alpha = 0.01$.
  a. State the null and alternative hypotheses.
  b. Find the theoretical frequencies.
  c. Compute the χ^2 value using Formula 9.1.
  d. Determine the degrees of freedom and the critical χ^2 value.
  e. If the computed χ^2 is greater than the critical χ^2 value, reject h_0.

9–12. Gregor Mendel stated that the number of sweet peas of a certain type falling into the classifications round and yellow, wrinkled and yellow, round and green, and wrinkled and green should be in the ratio $9:3:3:1$. Suppose that 200 such sweet peas revealed 112, 38, 34, and 16 in the respective classes. Are these data consistent with the Mendel model? Use $\alpha = 0.05$.
  a. State the null and alternative hypotheses.
  b. Determine the theoretical frequencies.
  c. Compute the χ^2 value using Formula 9.1.
  d. Determine the degrees of freedom and the critical χ^2 value.
  e. If the computed χ^2 is greater than the critical χ^2 value, reject h_0.

9–13. A market researcher hypothesized that consumers would choose item A 38% of the time, while items B and C are chosen 27% and 35%, respectively. To test this hypothesis, 200 consumers are recruited. Each consumer is asked to choose item A, B, or C. The results are:

Choice	Frequency
A	80
B	50
C	70
Total	200

Do these data fit the hypothesized distribution of choices? Use $\alpha = 0.01$.
  a. State the null and alternative hypotheses.
  b. Find the theoretical frequencies.
  c. Compute the χ^2 value using Formula 9.1.
  d. Determine the degrees of freedom and the critical χ^2 value.
  e. If the computed χ^2 value is greater than the critical χ^2 value, reject h_0.

9–14. Two groups consisting of 100 mental patients are chosen in a study concerning the effectiveness of therapy. One group received therapy while the other received a nontherapeutic task (placebo). After a period of time each patient is evaluated to determine if the patient improved or did not improve in functioning. The results are observed as:

	Improved	Not improved	Total
Therapy	75	25	100
Placebo	58	42	100
Total	133	67	200

Do these data present sufficient evidence that there is a relationship between treatment and improvement? Use $\alpha = 0.05$.
   a. State the null and alternative hypotheses.
   b. Find the theoretical frequencies.
   c. Compute the χ^2 value using Formula 9.1.
   d. Determine the degrees of freedom and the critical χ^2 value.
   e. If the computed χ^2 value is greater than the critical χ^2 value, reject h_0.

Use the following data for questions 9–15 through 9–18.

Student	Sex	Age	Grade	Affiliation	Height
1	M	10	A	D	S
2	F	11	B	R	S
3	F	12	A	R	S
4	M	10	B	D	S
5	F	10	F	D	M
6	M	11	C	R	T
7	F	12	D	R	S
8	F	10	B	R	M
9	M	11	C	D	T
10	F	11	C	D	T
11	F	12	B	R	S
12	M	10	A	D	S
13	M	10	B	D	T
14	M	12	C	D	T
15	F	11	C	R	M
16	F	12	D	R	S
17	M	10	F	R	M
18	F	11	C	D	M
19	M	11	C	D	T
20	M	12	B	R	T

9–15. Construct a contingency table between Sex and Age.
   a. Determine the number of different entries for sex and age.
   b. Create a Sex-by-Age table.
   c. Find the number of 10-year-old males, 11-year-old males, 12-year-old females, etc., and enter the frequencies in the table.

9-16.   a. Construct a contingency table between Sex and Height.
   b. Is there a relationship between Sex and Height? Use $\alpha = 0.05$.
      1. Determine the number of different entries for Sex and Height.
      2. Create a Sex-by-Height table.
      3. Find the number of short males, short females, tall females, etc., and enter frequencies into the table.
      4. State the null and alternative hypotheses.
      5. Compute the theoretical frequencies.
      6. Compute χ^2. Note: With theoretical frequencies less than 5, the χ^2 should not be computed.

9-17.   Construct a contingency table between Grade and Sex.
   a. Determine the number of different types of entries for Grade and Sex.
   b. Create a Grade-by-Sex table.
   c. Find the number of males and females with grade A, grade B, etc., and enter the counts into the appropriate cell in the table.

9-18.   State the proper rule for when
   a. Yates' correction is used.
   b. The χ^2 value should not be computed using Formula 9.1.

9-19.   The binomial distribution for $n = 15$ and $p = .30$ has the probabilities found in column "A" below.

	"A"		"B"	
X	Probability		X	Frequency
0	.005		0	6
1	.03		1	25
2	.092		2	99
3	.170		3	150
4	.218		4	230
5	.207		5	200
6	.147		6	150
7	.081		7	77
8	.035		8	40
9	.011		9	7
10	.003		10	10
11	.001		11	2
12	.000		12	3
13	.000		13	1
14	.000		14	0

In a sample of 1,000, the frequencies in column "B" were observed. Using only $X = 0$ through $X = 9$, determine whether the observed frequencies fit the probability distribution given by the binomial? Use $\alpha = 0.05$.
   a. State the null and alternative hypotheses.
   b. Find the theoretical frequencies ($N = 100$).
   c. Compute the χ^2 value using Formula 9.1.
   d. Determine the degrees of freedom and the critical χ^2 value.
   e. If the computed χ^2 value is greater than the critical χ^2 value, reject h_0.

9-20.   A psychologist wanted to determine if the type of therapy was related to the level (amount) of recovery for psychotic patients. The data collected are given below.

		Type of therapy		
		A	B	C
Level of	High	19	32	14
Recovery	Medium	20	6	11
	Low	10	12	18

Determine if there is a relationship between the type of therapy and the levels of recovery using $\alpha = 0.05$.
  a. State the null and alternative hypotheses.
  b. Compute the theoretical frequencies. Find the row and column totals first.
  c. Compute the χ^2 value using Formula 9.1.
  d. Determine the degrees of freedom and the critical χ^2 value.
  e. If the computed χ^2 value is greater than the critical χ^2 value, reject h_0.

design also
to evaluate
There can

The n
drawn fro
subgroups
must plan
patients c
attendant
kinds of tr
if found t
lations sa
treatmen
ment gro
experime
luctant to
need it, a
from the

Just
populatic
all the p
are supp
affect th
This typ
affected
affected
ticularly
crepant
the 0.01
on analy
erwise,
differen

In
imenta
one-wa
tion. A
curves,
nonnor
sult an
of trea
variati
scores
becom
in the
a cont

Analysis of Variance— One-Way Designs

ever
resu
mer
of subjects
the methoc
the effects
full discus
the scope
designs, h
where in t
those met
literature
    A cha
more intr
findings
specific d
with any
what me
to use ar
with that
to meet a
research
ples, and
is preser
t-test to
uation c
contrast
fects of

PROE

Test
of Ec

With t
group
howev
examp
effect
sugge
inject
tion,
is evi
than

THE ONE-WAY ANOVA TEST OF DIFFERENCES BETWEEN SUBGROUP MEANS

The one-way ANOVA test is applied to the situation where several subsamples of equal size are drawn at random from a common pool and subjected to different types of experimental treatment. The sum of scores and sum of squared scores for each subgroup are computed separately and then combined to give the totals for the entire group (combining the subgroups). These data are utilized to obtain two different estimates of the population variance, σ^2, for the experimental variable, a "within-groups" estimate and a "between-groups" estimate.

In some texts and the literature these variance estimates are referred to as "mean square." The between-group variance represents differences due to treatment and the within-group variance represents random or uncontrolled error. These two variance estimates are compared by means of the F ratio test. Unlike the F-test presented in Chapter 8, which tested homogeneity of variance, this F-test is interested in how large the between-group variance is from the within-group variance.

The null hypothesis for the one-way ANOVA states that there is no real difference between the group means. The alternative hypothesis is that there is at least one pair of group means not equal. In symbol form:

$$h_0 : \mu_i = \mu_j \text{ for all } i \text{ and } j$$

$$h_1 : \mu_i \neq \mu_j \text{ for some } i \text{ and } j \text{ where } i \neq j.$$

If F is significant, the null hypothesis (h_0) is rejected and the differences between means are presumed to exceed what is expected if the samples are drawn from populations with equal means. If F is not significant, h_0 is not rejected.

Within-Groups Variance Estimate

The variances of the population from which three samples are drawn at random is supposed to be the same for all three groups, σ^2. An estimate of σ^2 is obtained from each group (s_1^2, s_2^2, s_3^2), by the formula $\Sigma x^2/(n-1)$ where n is the number of cases in the subgroup. A better estimate of σ^2 is obtained by combining the data from all three groups to get one overall estimate of σ^2. This is done by pooling the sums of squares of deviation scores and pooling the degrees of freedom:

$$s^2 = \frac{\Sigma x_1^2 + \Sigma x_2^2 + \ldots + \Sigma x_k^2}{(n_1 - 1) + (n_2 - 1) + \ldots + (n_k - 1)} \tag{10.1}$$

An alternate way of pooling the sums of squares is to add all three groups together and take Σx^2 for the total group and divide by $N - 1$ where N is the sum of $n_1 + n_2 + n_3$. This sum of squared deviation scores is taken with respect to the overall mean ($x_i = X_i - M$), though in Formula 10.1 each deviation score is taken for its own subgroup mean. Formula 10.1 gives a better estimate of σ^2 because the means of the subgroup populations are not presumed to be equal, as is required to permit the use of deviation scores for the overall mean (**grand mean**) of the three groups

grand mean   The overall mean for all groups taken together.

STATISTICS IN THE WORLD AROUND YOU

In the April 14, 1980 issue of *Advertising Age,* an announcement was made about the battle between three cigarette manufacturers over which of their products yield the lowest tar content. The three brands were Carlton, NOW, and Cambridge Box. In a research study, data were gathered on these three different brands of cigarettes. Five cigarettes of each brand were measured for the amount of tar. The data are given below:

Brand	Tar per cigarette (milligrams)				
Carlton	.15	.13	.22	.15	.14
NOW	.20	.24	.17	.19	.18
Cambridge	.17	.18	.21	.23	.21

To compare the three brands for differences, a one-way ANOVA is applied to the data. The ANOVA tells us whether or not a difference exists between the three brands. In doing the calculations, a computed F value is 3.10. The critical value for the .05 level of significance using 2 and 12 degrees of freedom is 3.88. Since 3.10 is less than 3.88, there is insufficient evidence to claim any one of the brands is producing significantly lower tar than the others. The means do indicate that the mean for 5 cigarettes of Carlton (.158) is lower than either NOW (.196) or Cambridge (.200). However, the differences are not considered to be statistically significant.

Research studies on consumer products are common in the business world. Psychology is applied in the development of advertisement and in the planning of marketing programs. Manufacturers use the information generated by such studies to help form their advertising strategy and/or take corrective measures to become the "best" in that consumer product category.

combined. By taking deviations for subgroup means and pooling them across groups, the estimate of the population variance from the pooled data is not improperly inflated by any differences that exist between the means. Of course, if the differences between means are of negligible importance, the two different ways of obtaining the pooled variance estimate are roughly comparable. The method illustrated by Formula 10.1 is always used, however, to obtain the within-groups estimate of the population variance for the experimental variable being measured. The one-way ANOVA test of the difference between means makes use of this within-groups estimate of the population variance by comparing it to a second estimate of the population variance derived from information about the variability of the subgroup means.

**Between-Groups
Variance Estimate**

Formula 8.5 in Chapter 8 provides an estimate of the variability of the sampling distribution of means, the **standard error of the mean:**

standard error of the mean
The standard deviation of the
sampling distribution of
sample means.

$$s_M = \frac{s}{\sqrt{N}} \qquad (8.5)$$

Squaring both sides gives:

$$s_M{}^2 = \frac{s^2}{N} \qquad (10.2)$$

Another estimate of $s_M{}^2$ is obtained from the means of the subgroups in the experiment. This is done merely by calculating s^2, the variance estimate, based on the squared deviations of these subgroup means, M_i about the grand mean, M as follows:

$$s_M{}^2 = \frac{\sum_{i=1}^{k} (M_i - M)^2}{k - 1} \qquad (10.3)$$

where M_i = the mean of subgroup i,

M = the grand mean, and

k = the number of subgroups.

Substituting Formula 10.3 into Formula 10.2 gives:

$$\frac{s^2}{N} = \frac{\sum_{i=1}^{k} (M_i - M)^2}{k - 1}$$

Multiplying both sides by N gives:

$$s^2 = \frac{N \sum_{i=1}^{k} (M_i - M)^2}{k - 1} \qquad (10.4)$$

sum of squares   The sum of
the squared deviations of
sample values about the
sample mean.

Formula 10.4 gives another estimate of the population variance derived from information about the variability of subgroup means. In Formula 10.3, $k - 1$ is the number of degrees of freedom in estimating the true variance of the distribution of means, $\sigma_M{}^2$, and $k - 1$ also represents the number of degrees of freedom in Formula 10.4 for the between-groups estimate of the population variance, σ^2. Since a variance estimate is a ratio of a **sum of squares** of deviation scores to a degrees of freedom value, it follows that the numerator of the expression in Formula 10.4, $N \sum_{i=1}^{k} (M_i - M)^2$, represents the sum of squares for the between-groups estimate of σ^2, the true variance of the population with respect to the experimental variable.

The *F* Ratio Test in the One-Way ANOVA

Two ways of estimating σ^2, the true variance of the population for the experimental variable, are available from the two methods just described, one based on variability of subjects about their own subgroup means (within-groups estimate of σ^2) and the other based on the variability of the subgroup means about the grand mean (between-groups estimate of σ^2). If there are significant treatment effects causing the means of the various experimental subgroups to vary more than randomly about the grand mean, then the between-groups estimate of σ^2 will be unduly inflated. The more pronounced the differences between sample means, the more inflated this variance estimate becomes.

It follows, therefore, that if the between-groups estimate of σ^2 is significantly greater than the within-groups estimate of σ^2, the sample means must differ from each other more than is expected of random samples from populations all having the same population mean, μ.

To test whether the means of subgroups are significantly different from each other using the one-way ANOVA, an *F* ratio is obtained dividing the between-groups estimate of σ^2 by the within-groups estimate of σ^2. The significance of this *F* is evaluated in the same manner as before (see Problem 27, Chapter 8).

The only difference between this application of *F* and that in Problem 27 is that the between-groups estimate of σ^2 is *always* placed on top, whether it is larger or not. It almost always is larger. In case it is not, however, *F* becomes less than 1.0 and is automatically not significant. This merely means that the variability of the group means is even less than what might be expected as a result of random sampling variations for samples drawn from populations with the same mean.

Example   The application of the one-way ANOVA with samples of equal size is illustrated by means of a worked-out example. Table 10.1 shows the results of a hypothetical experiment in which an instructor tested the differential utility of three sets of self-instructional programmed materials. Thirty students are selected from the instructor's classes to participate in the experiment. These subjects are randomly assigned to groups 1, 2, and 3. Instructional materials sets are then numbered and randomly assigned to groups 1, 2, and 3. Each group works with its own set of instructional materials for one month. At the end of this period, all three groups take the same examination over the subject matter that all instructional materials sets were supposed to cover. Scores on this examination for the individuals in the three groups are given in table 10.1 along with the squares of those scores. The sums of the columns are given below the scores in the row labeled Σ. The remainder of the calculations for making this test are divided into getting the within-groups variance estimate, the between-groups variance estimate, and computing the *F* ratio.

TABLE 10.1
One-way ANOVA with samples of equal size

Subject number	Group 1		Group 2		Group 3	
	X	X^2	X	X^2	X	X^2
1	45	2025	10	100	50	2500
2	50	2500	55	3025	90	8100
3	70	4900	35	1225	40	1600
4	40	1600	50	2500	70	4900
5	40	1600	65	4225	80	6400
6	10	100	30	900	35	1225
7	30	900	45	2025	25	625
8	80	6400	40	1600	55	3025
9	35	1225	20	400	75	5625
10	20	400	25	625	45	2025
Σ	420	21650	375	16625	565	36025

$\Sigma\Sigma X = 1360$ $\qquad\qquad \Sigma\Sigma X^2 = 74300$

$N = n_1 + n_2 + n_3 = 10 + 10 + 10 = 30$. Since $n_1 = n_2 = n_3$, $n = 10$.

Within-groups variance estimate:

$$\Sigma x^2 = \Sigma X^2 - \frac{(\Sigma X)^2}{n}$$

$$\Sigma x_1^2 = 21650 - \frac{(420)^2}{10} = 21650 - 17640 = 4010.0$$

$$\Sigma x_2^2 = 16625 - \frac{(375)^2}{10} = 16625 - 14062.5 = 2562.5$$

$$\Sigma x_3^2 = 36025 - \frac{(565)^2}{10} = 36025 - 31922.5 = 4102.5$$

Pooled sum of squares $= \Sigma x_1^2 + \Sigma x_2^2 + \Sigma x_3^2$
$= 4010.0 + 2562.5 + 4102.5$
$= 10675.0$

Pooled sum of degrees of freedom $= (n_1 - 1) + (n_2 - 1) + (n_3 - 1)$
$= 9 + 9 + 9 = 27$

Within-groups variance estimate $= s_W^2 = \dfrac{10675.0}{27} = 395.4$

Between-groups variance estimate:

$$M_1 = \frac{420}{10} = 42.0 \qquad\qquad\qquad M_1^2 = 1764.00$$

$$M_2 = \frac{375}{10} = 37.5 \qquad\qquad\qquad M_2^2 = 1406.25$$

$$M_3 = \frac{565}{10} = 56.5 \qquad\qquad\qquad M_3^2 = 3192.25$$

$$\Sigma \ldots \ldots \ldots 136.0 \qquad\qquad\qquad \Sigma = 6362.50$$

> **TABLE 10.1**
>
> **Continued**
>
> ---
>
> $$\sum_{i=1}^{3} (M_i - M)^2 = \Sigma M_i^2 - \frac{(\Sigma M_i)^2}{K} = 6362.5 - \frac{(136)^2}{3} = 6362.5 - 6165.3$$
> $$= 197.2$$
>
> Pooled sum of squares $= SS_B = n\sum_{i=1}^{3} (M_i - M)^2 = 10(197.2) = 1972.0$
>
> degrees of freedom $= k - 1 = 2$
>
> between-groups variance estimate $= s_B^2 = \dfrac{1972.0}{2} = 986.0$
>
> *F Ratio:*
>
> $$F = \frac{s_B^2}{s_W^2} = \frac{986.0}{395.4} = 2.49 \text{ N.S.}$$
>
> $F_{.05} = 3.35 \qquad F_{.01} = 5.49$
>
> ---

To obtain the within-groups variance estimate, first compute the sum of squares of deviation scores for each separate subgroup for its own subgroup mean, using the now familiar formula $\Sigma x^2 = \Sigma X^2 - (\Sigma X)^2/N$, where N in this particular case indicates n, the size of the subgroup. These computed values Σx_1^2, Σx_2^2, Σx_3^2, are given below table 10.1 as 4010.0, 2562.5, and 4102.5. The values are summed to obtain the within-groups sum of squares. The degrees of freedom are 9 in each group, so the pooled df for the within-groups sum of squares is 27. Dividing the sum of squares by the df gives $s_W^2 = 10675.0/27 = 395.4$ as the within-groups estimate of σ^2.

To obtain the between-groups variance estimate, the means of the subsamples and their squares are computed. The sums of these values are used to obtain $\sum_{i=1}^{3} (M_i - M)^2$, the sum of the squared deviations of the subsample means, M_i, from the grand mean, M. This value, 197.2, must be multiplied by 10, the subsample size (n), to get the sum of squares for the between-groups estimate of σ^2, as shown in Formula 10.4. The sum of squares between-groups, 1972.0, is divided by $(k - 1) = 2$, the number of df, to obtain the between-groups variance estimate, $s_B^2 = 986.0$.

Finally, to obtain F, the between-groups variance estimate is divided by the within-groups variance estimate to give $F = s_B^2/s_W^2 = 2.49$. Entering table E, Appendix B with 2 df for the larger variance estimate and 27 df for the smaller variance estimate, $F_{.05} = 3.35$ and $F_{.01} = 5.49$ are obtained as the values of F required for significance at the 5% and 1% levels, respectively. Since the obtained value of F is only 2.49, it is labeled N.S. for not significant. As a consequence of the insignificant F ratio, h_0, the null hypothesis, cannot be rejected on the basis of these results. If there are differences between means induced as a consequence of the treatment effects, this test does not show them to be statistically significant. The fairly large numerical difference between the means and size of F obtained in relation to $F_{.05}$

TABLE 10.2

One-way ANOVA summary for data in table 10.1

Source of variation	Sum of squares	Degrees of freedom	Mean square (variance estimate)	F
Between-groups	1972.0	2	986.0	2.49
Within-groups	10675.0	27	395.4	
Total	12647.0	29		

$F_{.05} = 3.35$; $F_{.01} = 5.49$

might lead the investigator to hope that a repetition of the experiment with larger samples will yield significant results. With sample sizes of only 10, however, the differences between the teaching materials are not large enough to yield a statistically significant effect.

The results of a one-way ANOVA are typically summarized in a table similar to that shown in table 10.2. The total sum of squares and degrees of freedom, obtained in table 10.2 by adding the values for between-groups and within-groups, should check with the values obtained independently for all three groups combined. The df of 29 checks because the df for the total group is $N - 1$ or 29 since the three subgroups with $n = 10$ add up to $N = 30$. The sum of squares for the total group is obtained using sums of the columns in table 10.1 as follows:

$$\Sigma X = 420 + 375 + 565 = 1360$$

$$\Sigma X^2 = 21650 + 16625 + 36025 = 74300$$

$$\Sigma x^2 = \Sigma X^2 - \frac{(\Sigma X)^2}{N} = 74300 - \frac{(1360)^2}{30}$$

$$\Sigma x^2 = 74300 - 61653.3 = 12646.7$$

This value differs from that in table 10.2 by only 0.3, a discrepancy attributable to rounding errors. In doing a one-way ANOVA, always make this check to guard against errors.

SUMMARY OF STEPS

Testing the Difference between Means of Several Samples of Equal Size Using One-Way ANOVA

Step 1 Compute the sum of raw scores, ΣX, and the sum of raw scores squared, ΣX^2, for each sample separately.

Step 2 Obtain the overall sum of raw scores, $\Sigma\Sigma X$, and the overall sum of raw scores squared, $\Sigma\Sigma X^2$.

Step 3 Compute the mean of each sample, M_i, and the overall mean, M.

T test = Significant diff. in means of 2 variables

Step 4 Find the sum of squares of deviation scores within each sample separately using the totals obtained in Step 1 and the formula $\Sigma x^2 = \Sigma X^2 - (\Sigma X)^2 / n$ where n is the number of cases in each sample.

Step 5 Obtain the between-groups sum of squares by summing the squared differences between sample means and the overall mean, $\Sigma(M_i - M)^2$, and multiplying by n, the common sample size.

Step 6 Obtain the between-groups variance estimate, s_B^2, by dividing the between-groups sum of squares from Step 5 by $k - 1$ where k is the number of samples.

Step 7 Obtain the within-groups sum of squares by adding up the sum of squared deviation scores for the separate samples, computed in Step 4.

Step 8 Obtain the within-groups variance estimate, s_W^2, by dividing the within-groups sum of squares from Step 7 by $k(n - 1)$, the sum of the degrees of freedom for the separate samples.

Step 9 Verify that the between-groups sum of squares from Step 5 plus the within-groups sum of squares from Step 7 equals the total group sum of squares within the limits of rounding error. The total group sum of squares is obtained from the totals computed in Step 2 by the formula $\Sigma\Sigma X^2 - (\Sigma\Sigma X)^2 / N$ where $N = k \times n$.

Step 10 Divide the between-groups variance estimate, s_B^2, by the within-groups variance estimate, s_W^2, to obtain the F ratio or F value.

Step 11 Enter the Table of the F Distribution (Appendix B, table E) with $k - 1$ and $k(n - 1)$ degrees of freedom for the between-groups and within-groups variance estimates, respectively, to obtain the critical values of F, $F_{.05}$ and $F_{.01}$.

Step 12 If the value of F computed in Step 10 is greater than $F_{.05}$ but less than $F_{.01}$, reject the null hypothesis at the 5% level of confidence. If the value of F computed in Step 10 is greater than $F_{.01}$, reject the null hypothesis at the 1% level of confidence. If the obtained value of F is less than $F_{.05}$, do not reject the null hypothesis.

Step 13 If the null hypothesis is rejected in Step 10, t-tests can be conducted to determine which pairs of sample means are significantly different from each other (see problems 40 and 41).

MATHEMATICAL NOTES

That the total sum of squares must be equal to the within-groups plus the between-groups sums of squares is proved in this section. Begin with the formula for a deviation score about the grand mean, M:

$$x = X_{ij} - M$$

$$x = (X_{ij} - M) + (M_i - M)$$

$$x = x_{ij} + d_i$$

$$x^2 = x_{ij}^2 + d_i^2 + 2x_i d_i$$

Summing both sides over the entire total group, from 1 to N while breaking up the sums on the right side in the subgroups separately gives:

$$\sum_{1}^{N} x^2 = \sum_{j=1}^{n} x_{1j}^2 + \sum_{j=1}^{n} x_{2j}^2 + \sum_{j=1}^{n} x_{3j}^2 + \cdots \sum_{j=1}^{n} x_{kj}^2 + nd_1^2 + nd_2^2 +$$ (10.5)

$$nd_3^2 + 2 \sum_{j=1}^{n} x_{1j}d_1 + 2 \sum_{j=1}^{n} x_{2j}d_2 + \cdots + 2 \sum_{j=1}^{n} x_{kj}d_k$$

$$\sum_{1}^{N} x^2 = \text{within-groups sum of squares} + \text{between-groups sum of squares} +$$

$$2d_1 \sum_{j=1}^{n} x_{1j} + 2d_2 \sum_{j=1}^{n} x_{2j} + \cdots + 2d_k \sum_{j=1}^{n} x_{kj}$$ (10.6)

The first row of terms on the right of Formula 10.5 is just the summation of the sums of squares of the deviation scores within groups, the within-groups sum of squares. The second row of terms in Formula 10.5 is $n\Sigma d_1^2$ or $n\Sigma(M_i - M)^2$, which by Formula 10.4 is the between-groups sum of squares. The third row of terms in Formula 10.5 reduces to the second row of terms in Formula 10.6 because the d values are constants with respect to the sums over subgroups and hence can be taken outside the summation signs. What is left is the sum of deviation scores, which is zero in each case so the second row of terms in Formula 10.6 vanishes, leaving:

$$\sum_{1}^{N} x^2 = \text{within-groups } SS_q + \text{between-groups } SS_q$$

PROBLEM 39

Test the Difference between Means of Several Samples of Unequal Size Using One-Way ANOVA

Many research situations occur where it is desirable to test the significance of differences among means of several samples of unequal size. This occurs in a planned experiment where the subgroup sizes are originally equal but become unequal due to dropouts. Dropouts should be replaced, but this is not always possible. In such cases, the method of Problem 39 is used to test the mean differences instead of the method illustrated in Problem 38.

An even more common application of Problem 39 is for data that are naturally occurring rather than experimentally obtained. The different subsamples might not consist of subjects randomly assigned from a common pool who are then to be subjected to different experimental treatments. Rather, they might represent samples from basically different populations that are being compared on some variable of interest. For example, an investigator is interested in the relationship of blood levels of potassium and the area of the country where the individual resides. The results of blood tests made on persons inducted at military centers throughout the country are obtained. By sifting these records, several samples of individuals from particular localities are obtained. The available data on potassium blood level is then used as the "experimental" variable in a one-way ANOVA. These samples very likely will be of different size, however. The investigator can throw out data at random to make all samples of equal size in order to use the method of Problem 38, but it is preferable to use the method of Problem 39, which permits the subsamples to vary in size.

FORMULAS FOR CALCULATING VARIANCE ESTIMATES FOR ONE-WAY ANOVA WITH SUBGROUPS OF UNEQUAL SIZE

Formula 10.1 gives the method of determining the within-groups estimate of the population variance, σ^2, for the one-way ANOVA with samples of equal size. This same formula also applies when the samples are of unequal size:

$$s^2 = \frac{\Sigma x_1^2 + \Sigma x_2^2 + \cdots + \Sigma x_k^2}{(n_1 - 1) + (n_2 - 1) + \cdots + (n_k - 1)} \qquad (10.7)$$

The only difference is that the values of the samples sizes, $n_1, n_2, \ldots, n_k$ are different when Problem 39 is being worked out though they are all equal for Formula 10.1 and Problem 38. In Formula 10.7 for Problem 39, the implied limits on the summations are from 1 to n_i, so these change with the sample size. In Formula 10.1 for Problem 38, the limits on the summations are all the same because all the sample sizes are equal.

In both Problems 38 and 39, then, the sums of squares for the within-groups variance estimate are obtained by pooling the sums of squares of deviations from the various subgroups. The degrees of freedom are also pooled. The population variance estimate in both problems is the ratio of the pooled sum of squares to the pooled degrees of freedom.

Between-Groups Variance Estimate

The formula for the between-groups variance estimate in the one-way ANOVA with subgroups of unequal size varies slightly from Formula 10.4, which applies to subgroups of equal size. In Formula 10.4, each squared discrepancy between a subgroup mean and the grand mean is effectively weighted by its sample size, n. Since n is the same for all groups, it becomes a constant and moves outside the summation sign. When the samples are of unequal size, however, each squared discrepancy is weighted by its own subsample size, n_i, which puts the n back inside the summation sign since it varies with the sample as shown in Formula 10.8:

$$s^2 = \frac{\sum_{i=1}^{k} n_i(M_i - M)^2}{k - 1} \qquad (10.8)$$

Each squared discrepancy of subgroup mean to grand mean is weighted according to the subgroup sample size. Those with larger sample sizes receive more weight. The degrees of freedom are the same for the between-groups variance estimate in the one-way ANOVA both for samples of equal size and samples of unequal size, namely, $k - 1$, where k is the number of groups (or levels of the independent variable).

Example   The procedure for doing a one-way ANOVA with samples of unequal size is illustrated by a worked-out example. Table 10.3 gives the raw scores and squared raw scores for four samples of male servicemen from four widely scattered regions of the country. The raw scores represent scaled potassium blood levels for these recruits. The question is whether these subgroups can be considered to be random samples from populations that have equal population mean potassium blood levels. This question is answered by doing a one-way ANOVA for samples of unequal size. The null hypothesis, therefore, is a statement that the group means are all statistically equal. The alternative hypothesis states that at least one pair of group means is unequal. As with the one-way ANOVA for groups of equal size, this test assumes normally distributed populations with equal variances for the different groups for the experimental variable while entertaining the possibility that the population means will be different.

The sums of the columns in table 10.3 give ΣX and ΣX^2 for the subgroups. The overall sums for the groups combined are given as $\Sigma\Sigma X$ and $\Sigma\Sigma X^2$. The computations for the variance estimates are given at the bottom of table 10.3. The within-groups variance estimate is obtained in the same way that it is obtained in table 10.1 for the one-way ANOVA with groups of equal size (Problem 38).

TABLE 10.3

One-way ANOVA with samples of unequal size

Subject number	Group 1		Group 2		Group 3		Group 4	
	X	X^2	X	X^2	X	X^2	X	X^2
1	8	64	16	256	14	196	27	729
2	13	169	19	361	10	100	15	225
3	6	36	8	64	10	100	20	400
4	16	256	16	256	2	4	19	361
5	1	1	18	324	9	81	10	100
6	14	196	20	400	3	9	17	289
7	12	144	17	289	16	256	21	441
8	14	196	30	900	4	16	20	400
9	21	441	18	324	17	289	32	1024
10	15	225	24	576	9	81	24	576
11	13	169	11	121	20	400		
12	25	625	20	400	8	64		
13	10	100	26	676	12	144		
14	19	361	19	361	11	121		
15	14	196	15	225				
Σ	201	3179	277	5533	145	1861	205	4545

TABLE 10.3
Continued

$\Sigma\Sigma X = 828 \qquad \Sigma\Sigma X^2 = 15118$

$N = n_1 + n_2 + n_3 + n_4 = 15 + 15 + 14 + 10 = 54$

Within-groups variance estimate:

$$\Sigma x^2 = \Sigma X^2 - \frac{(\Sigma X)^2}{n}$$

$$\Sigma x_1^2 = 3179 - \frac{(201)^2}{15} = 3179 - 2693.4 = 485.6$$

$$\Sigma x_2^2 = 5533 - \frac{(277)^2}{15} = 5533 - 5115.3 = 417.7$$

$$\Sigma x_3^2 = 1861 - \frac{(145)^2}{14} = 1861 - 1501.8 = 359.2$$

$$\Sigma x_4^1 = 4545 - \frac{(205)^2}{10} = 4545 - 4202.2 = 342.5$$

pooled sum of squares $= 485.6 + 417.7 + 359.2 + 342.5 = 1605.0$

pooled sum of degrees of freedom $= 14 + 14 + 13 + 9 = 50$

Within-groups variance estimate $= s_w^2 = \dfrac{1605.0}{50} = 32.1$

Between-groups variance estimate:

$$M_1 = \frac{201}{15} = 13.40; \ M_2 = \frac{277}{15} = 18.47; \ M_2 = \frac{145}{14} = 10.36;$$

$$M_4 = \frac{205}{10} = 20.50; \ M = \frac{828}{54} = 15.3$$

$n_1(M_1 - M)^2 = 15(13.40 - 15.3)^2 = 15(-1.9)^2 = 15(3.61) = 54.15$

$n_2(M_2 - M)^2 = 15(18.47 - 15.3)^2 = 15(3.17)^2 = 15(10.05) = 150.75$

$n_3(M_3 - M)^2 = 14(10.36 - 15.3)^2 = 14(-4.94)^2 = 14(24.40) = 341.6$

$n_4(M_4 - M)^2 = 10(20.50 - 15.3)^2 = 10(5.2)^2 = 10(27.04) = 270.4$

$$\sum_{i=1}^{4} n_i(M_i - M)^2 = 54.15 + 150.75 + 341.6 + 270.4 = 816.9$$

$$\text{Degrees of freedom} = (k - 1) = (4 - 1) = 3$$

Between-groups variance estimate $= s_B^2 = \dfrac{816.9}{3} = 272.3$

F ratio:

$$F = \frac{s_B^2}{s_W^2} = \frac{272.3}{32.1} = 8.48**$$

$F_{.05} = 2.79 \qquad F_{.01} = 4.20$

TABLE 10.4

One-way ANOVA summary for data of table 10.3

Source of variation	Sum of squares	Degrees of freedom	Mean square (variance estimate)	F
Between-groups	816.9	3	272.3	8.48**
Within-groups	1605.0	50	32.1	
Total	2421.9	53		

For the between-groups variance estimate each deviation score of a subgroup mean from the grand mean is calculated and multiplied by the subgroup sample size, n_i. These figures are summed to obtain the between-groups sum of squares according to Formula 10.8. The between-groups sum of squares is then divided by the between-groups degrees of freedom, one less than the number of groups. The results of this one-way ANOVA are summarized in table 10.4. The results show that the null hypothesis is rejected. There are differences between the group means.

The total sum of squares must be checked against the value obtained using ΣX and ΣX^2 for the entire group as follows:

$$\Sigma x^2 = \Sigma X^2 - \frac{(\Sigma X)^2}{N} = 15118 - \frac{(828)^2}{54} = 15118 - 12696 = 2422.0$$

The difference between 2422 and the value of 2421.9 shown in table 10.4 is due to rounding errors.

The total degrees of freedom shown in table 10.4 checks since 53 equals $N - 1$. The F ratio has two asterisks (**) since it exceeds the $F_{.01}$ value with 3 and 50 df, respectively, for the larger and smaller variance estimates.

Checking Subgroup Variance Estimates

Since F is significant, it is necessary to examine the s^2 values for the various subgroups to see if they are close enough to each other in size to make the F-test appropriate. Using the Σx_i^2 values from table 10.3, these variance estimates are obtained as follows:

$$s_1^2 = \frac{\Sigma x_1^2}{n_1 - 1} = \frac{485.6}{14} = 34.7$$

$$s_2^2 = \frac{\Sigma x_2^2}{n_2 - 1} = \frac{417.7}{14} = 29.8$$

$$s_3^2 = \frac{\Sigma x_3^2}{n_3 - 1} = \frac{359.2}{13} = 27.6$$

$$s_4^2 = \frac{\Sigma x_4^2}{n_4 - 1} = \frac{342.5}{9} = 38.1$$

The largest F is $s_4{}^2/s_3{}^2 = 1.4$. For 9 and 13 *df*, respectively, for the larger and smaller variance estimate, $F_{.05}$ is 2.72 so the null hypothesis, h_0, cannot be rejected as far as variances are concerned (see Appendix B, table E). Since there is no proof that the population variances are unequal, the one-way ANOVA to test differences between means is appropriate as far as the assumption of equal variances is concerned. In this case with an F twice as large as that required for significance at the 0.01 level, differences in $s_1{}^2$ values have to be very substantial to cast doubt on the conclusion that the means differ significantly from each other.

SUMMARY OF STEPS

Testing the Difference between Means of Several Samples of Unequal Size Using One-Way ANOVA

Step 1   Compute the sum of raw scores, ΣX, and the sum of raw scores squared, ΣX^2, for each sample separately.

Step 2   Obtain the overall sum of raw scores, $\Sigma\Sigma X$, and the overall sum of raw scores squared, $\Sigma\Sigma X^2$.

Step 3   Compute the mean of each sample, M_i, and the overall mean, M.

Step 4   Find the sum of squares of deviation scores within each sample separately using the totals obtained in Step 1 and the formula $\Sigma x^2 = \Sigma X^2 - (\Sigma X)^2/n$ where n is the sample size.

Step 5   For each sample, obtain the between-groups sum of squares by computing $n_i (M_i - M)^2$, where n_i is the sample size. Add these values over the k samples.

Step 6   Obtain the between-groups variance estimate, $s_B{}^2$, by dividing the between-groups sum of squares from Step 5 by $k - 1$ where k is the number of samples.

Step 7   Obtain the within-groups sum of squares by adding up the sum of squared deviation scores for the separate samples, computed in Step 4.

Step 8   Obtain the within-groups variance estimate, $s_W{}^2$, by dividing the within-groups sum of squares from Step 7 by $\sum_{i=1}^{k} (n_i - 1)$, the sum of the degrees of freedom for the k separate samples.

Step 9   Verify that the between-groups sum of squares from Step 5 plus the within-groups sum of squares from Step 7 equals the total group sum of squares within the limits of rounding error. The total group sum of squares is obtained from the totals computed in Step 2 by the formula $\Sigma\Sigma X^2 - (\Sigma\Sigma X)^2/N$ where N is the sum of the separate sample sizes, or the overall sample size.

Step 10   Divide the between-groups variance estimate, $s_B{}^2$, by the within-groups variance estimate, $s_W{}^2$, to obtain the F value.

Step 11 Enter the Table of the F Distribution (Appendix B, table E) with $k - 1$ and $\sum_{i=1}^{k} (n_i - 1)$ degrees of freedom for the between-groups and within-groups variance estimates, respectively, to obtain the critical values of F, $F_{.05}$ and $F_{.01}$.

Step 12 If the value of F computed in Step 10 is greater than $F_{.05}$ but less than $F_{.01}$, reject the null hypothesis at the 5% level of confidence. If the value of F computed in Step 10 is greater than $F_{.01}$, reject the null hypothesis at the 1% level of confidence. If the obtained value of F is less than $F_{.05}$, do not reject the null hypothesis.

Step 13 If the null hypothesis is rejected in Step 10, conduct multiple-comparison tests to determine which pairs of sample means are significantly different from each other.

MATHEMATICAL NOTES

The representation of the total sum of squares as the sum of the within-groups sum of squares plus the between-groups sum of squares is shown in the Mathematical Notes section of Problem 38 for the case with equal size subsamples. Inspection of Formula 10.5 readily reveals that a similar mathematical development can be used to show that these sums of squares also add up for the case of subgroups of unequal size. The between-groups sum of squares is calculated as the numerator of Formula 10.8, however, instead of Formula 10.4, which requires equal subgroup sample sizes. In Formula 10.5 each n is subscripted to represent the n of the given subgroup, and these values may differ in size. The within-groups sum of squares is obtained in the same way as with subsamples of equal size although the upper limit on the summation differs from subgroup to subgroup ($n_1, n_2, \ldots, n_k$) instead of being just n for every group.

OTHER APPLICATIONS OF THE ONE-WAY ANOVA

The one-way ANOVA can also be applied to certain nonexperimental studies in which subjects are not randomly assigned to the different subgroups from the same pool. For example, it is used to compare different populations to see if their means vary with respect to some variable of interest. Random samples of males are drawn from the general population within several widely separated states of the United States. Measurements of height are taken and the one-way ANOVA is used to investigate the variability of means in height of the subgroups. If F proves to be significant, an investigation will be launched into the reasons for mean differences in height of the male populations of the different states. In this application of the one-way ANOVA, however, there is no experimental intervention, giving different treatment to the various subgroups followed by measurement of the posttreatment state. If any such variation in treatment takes place, with the expectation of evaluating the effect of the different treatment by the one-way ANOVA, random assignment of the subjects from a common pool to the different subgroups is an absolute necessity.

t-Tests Following the One-Way ANOVA F-Test

It is possible to compare the mean of each subsample with the mean of each other subsample in the previous example using the t-test (see Problem 29). This practice is discouraged in favor of the one-way ANOVA approach because in making many t-tests a certain percentage of them can be expected to achieve significance by sampling at random from populations with equal means and variances. In fact, out of 100 t-tests, about 5 achieve significance at the 5% level by chance alone. With seven subgroups, for example, one significant t-test at the 0.05 level can be expected by chance comparing each subgroup mean with each other subgroup mean since there are $(7 \times 6)/2 = 21$ such comparisons.

It is most appropriate to use the one-way ANOVA to determine if there is an overall significant F ratio before examining differences between individual pairs of subgroup means for significance. If F is not significant, it is not statistically acceptable to claim significance for individual pair subgroup mean comparisons that exceed $t_{.05}$ or some other chosen level of confidence. If F is significant, then it is appropriate to make individual t-tests to determine the significance of differences between pairs of subgroup means. Remember that the more t-tests are made, the greater the possibility that a spuriously significant t will be obtained. The investigator must exercise suitable caution in interpreting the results. In fact, where many such t-tests are being made, higher values than the usual $t_{.05}$ and $t_{.01}$ critical values are required for significance. Procedures for determining the higher critical values are described in connection with Problem 40.

Where F is not significant, some investigators proceed to make individual t-tests for the large subgroup mean pair differences even though it is not strictly correct to do so statistically. However, some hypotheses for further investigation can be obtained in this manner. For example, if one pair of means out of a large number of mean pairs gives a significant t while the other pairs do not, the investigator might be encouraged to set up another experiment using just those two treatments to see if the phenomenon is replicable, that is, whether or not a statistically significant t-test will be obtained again with new samples.

Multiple Comparisons

In a one-way analysis of variance involving more than two groups, a significant F ratio leads to a rejection of the overall hypothesis of equality of means. A significant F ratio indicates that at least one pair of means is statistically different, but it does not point to which pair or pairs of group means that are significant. The F-test in analysis of variance is an overall test. As a result, the overall F-test is not a very interesting or useful test to most researchers. Generally, the interest of researchers lies in the difference between certain, or particular, group means. For example, a market researcher wants to compare the increases in sales generated by two promotional plans: (1) buy one item and get the second item at half price and (2) buy two items at regular price and get the third item free over the existing sales strategy, which is (3) the item at regular price. An overall F-test in ANOVA tells the researcher whether a difference between the three plans (conditions) exists. If the test is significant, it does *not* tell where the difference lies.

Thus, the researcher needs additional tools to look "deeper into the data." As previously mentioned, the t-test can be used to make comparisons between the group means. The use of this method for too many comparisons encounters the same problems as mentioned in the previous section, spuriously significant t-tests. In Problem 40, the use of the t-test for purposes of comparison after a significant F-test is obtained is discussed. Additionally, to handle the problems that arise from using too many t-tests, the Scheffé test is introduced. In Problem 41, the Tukey HSD test is presented as an alternative to the t-test and the Scheffé test.

PROBLEM 40

Multiple Comparisons Using the t-Test

In Chapter 8, the t-test for comparing two group means is introduced and developed. The t-test is used to compare individual pairs of means following a significant F ratio test in an analysis of variance. To use this t-test for multiple comparisons, a slight change to the t-test formula in Chapter 8 is required. The change occurs in the calculation of the standard error term s_{DM}. The term s_W^2 replaces $(\Sigma x_1^2 + \Sigma x_2^2)/(n_1 + n_2 - 2)$ in the formula for s_{DM}. The formula for the t-test in multiple comparisons is:

$$t_{ij} = \frac{M_i - M_j - 0}{\sqrt{s_W^2 \left[(n_i + n_j)/n_i n_j \right]}} \qquad (10.9)$$

where

t_{ij} = the t value computed for comparing group i's mean with group j's mean,

M_i, M_j = the means for groups i and j, respectively,

s_W^2 = the mean square for within-groups, and

n_i, n_j = the sample size for groups i and j, respectively.

The s_W^2 value is found by examining the ANOVA summary table 10.4 under the column labeled "Mean Square" and the intersecting row labeled "Within-groups". This estimate is obtained by pooling all the sums of squares and dividing by the pooled degrees of freedom. The degrees of freedom are also found by examining the ANOVA summary table using the same row, but under the column labeled "Degrees of Freedom." This t value is evaluated by specifying the appropriate α level, determining the degrees of freedom and finding the critical value using table C in Appendix B. For k groups, there are $k(k - 1)/2$ possible comparisons. Since multiple t-tests use the same denominator, s_W^2, the tests of significance are *not* statistically independent, although the comparisons among means are independent for normally distributed populations. For a large number of degrees of freedom, the significance tests are regarded as independent. In the following example, the computation of the t-test is illustrated. Each possible comparison is given. In an actual problem, only those groups that are hypothesized as different are used in the t-test comparison.

> **TABLE 10.5**

t-tests of difference between subgroup means for the data in table 10.3

$$t_{12} = \frac{13.4 - 18.5 - 0}{\sqrt{32.1\left(\dfrac{15 + 15}{15 \cdot 15}\right)}} = \frac{-5.1}{\sqrt{4.27}} = -2.46*$$

$$t_{13} = \frac{13.4 - 10.4 - 0}{\sqrt{32.1\left(\dfrac{15 + 14}{15 \cdot 14}\right)}} = \frac{3.0}{\sqrt{4.43}} = 1.43$$

$$t_{14} = \frac{13.4 - 20.5 - 0}{\sqrt{32.1\left(\dfrac{15 + 10}{15 \cdot 10}\right)}} = \frac{-7.1}{\sqrt{5.35}} = -3.07**$$

$$t_{23} = \frac{18.5 - 10.4 - 0}{\sqrt{32.1\left(\dfrac{15 + 14}{15 \cdot 14}\right)}} = \frac{8.1}{\sqrt{4.43}} = 3.86**$$

$$t_{24} = \frac{18.5 - 20.5 - 0}{\sqrt{32.1\left(\dfrac{15 + 10}{15 \cdot 10}\right)}} = \frac{-2.0}{\sqrt{5.35}} = -0.87$$

$$t_{34} = \frac{10.4 - 20.5 - 0}{\sqrt{32.1\left(\dfrac{14 + 10}{14 \cdot 10}\right)}} = \frac{-10.1}{\sqrt{5.50}} = -4.30**$$

$df = 22,\ t_{.05} = 2.07,\ t_{.01} = 2.82$

$df = 23,\ t_{.05} = 2.07,\ t_{.01} = 2.81$

$df = 27,\ t_{.05} = 2.05,\ t_{.01} = 2.77$

$df = 28,\ t_{.05} = 2.05,\ t_{.01} = 2.76$

Table 10.5 gives an example of t values computed using Formula 10.9. The data used in table 10.5 are from table 10.3.

Three of the t-test comparisons in table 10.5, 1–4, 2–3, and 3–4, yield differences between means significant at the 0.01 level. These are indicated by two asterisks (**) after the t value. The sign on the t value is ignored, since the t distribution is symmetrical. One additional comparison, 1–2, gives a statistically significant difference between means, but at the 0.05 level. This is indicated in table 10.5 by one asterisk after the t value. The two remaining comparisons failed to yield statistically significant differences between means.

Unfortunately, these t-tests and their apparent levels of significance cannot be taken at face value. The t-test is basically designed for the situation where only one comparison between two available means is to be made. In this particular case, with four treatment groups, a total of six t-tests are made. The more t-tests that are made, the greater the likelihood of getting one that exceeds the usual significance level by chance. With several groups, then, just making all the possible t-tests and reporting those that achieved statistical significance at the 0.05 and/or the 0.01 level is not acceptable.

If an investigator decides in advance to test only one particular pair of means out of several possible pairs, and if the resulting F is significant, the t-test is used in the usual way. This does not involve multiplying the chances of getting a significant t when the null hypothesis is true. If the plan is to test all the possible mean differences, however, the ordinary significance levels for t are too low. This underestimation is a function of the number of t-tests to be made. As the number of groups increases, the possible comparisons multiply rapidly, and as a consequence, so do the opportunities to obtain significant results by chance.

A method for dealing with this problem, which determines how big t must be to achieve significance where all possible comparisons are being made has been devised by Henry Scheffé. The **Scheffé test** also makes allowances for comparisons involving averages of subgroups. This multiplies the number of comparisons requiring a value of t for significance that might be higher than it need be if the investigator had no idea of investigating means of combined groups. If an obtained value exceeds the critical Scheffé t' value, however, it is certainly safe to treat the difference as statistically significant.

Scheffé test A very conservative post-hoc test used to compare pairs or combinations of group means following a significant F-test in ANOVA.

The Scheffé Method

To determine a value $t'_{.05}$, which must be exceeded by an obtained t for it to be considered significant at the .05 level when all possible comparisons are being made in the one-way ANOVA, define t' as follows:

$$t'_{.05} = \sqrt{(k-1)F_{.05}} \qquad (10.10)$$

where $t'_{.05} = $ the Scheffé critical value of t,

$k = $ the number of groups in the one-way ANOVA, and

$F_{.05} = $ the F ratio required for significance with $k-1$ as the df for the larger variance estimate and $N-k$ as the df for the smaller variance estimate. ($N = $ total sample size)

Any and all obtained t ratios that exceed $t'_{.05}$ are regarded as significant at the 0.05 level at least. This test applies to the one-way ANOVA with equal size subgroups as well. If a 1% level of t is desired, use the following formula:

$$t'_{.01} = \sqrt{(k-1)F_{.01}} \qquad (10.11)$$

The definition of terms in Formula 10.11 parallel those for Formula 10.10.

For the data in tables 10.3, 10.4, and 10.5, the $F_{.05}$ and $F_{.01}$ values for Formulas 10.10 and 10.11 are 2.79 and 4.20 using 3 and 50 degrees of freedom for the larger and smaller variance estimates, respectively. Substituting these values of F in Formulas 10.10 and 10.11 gives:

$$t'_{.05} = \sqrt{3(2.79)} = \sqrt{8.37} = 2.89$$

$$t'_{.01} = \sqrt{3(4.20)} = \sqrt{12.60} = 3.55$$

Comparing these values of $t'_{.05}$ and $t'_{.01}$ with the $t_{.05}$ and $t_{.01}$ values required for significance at the 0.05 and 0.01 level for the t-test as shown in table 10.5, it is apparent that with $k = 4$, the Scheffé $t'_{.05}$ value is not dissimilar to the t-test $t_{.01}$ value. The Scheffé $t'_{.01}$ value, of course, is considerably larger than the t-test $t_{.01}$ value.

In table 10.5, the comparisons 2–3 and 3–4 still remain significant at the 0.01 level when compared against $t'_{.01} = 3.55$, but the 1–4 comparison t value of 3.07 drops down to being significant at only the 0.05 level. The 1–2 comparison value of $t = -2.46$ drops from being significant at the 0.05 level to not being significant at all.

To summarize, if only one t-test is to be made following a significant ANOVA F-test, the significance levels shown in table 10.5 represent appropriate standards for making that evaluation. If all possible t-tests are to be made, including those for averages of all possible combined subgroups, the Scheffé $t'_{.05}$ and $t'_{.01}$ values give a fair criteria of significance for the obtained t values. If only some of the comparisons are actually to be considered, the appropriate significance levels for t lie somewhere in between (see Problem 41). Using Scheffé values is a conservative approach that will guard against being too aggressive in rejecting the null hypothesis at a specified level of significance.

PROBLEM 41

The Tukey HSD Multiple-Comparison Test

The Scheffé test is more rigorous with respect to Type I error than other multiple-comparison methods. The Scheffé test leads to fewer significant differences than other methods. A general rule is that if a Scheffé comparison is statistically significant, all other multiple-comparison methods will also be significant. The Scheffé test as mentioned earlier, is a conservative test and assumes the researcher is going to make all possible comparisons between the group means and combinations of group means. If in reality the researcher wishes only to make several comparisons, but not all, the Scheffé test might be too conservative and lead to too many Type II errors. The researcher might consider the Tukey HSD (Honestly Significant Difference) test, developed by John Tukey, as an alternative to the Scheffé test.

Tukey HSD test   A post-hoc test used to compare pairs of group means following a significant F-test in ANOVA.

The **Tukey HSD test** is based upon the same assumptions underlying the t-test. The Tukey test is not as liberal as the t-test and leads to fewer significant differences. It does, however, lead to more significant differences than the Scheffé test. The Tukey method involves finding a certain value required for significance at the 0.05 or 0.01 levels for k and df, where k is the total number of groups in the one-way ANOVA and df is the degrees of freedom associated with the within-group mean square, s_W^2.

A comparison between two group means is statistically significant if the absolute value of the difference between means exceeds the value of HSD, where HSD is defined as:

$$\text{HSD} = q_{k,v} \sqrt{s_W^2/n} \qquad (10.12)$$

where $q_{k,v}$ = the studentized range statistic value,

s_W^2 = the within-group variance estimate, and

n = the number of subjects in each group.

The value of q is obtained from table M in Appendix B, the Distribution of the Studentized Range Statistic. Given k means based on equal n's, the studentized range q is simply the difference between the largest and smallest means divided by an estimate of the standard error. With different values of k and df associated with s_W^2, the sampling distribution of q, the studentized range is defined. To use table M, three values are necessary: the level of significance (α), the within-group degrees of freedom $v = (N_T - k)$, and the number of groups (k). N_T is the total number of observations and n is the sample size for each group. Ideally, in order for the studentized range to work properly, the groups to be compared should be of equal size. The q statistic and HSD will still work if the sample size of the groups are not too different from one another, provided that n is replaced with n_h, the harmonic mean of the number of observations. That is, n_h is computed using the following formula:

$$n_h = \frac{k}{[(1/n_1) + (1/n_2) + \cdots + (1/n_k)]} \qquad (10.13)$$

where k = the number of groups, and

$n_1, n_2, \ldots, n_k$ = the sample sizes for each group.

The value of n_h can be used in place of n in Formula 10.12. The use of n_h is *not* appropriate if there are large differences between the sample sizes.

Using the data in table 10.5, first compute the absolute differences between the group means:

$D_{12} = 5.1$	$D_{23} = 8.1$
$D_{13} = 3.0$	$D_{24} = 2.0$
$D_{14} = 7.1$	$D_{34} = 10.1$

Next, the degrees of freedom for s_W^2 and k are determined:

$$df = N_T - k = n_1 + n_2 + n_3 + n_4 - k$$

$$= 15 + 15 + 14 + 10 - k = 54 - k$$

$$k = \text{number of groups} = 4, \text{ so } df = 54 - 4 = 50.$$

To find q, enter table M of Appendix B with df and k.

$$q_{.05,50} = 3.77 \text{ (interpolated value)}$$

$$q_{.01,50} = 4.65 \text{ (interpolated value)}$$

From table 10.4,

$$s_W^2 = 32.1$$

Using Formula 10.13,

$$n_h = 4/\left[(1/15) + (1/15) + (1/14) + (1/10)\right]$$

$$= 4/.3048 = 13.12$$

Hence,

$$HSD_{.05} = 3.77 \sqrt{(32.1/13.125)} = 5.90$$

$$HSD_{.01} = 4.65 \sqrt{(32.1/13.125)} = 7.27$$

If any of the absolute differences is greater than $HSD_{.01}$, the difference is statistically significant at the 0.01 level. If an absolute difference is greater than $HSD_{.05}$ but not greater than $HSD_{.01}$, then the difference is significant at the 0.05 level. If the absolute difference is less than $HSD_{.05}$, the difference is not significant.

For the data in tables 10.3 and 10.4, the differences that reached significance at the 0.01 level were 2–3 and 3–4. These comparisons are consistent with the Scheffé test. The comparison between 1–4, which reached significance at the 0.05 level, is also consistent with the Scheffé test. Comparison 1–2, which was significant at the 0.05 level with the t-test is not significant using the Tukey HSD test.

Only three methods of comparison are discussed in this chapter. There are many other multiple-comparison methods, but a thorough discussion of all or most of those methods goes beyond the scope of this text.

SUMMARY OF STEPS

t-Test

Step 1   Compute all group means.

Step 2   Perform one-way ANOVA calculations.

Step 3   Specify one between-groups mean comparison.

Step 4   Using Formula 10.9, find the t value for the pair of means specified in Step 3. Note: The s_W^2 is the mean square for within-groups found in the ANOVA summary table.

Step 5   With degrees of freedom $N_T - k$ and level of significance, α, find critical value of t in table C of Appendix B.

Step 6   If the t value computed in Step 4 exceeds the critical value, reject the null hypothesis of equality. Otherwise, do not reject the hypothesis. If there are several comparisons, either planned or unplanned, (i.e., *post hoc*), use the Scheffé test or the Tukey test.

SUMMARY OF STEPS

Scheffé Test

Step 1   Compute all group means.

Step 2   Perform one-way ANOVA calculations.

Step 3   Taking each pair of group means, compute the t_{ij} value given by Formula 10.9.

Step 4   Compute the critical value $t'_{.05} = \sqrt{(k - 1)\ F_{.05}}$ and $t'_{.01} = \sqrt{(k - 1)F_{.01}}$.

Step 5   Compare each t_{ij} value computed in Step 3 with $t'_{.05}$ and $t'_{.01}$.

Step 6   If t_{ij} exceeds $t'_{.01}$, the difference between that pair of means is statistically significant at the 0.01 level. If t_{ij} exceeds $t'_{.05}$ but does not exceed $t'_{.01}$, that pair of means is statistically different at the 0.05 level.

SUMMARY OF STEPS

Tukey HSD Test

Step 1   Compute all group means.

Step 2   Perform one-way ANOVA calculations.

Step 3   Taking each pair of group means, compute the absolute difference, D_{ij}.

Step 4   Compute the critical value for 0.05 and 0.01 using Formula 10.12. Find the value of $q_{k,v}$ by specifying the level of significance, α, the degrees of freedom, and the number of groups and entering table M in Appendix B.

Step 5   If the differences computed in Step 3 exceed the critical value $HSD_{.01}$, that pair of group means is statistically significant at the 0.01 level. If the difference between any pair of means is greater than $HSD_{.05}$ but less than $HSD_{.01}$, that pair is statistically different at the 0.05 level of significance.

Exercises

10–1.   The data given below represent *ad libitum* consumption totals in cc of 10% alcohol solution, 20% alcohol solution, and distilled water, respectively, in a specified time period by three groups of rats maintained on the same deficiency diet. Test to see if there are differences among these means at the 0.05 level using the one-way ANOVA.

Individual	10% Alcohol	20% Alcohol	Water
1	40	26	13
2	45	20	16
3	38	30	14
4	37	24	20
5	36	25	15
6	30	19	17
7	33	27	10
8	38	25	14
9	39	26	16
10	35	26	15

   a. State the null and alternative hypotheses.
   b. Compute ΣX (sum of all scores), ΣX^2 (sum of all squared scores) and Σx^2 (total sum of squares).
   c. Compute N (total number of observations).
   d. Compute ΣX and ΣX^2 for each group.
   e. Find the mean (M_i) for each group and the grand mean (M).
   f. Find Σx_i^2 (sum of squares) for each group.
   g. Compute the pooled sum of squares and degrees of freedom for within-groups.
   h. Compute the within-groups variance estimate, s_W^2.
   i. For each group compute $n_i(M_i - M)^2$.
   j. Compute the pooled sum of squares for between-groups, $\Sigma n_i(M_i - M)^2$.
   k. Find the between-groups degrees of freedom $[k - 1]$ and compute the between-groups variance estimate, s_B^2.
   l. Compute the F ratio, $F = s_B^2 / s_W^2$.
   m. Enter table E of Appendix B with $\alpha = 0.05$ and 0.01 with numerator degrees of freedom = between-groups degrees of freedom, and denominator degrees of freedom = within-groups degrees of freedom to find $F_{.05}$ and $F_{.01}$.
   n. If the F ratio found in (l) is greater than the critical value, $F_{.05}$, a significant difference exists between group means, reject h_0.
   o. Complete the ANOVA summary table:

Source	Sum of Squares	df	Mean Square	F-Ratio
Between-groups				
Within-groups				
Total				

10-2. A personality test is administered to samples of male (Group AM) and female (Group AF) applicants for employment in a given firm as well as to volunteer male (Group EM) and female (Group EF) employees, who are promised that the test results will not affect their present or future opportunities with the company. Emotional stability scores on the test are converted to Stanine Scores. These scores for the four groups are given below. Perform a one-way ANOVA to test for differences between means.

Individual	AM	AF	EM	EF
1	8	6	7	5
2	4	8	5	2
3	9	6	4	6
4	7	7	3	4
5	9	4	4	9
6	2	6	6	1
7	5	3	2	4
8	8	5	1	3
9	6	8	8	3
10	7	7	5	7
11	4	2	6	4
12	7	7	5	4

a. State the null and alternative hypotheses.
b. Compute ΣX, ΣX^2, and Σx^2.
c. Compute N.
d. Compute ΣX and ΣX^2 for each group.
e. Find the mean (M_i) for each group and the grand mean (M).
f. Find Σx_i^2 for each group.
g. Compute the pooled sum of squares and degrees of freedom for within-groups.
h. Compute the within-groups variance estimate, s_W^2.
i. For each group compute $n_i(M_i - M)^2$.
j. Compute the pooled sum of squares for between-groups, $\Sigma n_i(M_i - M)^2$.
k. Find the between-groups degrees of freedom $[k - 1]$ and compute the between-groups variance estimate, s_B^2.
l. Compute the F ratio, $F = s_B^2 / s_W^2$.
m. Enter table E of Appendix B with $\alpha = 0.05$ and 0.01 with numerator degrees of freedom = between-groups degrees of freedom, and denominator degrees of freedom = within-groups degrees of freedom to find $F_{.05}$ and $F_{.01}$.
n. If the F ratio found in (l) is greater than the critical value, $F_{.05}$, a significant difference exists between group means, reject h_0.
o. Complete the ANOVA summary table:

Source	Sum of Squares	df	Mean Square	F Ratio
Between-groups				
Within-groups				
Total				

10–3.   Check to see if the assumption of equal variances for the F-test in question 10–1 is tenable.

  a. Note the number of groups, k.
  b. Using X and X^2, compute the variance of each group.
  c. Form all possible ratios between each group variance. Ratios are formed by dividing the larger variance by the smaller. Note the largest of such ratios.
  d. Find the critical value in table E of Appendix B. The degrees of freedom are $n_i - 1$ and $n_j - 1$.
  e. Compare the largest ratio found in (c) with the critical value found in (d). If the ratio exceeds the critical value, the assumption of equal variance is not tenable.

10–4.   Check to see if the assumption of equal variances for the F-test in question 10–2 is tenable.

  a. Note the number of groups, k.
  b. Using X and X^2, compute the variance of each group.
  c. Form all possible ratios between each group variance. Ratios are formed by dividing the larger variance by the smaller. Note the largest of such ratios.
  d. Find the critical value in table E of Appendix B. The degrees of freedom are $n_i - 1$ and $n_j - 1$.
  e. Compare the largest ratio found in (c) with the critical value found in (d). If the ratio exceeds the critical value, the assumption of equal variance is not tenable.

10–5.   If it is appropriate, make individual t-tests for pairs of means in question 10–1. What are the Scheffé $t'_{.05}$ and $t'_{.01}$ values?

  a. Determine if the F-test in the ANOVA is significant. If it is significant, then proceed with (b), otherwise, stop at this point and indicate that the F-test is not significant.
  b. Using Formula 10.9, compute the t_{ij} values for all possible pairs of group means.
  c. Using Formulas 10.10 and 10.11, find the Scheffé $t'_{.05}$ and $t'_{.01}$.
  d. Compare the computed t_{ij} value with the Scheffé t' values and determine which pair or pairs of means are significantly different.

10–6.   If it is appropriate, make individual t-tests for pairs of means in question 10–2. What are the Scheffé $t'_{.05}$ and $t'_{.01}$ values?

  a. Determine if the F-test in the ANOVA is significant. If it is significant, then proceed with (b), otherwise, stop at this point and indicate that the F-test is not significant.
  b. Using Formula 10.9, compute the t_{ij} values for all possible pairs of group means.
  c. Using Formulas 10.10 and 10.11, find the Scheffé $t'_{.05}$ and $t'_{.01}$.
  d. Compare the computed t_{ij} value with the Scheffé t' values and determine which pair or pairs of means are significantly different.

10-7.   The data below represent scaled and rounded off blood levels of a certain hormone for normal individuals, (A); nonpsychiatric patients, (B); and hospitalized schizophrenics, (C). Using the one-way ANOVA test for differences among the means.

Individual	A	B	C
1	13	12	9
2	12	12	3
3	12	10	8
4	14	11	6
5	16	14	11
6	13	13	7
7	9	10	5
8	11	8	9
9	12	13	8
10	13	11	9
11	14	12	
12	13		

a. State the null and alternative hypotheses.
b. Compute ΣX, ΣX^2, and Σx^2.
c. Compute N.
d. Compute ΣX and ΣX^2 for each group.
e. Find the mean (M_i) for each group and the grand mean (M).
f. Find Σx_i^2 for each group.
g. Compute the pooled sum of squares and degrees of freedom for within-groups.
h. Compute the within-groups variance estimate, s_W^2.
i. For each group compute $n_i(M_i - M)^2$.
j. Compute the pooled sum of squares for between-groups, $\Sigma n_i(M_i - M)^2$.
k. Find the between-groups degrees of freedom $[k - 1]$ and compute the between-groups variance estimate, s_B^2.
l. Compute the F ratio, $F = s_B^2 / s_W^2$.
m. Enter table E of Appendix B with $\alpha = 0.05$ and 0.01 with numerator degrees of freedom = between-groups degrees of freedom, and denominator degrees of freedom = within-groups degrees of freedom to find $F_{.05}$ and $F_{.01}$.
n. If the F ratio found in (l) is greater than the critical value, $F_{.05}$, a significant difference exists between group means, reject h_0.
o. Complete the ANOVA summary table:

Source	Sum of Squares	df	Mean Square	F Ratio
Between-groups				
Within-groups				
Total				

10-8.   Rats are randomly assigned to four conditions A, B, C, D. The rats are placed in Skinner boxes. In group A, the rat receives a food pellet every time it presses the bar. In Group B, the rat receives a food pellet every other time it presses the bar; in groups C and D, the reinforcement rate drops to one in three and one in four, respectively. The data below give the number of bar presses in a given time period after the start of the experiment. Two rats in group B and one rat in group D died before completion of the experiment, leading to unequal numbers of subjects in each group. Test for differences among means using the one-way ANOVA.

Individual	A	B	C	D
1	8	13	14	18
2	10	13	16	16
3	14	16	20	20
4	7	14	17	20
5	3	12	10	19
6	9	14	13	21
7	10	6	16	19
8	10	10	15	25
9	11		15	13
10	8		16	

a. State the null and alternative hypotheses.
b. Compute ΣX, ΣX^2, and Σx^2.
c. Compute N.
d. Compute ΣX and ΣX^2 for each group.
e. Find the mean (M_i) for each group and the grand mean (M).
f. Find Σx_i^2 for each group.
g. Compute the pooled sum of squares and degrees of freedom for within-groups.
h. Compute the within-groups variance estimate, s_W^2.
i. For each group compute $n_i(M_i - M)^2$.
j. Compute the pooled sum of squares for between-groups, $\Sigma n_i(M_i - M)^2$.
k. Find the between-groups degrees of freedom $[k - 1]$ and compute the between-groups variance estimate, s_B^2.
l. Compute the F ratio, $F = s_B^2 / s_W^2$.
m. Enter table E of Appendix B with $\alpha = 0.05$ and 0.01 with numerator degrees of freedom = between-groups degrees of freedom, and denominator degrees of freedom = within-groups degrees of freedom to find $F_{.05}$ and $F_{.01}$.
n. If the F ratio found in (l) is greater than the critical value, $F_{.05}$, a significant difference exists between group means, reject h_0.
o. Complete the ANOVA summary table:

Source	Sum of Squares	df	Mean Square	F Ratio
Between-groups				
Within-groups				
Total				

10–9.   If it is appropriate, carry out individual t-tests for pairs of means in question 10–7. What are the Scheffé $t'_{.05}$ and $t'_{.01}$ values?

    a. Determine if the F-test in the ANOVA is significant. If it is significant, then proceed with (b), otherwise, stop at this point and indicate that the F-test is not significant.

    b. Using Formula 10.9, compute the t_{ij} values for all possible pairs of group means.

    c. Using Formulas 10.10 and 10.11, find the Scheffé $t'_{.05}$ and $t'_{.01}$.

    d. Compare the computed t_{ij} value with the Scheffé t values and determine which pair or pairs of means were significantly different.

10–10.   If it is appropriate, carry out individual t-tests for pairs of means in question 10–8. What are the Scheffé $t'_{.05}$ and $t'_{.01}$ values?

    a. Determine if the F-test in the ANOVA is significant. If it is significant, then proceed with (b), otherwise, stop at this point and indicate that the F-test is not significant.

    b. Using Formula 10.9, compute the t_{ij} values for all possible pairs of group means.

    c. Using Formulas 10.10 and 10.11, find the Scheffé $t'_{.05}$ and $t'_{.01}$.

    d. Compare the computed t_{ij} value with the Scheffé t values and determine which pair or pairs of means are significantly different.

10–11.   Perform the Tukey HSD test on the pairs of means in question 10–1.

    a. Determine if the F-test in the ANOVA is significant. If the F-test for the overall difference between means is significant, proceed with (b), otherwise stop at this point and indicate that multiple comparisons cannot be performed on a nonsignificant F.

    b. Form D_{ij}, the absolute differences between the means of groups i and j.

    c. Compute the Tukey HSD critical value for 0.05 and 0.01 using Formula 10.12 and table M of Appendix B.

    d. Compare the D_{ij} values found in (b) with the critical values found in (c) and determine which pairs of group means are statistically different.

10–12.   Perform the Tukey HSD test on the pairs of means in question 10–2.

    a. Determine if the F-test in the ANOVA is significant. If the F-test for the overall difference between means is significant, proceed with (b), otherwise stop at this point and indicate that multiple comparisons cannot be performed on a nonsignificant F.

    b. Form D_{ij}, the absolute differences between the means of groups i and j.

    c. Compute the Tukey HSD critical value for 0.05 and 0.01 using Formula 10.12 and table M of Appendix B.

    d. Compare the D_{ij} values found in (b) with the critical values found in (c) and determine which pairs of group means are statistically different.

10–13. Perform the Tukey HSD test on the pairs of means in question 10–7.
   a. Determine if the F-test in the ANOVA is significant. If the F-test for the overall difference between means is significant, proceed with (b), otherwise stop at this point, and indicate that multiple comparisons cannot be performed on a nonsignificant F.
   b. Form D_{ij}, the absolute differences between the means of groups i and j.
   c. Compute the Tukey HSD critical value for 0.05 and 0.01 using Formula 10.12 and table M of Appendix B.
   d. Compare the D_{ij} values found in (b) with the critical values found in (c) and determine which pairs of group means are statistically different.

10–14. Perform the Tukey test on the pairs of means in question 10–8.
   a. Determine if the F-test in the ANOVA is significant. If the F-test for the overall difference between means is significant, proceed with (b), otherwise stop at this point and indicate that multiple comparisons cannot be performed on a nonsignificant F.
   b. Form D_{ij}, the absolute differences between the means of groups i and j.
   c. Compute the Tukey HSD critical value for 0.05 and 0.01 using Formula 10.12 and table M of Appendix B.
   d. Compare the D_{ij} values found in (b) with the critical values found in (c) and determine which pairs of group means are statistically different.

10–15. An experiment is conducted to examine the overall effectiveness of three training programs X, Y, Z, in training the soldering of electronic components to a printed circuit board. Fifteen trainees are randomly assigned to the three programs (five in each). After completing the course, each trainee is required to solder components onto four P.C. boards. The number of errors are recorded. One trainee withdrew from group X and two trainees dropped out from group Y. The data are presented below.

Training program	Number of errors
X	6 7 6 5
Y	5 6 5
Z	6 6 7 6 6

Perform an analysis of variance and significance test at $\alpha = 0.05$ to see if there is any difference between the programs in terms of average errors.
   a. State the null and alternative hypotheses.
   b. Compute ΣX, ΣX^2, and Σx^2.
   c. Compute N.
   d. Compute ΣX and ΣX^2 for each group.
   e. Find the mean (M_i) for each group and the grand mean (M).
   f. Find Σx_i^2 for each group.
   g. Compute the pooled sum of squares and degrees of freedom for within-groups.

h. Compute the within-groups variance estimate, s_W^2.
i. For each group compute $n_i(M_i - M)^2$.
j. Compute the pooled sum of squares for between-groups, $\Sigma n_i(M_i - M)^2$.
k. Find the between-groups degrees of freedom $[k - 1]$ and compute the between-groups variance estimate, s_B^2.
l. Compute the F ratio, $F = s_B^2 / s_W^2$.
m. Enter table E of Appendix B with $\alpha = 0.05$ and 0.01 with numerator degrees of freedom = between-groups degrees of freedom, and denominator degrees of freedom = within-groups degrees of freedom to find $F_{.05}$ and $F_{.01}$.
n. If the F ratio found in (l) is greater than the critical value, $F_{.05}$, a significant difference exists between group means, reject h_0.
o. Complete the ANOVA summary table:

Source	Sum of Squares	df	Mean Square	F Ratio
Between-groups				
Within-groups				
Total				

10–16. Perform a Tukey test for the analysis of variance given in question 10–15 to determine if any pair of means are significant. Use $\alpha = 0.05$.

a. Determine if the F-test in the ANOVA is significant. If the F-test for the overall difference between means is significant, proceed with (b), otherwise stop at this point and indicate that multiple comparisons cannot be performed on a nonsignificant F.
b. Form D_{ij}, the absolute differences between the means of groups i and j.
c. Compute the Tukey HSD critical value for 0.05 and 0.01 using Formula 10.12 and table M of Appendix B.
d. Compare the D_{ij} values found in (b) with the critical values found in (c) and determine which pairs of group means are statistically different.

10–17. Given the same problem as in question 10–15, instead of the number of errors, the average completion times are used:

Training program	Mean completion time (mins.)				
X	57	64	61	60	
Y	50	59	35		
Z	64	59	72	62	63

Perform an analysis of variance on these data to determine if there is any significant difference between the mean completion time. Use $\alpha = 0.01$.

a. State the null and alternative hypotheses.
b. Compute ΣX, ΣX^2, and Σx^2.
c. Compute N.
d. Compute ΣX and ΣX^2 for each group.

e. Find the mean (M_i) for each group and the grand mean (M).

f. Find Σx_i^2 for each group.

g. Compute the pooled sum of squares and degrees of freedom for within-groups.

h. Compute the within-groups variance estimate, s_W^2.

i. For each group compute $n_i(M_i - M)^2$.

j. Compute the pooled sum of squares for between-groups, $\Sigma n_i(M_i - M)^2$.

k. Find the between-groups degrees of freedom $[k - 1]$ and compute the between-groups variance estimate, s_B^2.

l. Compute the F ratio, $F = s_B^2 / s_W^2$.

m. Enter table E of Appendix B with $\alpha = 0.05$ and 0.01 with numerator degrees of freedom = between-groups degrees of freedom, and denominator degrees of freedom = within-group degrees of freedom to find $F_{.05}$ and $F_{.01}$.

n. If the F ratio found in (l) is greater than the critical value, $F_{.05}$, a significant difference exists between group means, reject h_0.

o. Complete the ANOVA summary table:

Source	Sum of Squares	df	Mean Square	F Ratio
Between-groups				
Within-groups				
Total				

10–18. Four groups of fifth graders are randomly selected and assigned to a different educational program in computer literacy. At the end of the educational program, each student is examined on a standardized examination. Twenty-eight students participated, but only 21 finished the course. The data are given below.

Educ. Program	Score on exam
A	85 80 62 45 81 70
B	73 72 55 76 69
C	86 83 40 90 62 71
D	66 79 65 80

Perform an analysis of variance of these data using $\alpha = 0.05$ to determine if there are any differences between the different educational programs.

a. State the null and alternative hypotheses.

b. Compute ΣX, ΣX^2, and Σx^2.

c. Compute N.

d. Compute ΣX and ΣX^2 for each group.

e. Find the mean (M_i) for each group and the grand mean (M).

f. Find Σx_i^2 for each group.

g. Compute the pooled sum of squares and degrees of freedom for within-groups.

h. Compute the within-groups variance estimate, s_W^2.

i. For each group compute $n_i (M_i - M)^2$.
j. Compute the pooled sum of squares for between-groups, $\Sigma n_i (M_i - M)^2$.
k. Find the between-groups degrees of freedom $[k - 1]$ and compute the between-groups variance estimate, s_B^2.
l. Compute the F ratio, $F = s_B^2 / s_W^2$.
m. Enter table E of Appendix B with $\alpha = 0.05$ and 0.01 with numerator degrees of freedom = between-groups degrees of freedom, and denominator degrees of freedom = within-groups degrees of freedom to find $F_{.05}$ and $F_{.01}$.
n. If the F ratio found in (l) is greater than the critical value, $F_{.05}$, a significant difference exists between group means, reject h_0.
o. Complete the ANOVA summary table:

Source	Sum of Squares	df	Mean Square	F Ratio
Between-groups				
Within-groups				
Total				

10–19. A consumer research company is hired to see if three additives used in automobile gasoline improve mileage. Twelve cars of the same make, year, and model are used in the study. The data after three tanks of gasoline are given below.

Additive	Average miles per gallon			
S	23	25	26	22
V	20	19	21	17
T	21	24	18	19

Perform an analysis of variance at $\alpha = 0.05$ to determine if there is any difference in average miles per gallon between the different additives.

a. State the null and alternative hypotheses.
b. Compute ΣX, ΣX^2, and Σx^2.
c. Compute N.
d. Compute ΣX and ΣX^2 for each group.
e. Find the mean (M_i) for each group and the grand mean (M).
f. Find Σx_i^2 for each group.
g. Compute the pooled sum of squares and degrees of freedom for within-groups.
h. Compute the within-groups variance estimate, s_W^2.
i. For each group compute $n_i (M_i - M)^2$.
j. Compute the pooled sum of squares for between-groups, $\Sigma n_i (M_i - M)^2$.
k. Find the between-groups degrees of freedom $[k - 1]$ and compute the between-groups variance estimate, s_B^2.
l. Compute the F ratio, $F = s_B^2 / s_W^2$.

m. Enter table E of Appendix B with $\alpha = 0.05$ and 0.01 with numerator degrees of freedom = between-groups degrees of freedom, and denominator degrees of freedom = within-groups degrees of freedom to find $F_{.05}$ and $F_{.01}$.

n. If the F ratio found in (l) is greater than the critical value, $F_{.05}$, a significant difference exists between group means, reject h_0.

o. Complete the ANOVA summary table:

Source	Sum of Squares	df	Mean Square	F Ratio
Between-groups				
Within-groups				
Total				

10–20. Referring to question 10–19, perform all pairwise comparisons of means using the Scheffé test ($\alpha = .05$).

a. Determine if the F-test in the ANOVA is significant. If it is significant, then proceed with (b), otherwise, stop at this point and indicate that the F-test is not significant.

b. Using Formula 10.9, compute the t_{ij} values for all possible pairs of group means.

c. Using Formulas 10.10 and 10.11, find the Scheffé $t'_{.05}$ and $t'_{.01}$.

d. Compare the computed t_{ij} value with the Scheffé t' values and determine which pair or pairs of mean are significantly different.

10–21. Perform all pairwise comparisons using the Tukey HSD test for the data in question 10–19.

a. Determine if the F-test in the ANOVA is significant. If the F-test for the overall difference between means is significant, proceed with (b), otherwise stop at this point and indicate that multiple comparisons cannot be performed on a nonsignificant F.

b. Form D_{ij}, the absolute differences between the means of groups i and j.

c. Compute the Tukey HSD critical value for 0.05 and 0.01 using Formula 10.12 and table M of Appendix B.

d. Compare the D_{ij} values found in (b) with the critical values found in (c) and determine which pairs of group means are statistically different.

Chapter **11**

Analysis of Variance—
Two-Way Designs

T he one-way ANOVA of Chapter 10 and the *t*-test of Chapter 8 evaluated the effect of only one independent variable or experimental factor on a dependent variable. The major difference between the *t*-test and the one-way ANOVA is the number of groups or levels within the independent variable. The *t*-test uses only two levels and the one-way ANOVA uses more than two. The two-way ANOVA presented in this chapter examines the simultaneous effect of two independent variables or experimental factors on a dependent variable. Each independent variable has two or more levels (groups). There are several different variations of the two-way ANOVA. However, only one of those will be presented here. Perhaps the simplest and most often used **two-way ANOVA** is the *factorial design*. If there are *p* levels of one experimental factor and *q* levels of another experimental factor, the two-way factorial ANOVA design consists of *pq* experimental combinations. Additionally, to use this design properly, subjects are randomly assigned to *pq* experimental **factor** combinations, with each subject receiving only one combination. Problem 42 for example, contains two levels of one experimental factor (Type of teaching method: Discussion vs. Lecture) and three levels of another factor (Type of Instructor A, B, C). Problem 42 has 2 × 3 = 6 experimental combinations: Discussion–Instructor A, Discussion–Instructor B, Discussion–Instructor C, Lecture–Instructor A, Lecture–Instructor B, and Lecture–Instructor C.

However, unlike those designs in Chapter 10, the sources of variance are (1) each experimental factor, (2) the error factor (within-group) and (3) the interaction between experimental factors. Each experimental factor and the interaction are subjected to significance testing via the *F* ratio or *F*-test. If the interaction term is significant, then it is likely that one experimental factor behaves differently under different levels of the other experimental factor. Whenever a significant interaction occurs, the researcher must qualify the interpretation of each experimental factor.

Other variations of the factorial design and two-way ANOVAs can be found in more advanced texts such as Kirk (1982) and Winer (1971). For a very innovative treatise and use of multi-way ANOVAs (three or more experimental factors), the interested reader should consult Simon (1973, 1977), Box, Hunter, and Hunter (1978), or Daniel (1976). These manuscripts, however, presume that the reader has developed a strong understanding of ANOVA and experimental design.

two-way ANOVA   The case when a model is formed that has two different treatments.

factor   A variable of interest that is controlled in an analysis of variance.

PROBLEM 42

Test the Effects of Two Factors Simultaneously Using the Two-Way ANOVA

The head of a college department has three available instructors from whom to choose an individual to teach an important undergraduate course. The department head also needs to determine whether the lecture or discussion method produces greater student achievement on an end-of-course standardized examination. The examination scores are used by a national accreditation body to rate the department, so the scores are very important. The department head decides to conduct an experiment that will provide answers for both questions.

TABLE 11.1

Two-way analysis of variance

	Subject	Instructor A		Instructor B		Instructor C		$\Sigma\Sigma X$	$\Sigma\Sigma X^2$
		X	X^2	X	X^2	X	X^2		
Discussion	1	10	100	7	49	15	225		
	2	13	169	10	100	10	100		
	3	12	144	12	144	18	324		
	4	2	4	19	361	14	196		
	5	7	49	15	225	13	169		
	6	11	121	14	196	14	196		
	7	9	81	16	256	9	81		
	8	8	64	13	169	17	289		
	9	10	100	14	196	12	144		
	10	11	121	17	289	13	169		
	Σ	93	953	137	1985	135	1893	365	4831
Lecture	1	18	324	10	100	6	36		
	2	15	225	13	169	14	196		
	3	13	169	4	16	7	49		
	4	14	196	7	49	5	25		
	5	11	121	1	1	2	4		
	6	6	36	14	196	13	169		
	7	13	169	9	81	10	100		
	8	10	100	10	100	9	81		
	9	16	256	8	64	8	64		
	10	14	196	12	144	12	144		
	Σ	130	1792	88	920	86	868	304	3580
	$\Sigma\Sigma$	223	2745	225	2905	221	2761	669	8411

$$N = kn = 6 \cdot 10 = 60$$

Sixty students are selected to participate in the experiment. The students' names are placed in a hat. Individual names are drawn blindly from the hat for six separate experimental groups until all subjects have been assigned to one of the groups, giving 10 subjects per group. Each of the three instructors will teach two groups, one group by the lecture method and the other group by the discussion method, giving a total of six groups. The postcourse national examination scores for the students in these six groups are shown in table 11.1. These are scaled scores with a mean of 10 and a standard deviation of 3 based on national norms. No score higher than 19 or lower than 1 is given. The scaled scores themselves are labeled the X-scores and their squares are given under the heading X^2.

Computing Procedures for the Two-Way ANOVA

The sums of the columns in table 11.1 are given in the rows designated by the summation symbol Σ. Thus the sum of the X-scores for the first group is 93 and the sum of X^2 is 953.

In addition to showing ΣX and ΣX^2 for each of the six subgroups, table 11.1 also provides the double sums ($\Sigma\Sigma$) by rows and columns as well as the overall sum of X and X^2. For example, 365 and 4831 give ΣX and ΣX^2 for row 1, the sums of those values for the three groups that are using the discussion method. The sums for row 2 are $\Sigma X = 304$ and $\Sigma X^2 = 3580$, the result of adding the individual sums for the three groups using the lecture method. The column sums for the first column, for Instructor A, in table 11.1 are $\Sigma X = 223$ and $\Sigma X^2 = 2745$. For Instructor B, column 2, the sums are $\Sigma X = 225$ and $\Sigma X^2 = 2905$; for Instructor C, column 3, $\Sigma X = 221$ and $\Sigma X^2 = 2761$. Adding up the ΣX and ΣX^2 values by rows or by columns gives the total $\Sigma \Sigma X = 669$ and $\Sigma \Sigma X^2 = 8411$ for the entire sample comprising all six subgroups with $N = 60$.

Within-Groups Variance Estimate   Just as with the one-way ANOVA in Problems 38 and 39, a within-groups estimate of the population variance is computed with the two-way ANOVA. As with the one-way ANOVA, the sum of squared deviations relative to their own subgroup means is computed for each group separately. These sums of squares are pooled, $\Sigma \Sigma x_i^2$, to get the sum of squares within-groups. The $k - 1$ degrees of freedom from each group are also pooled to get $k(n - 1)$ df. The pooled sum of squares is divided by the pooled df to get s_W^2, the within-groups estimate of the population variance, σ^2. As with the one-way ANOVA, σ^2 is assumed to be equal for each subgroup. The computations leading to the determination that $s_W^2 = 11.90$ for the data in table 11.1 are:

Within-groups variance estimate:

$$\Sigma x^2 = \Sigma X^2 - \frac{(\Sigma X)^2}{n}$$

$$\Sigma x_1^2 = 953 - \frac{(93)^2}{10} = 953 - 864.9 = 88.1$$

$$\Sigma x_2^2 = 1985 - \frac{(137)^2}{10} = 1985 - 1876.9 = 108.1$$

$$\Sigma x_3^2 = 1893 - \frac{(135)^2}{10} = 1893 - 1822.5 = 70.5$$

$$\Sigma x_4^2 = 1792 - \frac{(130)^2}{10} = 1792 - 1690.0 = 102.0$$

$$\Sigma x_5^2 = 920 - \frac{(88)^2}{10} = 920 - 774.4 = 145.6$$

$$\Sigma x_6^2 = 868 - \frac{(86)^2}{10} = 868 - 739.6 = 128.4$$

Pooled sum of squares $= \Sigma \Sigma x_1^2 = 642.7$

Pooled sum of degrees of freedom $= k(n - 1) = 54$

Within-groups variance estimate $= s_W^2 = \dfrac{642.7}{54} = 11.90$

Between-Rows Variance Estimate   In the one-way ANOVA, variations of subgroup means from the grand mean provided the basis for another variance estimate, the between-groups variance estimate. In the two-way ANOVA, the groups in each row are pooled to provide two groups that give an estimate of the population variance by the same method used to get the between-groups variance estimate in the one-way ANOVA. The calculations for the between-rows variance estimate for the data in table 11.1 are on page 337.

STATISTICS IN THE WORLD AROUND YOU

Can aspirin prevent heart attacks? There is evidence that some people who suffer from heart disease (potential heart attack candidates) could benefit from taking an aspirin a day. There is on-going research taking place on whether or not aspirin can prevent heart attacks. The AMIS (Aspirin Myocardial Infarction Study) has completed at least one study on this topic. Philip Majerus at the Washington Medical School has conducted his own research. Essentially, such research involves using people who have already suffered a heart attack. The people are assigned to one or more treatment groups and a control group. The control group does not receive aspirin at all. The different treatment groups receive different doses of aspirin. To also test the hypothesis that heart attacks might be sex linked, the research study examines sex differences in the occurrence of heart disease. If there are three doses of aspirin to test and a control group, this study can be analyzed with a two-way ANOVA with four levels of one independent variable (dosage), and 2 levels of the sex variable (male and female). The design appears as follows:

Sex	Dosage 180 mg.	250 mg.	500 mg.	0 mg.
Male				
Female				

There have been similar studies for testing the hypothesis that vitamin C is effective in preventing the flu or that vitamin C is effective in the treatment against cancer.

Between-rows variance estimate:

$$M_{R_1} = \frac{365}{30} = 12.17 \qquad M_{R_1}^2 = 148.11$$

$$M_{R_2} = \frac{304}{30} = 10.13 \qquad M_{R_2}^2 = 102.62$$

$$\Sigma = \ldots \ldots 22.30 \qquad \Sigma = 250.73$$

$$\Sigma(M_{R_1} - M)^2 = 250.73 - \frac{(22.30)^2}{2} = 250.73 - 248.64 = 2.09$$

$$nC\Sigma(M_{R_i} - M)^2 = 10 \cdot 3 \cdot (2.09) = 62.70$$

$$df = R - 1 = 2 - 1 = 1$$

Between-rows variance estimate $= s_R^2 = \dfrac{62.70}{1} = 62.70$

When the three subgroups having the discussion method (row 1) are combined, $\Sigma X = 365$; this gives $M_{R_1} = 365/30 = 12.17$ for the mean of the first row. Since three groups are combined, the number of cases in the combined group is 30. The corresponding mean for the second row is $M_{R_2} = 304/30 = 10.13$. The sum of squares of deviations of these two means about their average, M, the grand mean, is found in the usual way, leading to a value of 2.09. This sum of squares of deviation scores must be multiplied by the number of cases in the combined group, $(n \times C) = 10 \times 3 = 30$, or the number of subjects per subgroup times the number of columns, to get the sum of squares for the between-rows variance estimate. This is analogous to the numerator for Formula 10.4, which gives the between-groups sum of squares for the one-way ANOVA. There, however, each unit group has n cases, so $\Sigma(M_i - M)^2$ is multiplied by n. Here, each unit group has nC cases, so $\Sigma(M_i - M)^2$ is multiplied by nC to get the between-rows sum of squares. The calculations to obtain this sum of squares are shown above, giving a value of 62.70. The degrees of freedom for this variance estimate are $(R - 1)$ where R is the number of rows, $(2 - 1)$ or 1. The between-rows variance estimate, therefore, obtained by dividing the sum of squares by the df, is $62.70/1 = 62.70$.

Between-Columns Variance Estimate   Just as the subgroups in each row are combined with the data in table 11.1 to form two larger groups to obtain an estimate of the population variance from the variation of row means about the grand mean, so are the subgroups in each column combined to form three larger groups. The variation in the means of these columns about the grand mean is used to provide yet another estimate of the population variance. As with the between-rows variance estimate, the basic method of obtaining this between-columns variance estimate is very similar to that used to obtain the between-groups variance estimate in the one-way ANOVA. The computations for the data in table 11.1 are shown below. The sum of squared deviations of the column means about the grand mean are obtained by computing column means and then using the usual formula $\Sigma x^2 = \Sigma X^2 - (\Sigma X)^2/N$.

Between-columns variance estimate:

$$M_{C_1} = \frac{223}{20} = 11.15 \qquad\qquad M_{C_1}^2 = 124.32$$

$$M_{C_2} = \frac{225}{20} = 11.25 \qquad\qquad M_{C_2}^2 = 126.56$$

$$M_{C_3} = \frac{221}{20} = 11.05 \qquad\qquad M_{C_3}^2 = 122.10$$

$$\Sigma = \ldots\ldots 33.45 \qquad\qquad \Sigma = 372.98$$

$$\Sigma(M_{C_1} - M)^2 = 372.98 - \frac{(33.45)^2}{3} = 372.98 - 372.97 = 0.01$$

$$nR\Sigma(M_{C_i} - M)^2 = 10 \cdot 2(0.01) = 0.20$$

$$df = C - 1 = 3 - 1 = 2$$

Between-columns variance estimate $= s_C^2 = \dfrac{0.20}{2} = 0.10$

In this case, for the data in table 11.1, Σx^2 for columns = 0.01. To obtain the between-columns sum of squares, the sum of squared deviations is then multiplied by the number of cases in each pooled group, $n \times R$, the number of cases per group by the number of rows, or 20. For the data in table 11.1, this value is 0.20. The number of degrees of freedom is the number of columns minus 1, or 2. The between-columns variance estimate, s_C^2, is obtained by dividing the between-columns sum of squares by the between-columns df, giving $s_C^2 = 0.20/2 = 0.10$.

Interaction Variance Estimate   In the one-way ANOVA, the total sum of squares equaled the sum of the within-groups sum of squares plus the between-groups sum of squares, except for rounding errors. This same principle holds for the two-way ANOVA. The total sum of squares and within-groups sum of squares are calculated in the same way for both the two-way and the one-way ANOVA. The between-groups sum of squares for the two-way ANOVA is calculated the same as for the one-way ANOVA if all subgroups are treated as they are in the one-way ANOVA, without regard to row and column position. For each subgroup, M_i and M_i^2 are obtained. The sums of these values are inserted in the usual formula for the sum of squared deviations and this total is multiplied by n, the subgroup size. The computations for these sums of squares for table 11.1 data yield the following:

$$
\begin{aligned}
\text{Within-groups sum of squares} &= 642.70 \\
\text{Between-groups sum of squares} &= \underline{308.95} \\
&\ 951.65
\end{aligned}
$$

Addition of these two sums of squares gives a figure that equals the independently calculated total sum of squares, 951.65, as shown below.

Total sum of squares:

$$
\sum_1^N x^2 = \Sigma X^2 - \frac{(\Sigma X)^2}{N} = 8411 - \frac{(669)^2}{60} = 951.65
$$

The between-groups sum of squares in the two-way ANOVA is not just due to one experimental factor, however. Differences between-column means as well

Between-groups sum of squares:

$$
M_{11} = M_1 = \frac{93}{10} = 9.3 \qquad M_1^2 = 86.49
$$

$$
M_{12} = M_2 = \frac{137}{10} = 13.7 \qquad M_2^2 = 187.69
$$

$$
M_{13} = M_3 = \frac{135}{10} = 13.5 \qquad M_3^2 = 182.25
$$

$$
M_{21} = M_4 = \frac{130}{10} = 13.0 \qquad M_4^2 = 169.00
$$

$$
M_{22} = M_5 = \frac{88}{10} = 8.8 \qquad M_5^2 = 77.44
$$

$$M_{23} = M_6 = \frac{86}{10} = 8.6 \qquad M_6^2 = 73.96$$

$$\Sigma = \dots \, 66.9 \qquad \Sigma = 776.83$$

$$\Sigma(M_i - M)^2 = 776.83 - \frac{(66.9)^2}{6} = 776.83 - 745.935 = 30.895$$

$$n\Sigma(M_i - M)^2 = 10(30.895) = 308.95$$

Within-SSq + between-SSq = 642.7 + 308.95 = 951.65

Between-SSq − row-SSq − column-SSq = 308.95 − 62.7 − 0.2 = 246.05

Interaction-SSq = 246.0

as differences between-row means contribute to the between-groups sum of squares. Adding the between-rows sum of squares, 62.70, and the between-columns sum of squares, 0.20, gives a total, however, of only 62.90 when the between-groups sum of squares is 308.95. Subtracting the between-rows and between-columns sums of squares from the between-groups sum of squares, 308.95 − 62.7 − 0.2, gives 246.05 as a "left over" sum of squares not attributable to row or column effects. In this particular experiment, this indicates that there are variations among the subgroup means that cannot be attributed to differences between teachers, or to differences between teaching methods. There must be some other factor.

The other factor is the *interaction* effect and this left over sum of squares is called the "interaction sum of squares." In this experiment, if there is a significant interaction effect, this means that some teachers do better with one method than they do with the other, or vice versa. If there were no interaction effect, the interaction sum of squares is zero and it does not matter which teacher is paired with which method. The interaction sum of squares seldom is zero, of course, but if it does not lead to a statistically significant result, no interaction effect is claimed.

The interaction sum of squares also leads to yet another estimate of the population variance, σ^2, when divided by the number of degrees of freedom. The df for the interaction sum of squares is $(R - 1)(C - 1)$, the product of the number of rows minus one times the number of columns minus one. Calculations for the interaction variance estimate, $s^2_{R \times C}$, are shown for the data in table 11.1.

Interaction variance estimate:

Interaction SSq = 246.0

Interaction $df = (R - 1)(C - 1) = (2 - 1)(3 - 1) = 2$

Interaction variance estimate $= s_{R \times C}^2 = \frac{246.0}{2} = 123.0$

Check on the Interaction Sum of Squares   The interaction sum of squares should also be calculated by another method to insure that the method shown for the data in table 11.1 has given correct results. The deviation of a subgroup mean from the grand mean, $(M_i - M)$, can be viewed as consisting of a row effect, $M_{R_i} - M$, a column effect, $M_{C_i} - M$, and an interaction effect. To get the part of the deviation that is due to interaction, subtract out the other two effects:

$$(M_i - M) - (M_{R_i} - M) - (M_{C_i} - M)$$

Removing parentheses and cancelling M's gives:

$$(M_i - M_{R_i} - M_{C_i} + M) \tag{11.1}$$

Formula 11.1 gives the deviation due to interaction. Squaring and summing over all such values for all subgroups gives:

$$\sum_{i=1}^{k} (M_i - M_{R_i} - M_{C_i} + M)^2 \tag{11.2}$$

As with other sums of squared deviation scores involving variations of means of groups, the sum of squared deviation must be multiplied by the number of cases in the group for which the mean is computed to obtain the sum of squares. This gives for the interaction sum of squares:

$$n \times \sum_{i=1}^{k} (M_i - M_{R_i} - M_{C_i} + M)^2 \tag{11.3}$$

Using Formula 11.2 on the six subgroups for the data in table 11.1 gives the results shown below.

Independent check on the interaction sum of squares:
$(M_i - M_{R_i} - M_{C_1} + M)^2$
$(9.30 - 12.17 - 11.15 + 11.15)^2 = (-2.87)^2 = \quad 8.24$
$(13.70 - 12.17 - 11.25 + 11.15)^2 = \quad (1.43)^2 = \quad 2.04$
$(13.50 - 12.17 - 11.05 + 11.15)^2 = \quad (1.43)^2 = \quad 2.04$
$(13.00 - 10.13 - 11.15 + 11.15)^2 = \quad (2.87)^2 = \quad 8.24$
$(8.80 - 10.13 - 11.25 + 11.15)^2 = (-1.43)^2 = \quad 2.04$
$(8.60 - 10.13 - 11.05 + 11.15)^2 = (-1.43)^2 = \quad 2.04$
$\Sigma = \quad 24.64$
$n\Sigma(M_i - M_{R_i} - M_{C_i} + M)^2 = 10(24.64) = 246.4$

Multiplying 24.64 by $n = 10$, as called for by Formula 11.3 gives 246.4 as the interaction sum of squares. This figure agrees within rounding error with the interaction sum of squares, 246.0, obtained by subtracting the between-columns and between-rows sum of squares from the between-groups sum of squares.

TABLE 11.2

Summary of two-way ANOVA for data in table 11.1

Source of variation	Sum of squares	Degrees of freedom	Mean square (variance estimate)	F
Between-rows	62.7	1	62.7	5.3*
Between-columns	0.2	2	0.1	0.0
Interaction	246.0	2	123.0	10.3**
Within-groups	642.7	54	11.9	
Total	951.6	59		

Rows: $F_{.05} = 4.02$; $F_{.01} = 7.12$

Columns: $F_{.05} = 3.17$; $F_{.01} = 5.01$

Interaction: $F_{.05} = 3.17$; $F_{.01} = 5.01$

Summary of the Two-Way ANOVA Results   As with the one-way ANOVA, it is customary to summarize the results of the two-way ANOVA in a table that shows the various sums of squares, degrees of freedom, variance estimates, and F ratio values. The summary of the data in table 11.1 is shown in table 11.2.

As with the one-way ANOVA, the within-groups variance estimate is used as the denominator for the F ratio tests. The between-rows, between-columns, and interaction variance estimates are divided, respectively, by the within-groups variance estimate to obtain the F ratios for testing these various effects. As table 11.2 shows, the between-rows F ratio is significant at the 0.05 level, the interaction F ratio is significant at the 0.01 level, and the between-columns F ratio is not significant. This suggests that: (1) the discussion method is generally more effective than the lecture method ($M_D = 12.17$, $M_L = 10.13$); (2) there are no overall differences among teachers ($M_A = 11.15$, $M_B = 11.25$, $M_C = 11.05$); and (3) some teachers do better with one teaching method than with the other method. Figure 11.1 shows Instructor A gets better performance from lecture than discussion. The opposite is true for Instructors B and C. This crossing pattern graphically depicts an interaction effect. The analysis of the data in this instance, then, suggests that the department head should select the instructor who does best using the discussion method to regularly teach the course.

If the overall F ratio for row and/or column is significant, means of rows or columns can be tested, pair by pair, to locate significant differences. If the interaction F ratio is significant, tests between individual cell means should be made. As with the one-way ANOVA, multiple comparisons increase the probability of obtaining a result significant by chance if the ordinary t-test is applied. For this reason, it is recommended that some procedure, such as the Scheffé method, be applied to decide how large the t ratio must be to be regarded as significant when several pairs of mean differences are being tested. The method described in connection with Problem 38 serves this purpose.

Figure 11.1
Subgroup (cell) means for data in table 11.1.

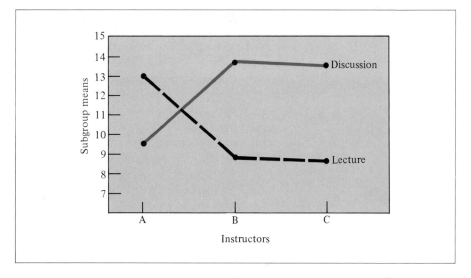

Applications of the Two-Way ANOVA

The two-way ANOVA just described is applicable to a wide variety of experiments, many of which do not involve random selection of cases for each subgroup from the same common pool. One of the factors, rows or columns, might involve distinct categories of individuals. For example, random samples of males and females might be given several different treatments, or people of high, medium, and low intelligence might each be given four different study methods. In this case, four random samples of size n are drawn from people who score high on intelligence tests, four samples of size n from people of medium intelligence, and four samples of size n from people of low intelligence.

It is also possible to have quantitative levels or categories both ways. People of high, medium, and low intelligence who are either high or low in weight can be tested on pain thresholds and a two-way ANOVA run on the scores. Or, brown-eyed and blue-eyed people, males and females, can be tested on a visual acuity task to provide data for a two-way ANOVA.

Special Cases of the Two-Way ANOVA   There are certain classes of experiments occasionally encountered that superficially appear to fit this model but which actually require a more complicated statistical treatment. If the conditions selected for the treatment in either rows or columns, or both, in the two-way ANOVA represent a random selection from a large number of possibilities, the data must be treated in a different manner. Suppose a random selection of six major university varsity broad jumpers each made ten jumps with each of two types of track shoes under properly randomized and spaced conditions. The jump lengths are the X values. This gives a 6 row, 2 column, two-way ANOVA, but the data should not be analyzed by the methods described in connection with Problem 39, since one of the two factors is based on a random selection of conditions (people, in this case). For experiments of this kind, the F ratios do not necessarily use the within-groups variance estimate in the denominator. Appropriate methods for dealing with this situation can be found in more advanced texts such as those found in the reference section.

Determining the Strength of an Effect in ANOVA

A significant statistical test (reject the null hypothesis) informs the researcher that something within the experiment has occurred that may not be attributed to chance variations alone. The significance level only gives the probability of making this claim incorrectly. In an analysis-of-variance, a significant F ratio only states that an independent variable has had a real effect on the dependent variable. If the level of significance was 0.05, this indicates that there is a 0.05 probability that the claim of a real effect is wrong. Despite statistical significance, there is really no assessment as to the strength of the effect of the independent variable on the dependent variable. A truly small effect can be made statistically significant by merely increasing the sample size. With all other factors held constant, the increase in the sample size will increase the degrees of freedom for the within-group source of variation and subsequently decrease the magnitude of the within-group mean square. With a smaller within-group mean square the F ratio becomes larger. The researcher as a result may conclude a real treatment effect, when in reality what is measured is a sample size effect. A significant statistical test only lets the researcher know that the independent variable may have a real effect on the dependent variable. It cannot account for the strength of the effect.

Dunnette (1966); Gallo, Jamieson, and Christian (1977); Linton and Gallo (1975); and Simon (1976, 1977) have examined a large number of published research experimental studies. Their findings show a very small number of studies reporting strength of effects. Additionally, in their own re-analysis of the studies using the published ANOVA summary tables, they showed that a majority of the major effects under study accounted for less than 5% of the total variance available. However, each major effect was statistically significant. In different words, most of the research studies reviewed claimed a major discovery between independent and dependent variables, even though the strength of the effect between independent and dependent variables was very weak.

For the analysis of variance methods presented in Chapters 10 and 11, there are two measures available for determining the strength of an effect. They are eta squared (η^2 : Greek letter eta) and omega squared (ω^2 : Greek letter omega). Eta squared can be computed directly from the information given on the ANOVA summary table and it provides an estimate as the strength of effect for the sample by finding the proportion of variance in the dependent variable that may be accounted for by the independent variable(s).

$$\eta^2 = \frac{\text{Sum of squares for treatment effect}}{\text{Sum of squares TOTAL}}$$

The sum of squares for treatment effect in a one-way ANOVA would be the sum of squares between-groups. In the two-way ANOVA it could be either the sum of squares associated with the first independent variable (between-rows), the second independent variable (between-columns) or the interaction between the two independent variables.

Omega squared is generally considered as a more precise estimate of strength. It is the strength of effect estimated for the population. Like eta squared, it is computed using information from the ANOVA summary table.

$$\omega^2 = \frac{(\text{Sum of squares treatment}) - df_{\text{treatment}} \times (\text{Mean square within-group})}{\text{Mean square within-group} + \text{Sum of squares TOTAL}}$$

The sum of squares treatment for a one-way ANOVA would be the between-group sum of squares. For the two-way ANOVA, sum of squares treatment can be either sum of squares between-rows, sum of squares between-columns, or their interaction.

To evaluate eta squared and omega squared, the range of values are zero to 1.00. The closer the value is to 1.00, the stronger the effect. Those effects with a eta or omega squared value near zero indicate a weak effect. Sample size effects are apparent when the effect is weak and the sample size is large.

SUMMARY OF STEPS

Testing the Effects of Two Factors Simultaneously Using the Two-Way ANOVA

Step 1 Compute the sum of the raw scores, ΣX, and the sum of the squared raw scores, ΣX^2, for each sample separately. There are $k = (R \times C)$ samples, where R is the number of rows and C is the number of columns.

Step 2 Using the totals from Step 1, find the sum of the squared deviation scores, Σx^2, for each sample separately by means of the formula $\Sigma x^2 = \Sigma X^2 - (\Sigma X)^2/n$, where n is the number of cases in each sample.

Step 3 Add up the sums of squared deviation scores obtained in Step 2 for all samples to get the within-groups sum of squares.

Step 4 Divide the within-groups sum of squares from Step 3 by the sum of the degrees of freedom for the k separate samples, $k(n - 1)$, to get the within-groups variance estimate, s_W^2.

Step 5 Using the totals from Step 1, combine the ΣX values for each row and compute a row mean for each row, M_{R_i}.

Step 6 Using the totals from Step 1, combine the ΣX values for each column and compute a column mean for each column, M_{C_i}

Step 7 Using the totals from Step 1, combine the ΣX values for all samples and compute an overall grand mean, M.

Step 8 Compute the row sum of squares by summing the values $(M_{R_i} - M)^2$ for each row and multiplying the total by nC, the sample size times the number of columns.

Step 9 Divide the row sum of squares obtained in Step 8 by $R - 1$, the degrees of freedom for rows, to obtain the between-rows variance estimate, s_R^2.

Step 10 Compute the column sum of squares by summing the values $(M_{C_i} - M)^2$ for each column and multiplying the total by nR, the sample size times the number of rows.

Step 11 Divide the column sum of squares obtained in Step 10 by $C - 1$, the degrees of freedom for columns, to obtain the between-columns variance estimate, s_C^2.

Step 12 Using the totals obtained in Step 1, calculate the mean of each sample, M_i.

Step 13 Obtain the between-groups sum of squares by adding up the $(M_i - M)^2$ values for all k samples and multiplying the total by n, the sample size.

Step 14 Using the totals from Step 1, calculate the total sum of squares by the formula $\Sigma x^2 = \Sigma\Sigma X^2 - (\Sigma\Sigma X)^2/N$, where N is the number of cases in all samples combined and the double summations are over all k samples.

Step 15 Add the within-groups sum of squares from Step 3 to the between-groups sum of squares obtained in Step 13 and check to see if this sum equals the total sum of squares obtained in Step 14.

Step 16 Calculate the interaction sum of squares by subtracting the between-rows sum of squares (Step 8) and the between-columns sum of squares (Step 10) from the between-groups sum of squares obtained in Step 13.

Step 17 Check the interaction sum of squares independently using Formula 11.3.

Step 18 Calculate the interaction variance estimate, $s^2_{R \times C}$, by dividing the interaction sum of squares by $(R - 1)(C - 1)$, the number of degrees of freedom for interaction.

Step 19 Calculate the F ratios by dividing the between-rows, between-columns, and interaction variance estimates by the within-groups variance estimate.

Step 20 Enter the Table of the F Distribution (table F, Appendix B) with the appropriate numbers of degrees of freedom to find the critical values of F, $F_{.05}$, and $F_{.01}$, for each of these F ratio tests.

Step 21 Determine the significance of each effect by comparing the obtained values of F with the critical values of F.

MATHEMATICAL NOTES

This section demonstrates that the total sum of squares for the two-way ANOVA can be obtained by adding up the within-groups, between-rows, between-columns, and interaction sums of squares. As with the one-way ANOVA, begin with a deviation score about the grand mean:

$$x = X_{ij} - M$$

$$x = X_{ij} + (M_i - M_i) + (M_{R_i} - M_{R_i}) + (M_{C_i} - M_{C_i}) - M$$

$$x = (X_{i_j} - M_i) + (M_{R_i} - M) + (M_{C_i} - M) + (M_i - M_{R_i} - M_{C_i} + M)$$

Squaring both sides and summing from 1 to N gives:

$$\sum_1^N x^2 = \sum_1^N (X_{ij} - M_i)^2 + \sum_1^N (M_{R_i} - M)^2 + \sum_1^N (M_{C_i} - M)^2$$

(11.4)

$$+ \sum_1^N (M_i - M_{R_i} - M_{C_i} + M)^2 + \text{cross-product terms}$$

The first term on the right of Formula 11.4 is the within-groups sum of squares, since it is just $\sum_1^N x_1^2 + \sum_1^N x_2^2 + \cdots + \sum_1^N x_k^2$. The second term on the right equals $nC \times \sum_1^R (M_{R_i} - M)^2$, since there are nC cases that have the same $(M_{R_i} - M)$ value and there are only R such values. This expression is the same as that for the between-rows sum of squares shown for the data in table 11.1. The third term on the right is rewritten as $nR \cdot \sum_1^C (M_{C_i} - M)^2$, since there are nR cases that have the same $(M_{C_i} - M)$ value and there are only C such values. This expression is the same as that for the between-columns sum of squares shown previously in table 11.1. The fourth term on the right of formula 11.4 is rewritten as $n \times \sum_1^k (M_i - M_R - M_C + M)^2$, since all the cases in the same subgroup have the same $(M_i - M_R - M_C - M)^2$ value. This expression is the same as Formula 11.3, the sum of squares for the interaction variance estimate.

The total sum of squares on the left of Formula 11.4 equals the sum of the sums of squares within-groups, between-rows, between-columns, and for interaction plus the cross-product terms. For the equation to hold as expected, the cross-product terms must be equal to zero. To see that these cross-product terms are zero, it is necessary to consider them within each subgroup. The first term with the others yields cross products that involve a constant times a deviation score, as with the one-way ANOVA. When the constant is removed outside the summation sign, a sum of deviation scores is left inside and this equals zero. The other cross products involve cross-product terms that are identical for every case within a subgroup, giving n times that cross product for the group. For the $2 \times \sum_1^N (M_{R_1} - M)(M_{C_1} - M)$ cross-product term this will give:

$$2[n(M_{R_1} - M)(M_{C_1} - M) + n(M_{R_1} - M)(M_{C_2} - M)$$

$$+ n(M_{R_1} - M)(M_{C_3} - M) + n(M_{R_2} - M)(M_{C_1} - M)$$

$$+ n(M_{R_2} - M)(M_{C_2} - M) + (M_{R_2} - M)(M_{C_3} - M)]$$

j. Find the degrees of freedom for rows and compute the between-rows variance estimate, s_R^2.

k. Compute the column sum of squares.

l. Find the degrees of freedom for columns and compute the between-column variance estimate, s_C^2.

m. Find the interaction sum of squares.

n. Find the degrees of freedom for interaction and compute the interaction variance estimate.

o. Calculate the F ratios for the row effect, the column effect, and interaction.

p. Find the critical F values, $F_{.05}$ and $F_{.01}$.

q. Determine the significance of each effect by comparing the critical value of F with the obtained value of F.

r. Compute the ANOVA summary table below:

Source	Sum of squares	df	Mean square	F value
Between-rows				
Between-columns				
Interaction				
Within-groups				
Total				

11–3.   If justified, make t-tests for the data in question 11–1, computing the Scheffé critical values $t'_{.05}$ and $t'_{.01}$. Interpret the results for the entire experiment.

a. Determine if the F-test in the ANOVA is significant for at least one effect. If one is significant, then proceed with (b), otherwise stop at this point and indicate that the F-test is not significant.

b. Using Formulas 11.5, 11.6 and 11.9, compute the t_{ij} values for all possible pairs of subgroup means.

c. Using Formulas 11.7 and 11.8, find the Scheffé $t'_{.05}$ and $t'_{.01}$.

d. Compare the computed t_{ij} value with the Scheffé t' values and determine which pair or pairs of means are significantly different.

11–4.   Perform the Tukey HSD test for the six group means of the data in question 11–1.

a. Determine if the F-test is significant in the ANOVA for at least one effect. If one F-test for the overall difference between means is significant, proceed with (b), otherwise stop at this point and indicate that multiple comparisons cannot be performed on a nonsignificant F.

b. Form D_{ij}, the absolute differences between the means of groups i and j, subgroups i and j, or cell means i and j.

c. Compute the Tukey HSD critical value for 0.05 and 0.01 using Formula 10.12 and table M of Appendix B.

d. Compare the D_{ij} values in (b) with the critical values found in (c) and determine which pairs of group means are statistically different.

TABLE 11.3

Two-way layout

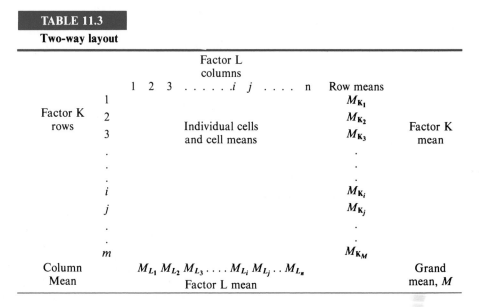

		Factor L columns								Row means	
		1	2	3	i	j	. . . .	n			
Factor K rows	1								M_{K_1}	
	2			Individual cells and cell means						M_{K_2}	Factor K mean
	3								M_{K_3}	
	.								.	
	i								M_{K_i}	
	j								M_{K_j}	
	m								M_{K_M}	
Column Mean		M_{L_1} M_{L_2} M_{L_3} . . . . M_{L_i} M_{L_j} . . M_{L_n}							Grand mean, M	
		Factor L mean								

In Problem 42, in table 11.2 the Rows source is statistically significant. Using the t ratio we can test the mean difference between lecture and discussion. Applying Formula 11.6,

where $M_{R_1} = 12.17$

$M_{R_2} = 10.13$

$s_W^2 = 11.90$

$df = 54$

$N_{R_1} = 30$

$N_{R_2} = 30$

gives:

$$t = \frac{12.17 - 10.13}{\sqrt{11.90\,[(30 + 30)/(30 \times 30)]}} = \frac{2.04}{\sqrt{(11.90)\,(0.0667)}}$$

$$t = \frac{2.04}{\sqrt{0.7937}} = \frac{2.04}{0.891} = 2.29$$

The critical value from table C of Appendix B for $t_{.05}$ and 54 df is 2.001 (interpolated). Since the obtained t of 2.29 is more than 2.001, reject the hypothesis of equal means. Discussion produces higher test scores than lecture ($M_D = 12.17$ and $M_L = 10.13$). Also, the interaction is significant. This indicates that lecture may be better for some instructors than others, while the same is true for discussion.

The cell means for this example can be computed by dividing the sums given in table 11.1 by 10. They are labeled, M_{11}, M_{12}, M_{13}, M_{21}, M_{22}, M_{23}. An alternative is to label them as M_1, M_2, M_3, M_4, M_5, M_6. M_{21} (lecture, Instructor A) is compared with M_{23} (lecture, Instructor C) below.

$$M_{21} = 13.0 \quad M_{23} = 8.6 \quad N_{21} = 10 \quad N_{23} = 10 \quad s_W{}^2 = 11.9$$

$$t = \frac{13 - 8.8}{\sqrt{11.9\,[(10 + 10)/(10 \times 10)]}} = \frac{4.4}{\sqrt{(11.9)\,(0.2)}}$$

$$= \frac{4.4}{\sqrt{2.38}} = \frac{4.4}{1.54} = 2.85$$

The critical value for $t_{.05}$ with 54 df is 2.001. The obtained t of 2.85 is more than 2.001, so the hypothesis of equal means is rejected. Lecture with Instructor A produced significantly higher scores than lecture with Instructor C.

Just as it was for Problem 40, the t-test is designed for the situation where only one comparison between two subgroup means is to be made. The more t-tests made using the same $s_W{}^2$ in the denominator of the t formula, the greater the likelihood of getting significance by chance. Some two-way ANOVA designs can have many levels that can be compared with each other. When there are many levels and the researcher's desire is to make all possible comparisons between the means, table C of Appendix B can be used to evaluate t values between all combinations of the subgroup means, but each should now be further evaluated with the Scheffé test. A description of this test is given in Problem 40. The form of the Scheffé test is the same here as it is for the one-way ANOVA. Formulas 11.7 and 11.8 give the Scheffé values used to compare the t values obtained from Formulas 11.5 and 11.6.

$$t' = \sqrt{(k - 1)\,F_{.05}} \tag{11.7}$$

$$t' = \sqrt{(k - 1)\,F_{.01}} \tag{11.8}$$

where k = the number of cell means (number of experimental factor combinations), $F_{.05}$ and $F_{.01}$ are found in table E of Appendix B for $k - 1$ and $N_T - k$ degrees of freedom. (N_T = total sample size.)

In the example in Problem 42, $k = 6$, $F_{.05} = 2.38$, $F_{.01} = 3.38$.

$$t'_{.05} = \sqrt{(6 - 1)2.38} = 3.45$$

$$t'_{.01} = \sqrt{(6 - 1)3.38} = 4.11$$

To achieve statistical significance at the 0.05 level, the computed t value must exceed 3.45. For the 0.01 level, the t value must be greater than 4.11.

SUMMARY OF STEPS

t Ratios Multiple-Comparison Method for Comparing Means in a Two-Way ANOVA

Step 1   Compute all cell, subgroup, and factor means. If factor 1 has n levels and factor 2 has m levels, then there are $n \times m$ cell means.

Step 2   Perform the two-way ANOVA calculations.

Step 3   From the hypothesis of the study, indicate which comparisons are to be made. Will it be a comparison between cell, subgroup, or factor means?

Step 4   Depending on the hypothesis, use Formulas 11.5 or 11.6 to compute the t-test statistic.

Step 5   Using a level of significance, the degrees of freedom $N_T - nm$ (N_T = total number of observations), and table C of Appendix B, find the critical value t_α.

Step 6   If the absolute value of the test statistic obtained in Step 4 exceeds the critical value, the hypothesis of equal means is rejected, otherwise it is not rejected.

SUMMARY OF STEPS

Scheffé Method of Multiple-Comparisons in a Two-Way ANOVA

Step 1   Compute the cell, subgroup, and factor means.

Step 2   Perform the two-way ANOVA calculations.

Step 3   Depending upon factor mean, subgroup mean, or cell mean comparisons, select the appropriate test statistic; either Formula 11.5 or 11.6 and compute the test statistic for each pair of means of interest.

Step 4   Compute the critical value t' using Formulas 11.7 and 11.8.

Step 5   If a t value computed in Step 3 is greater than t', reject the hypothesis of equal means. Otherwise do not reject the hypothesis.

PROBLEM 44

The Tukey HSD Multiple-Comparison Test for Comparing Means

The use of the Tukey HSD test for a two-way ANOVA is very similar to the Tukey test used in one-way ANOVAs. Like the Scheffé test, the major difference between the one-way and two-way ANOVA use of the Tukey HSD test is how the means are specified. That is, the comparisons can be between individual cell means, or between subgroup means (levels of independent variable). The comparisons take the form of difference scores between means. The difference scores are statistically significant if the absolute value of the difference between the means exceeds the value of HSD, the critical value. The Tukey HSD is defined in Chapter 10. Formula 10.12, which can be used for two-way ANOVA comparisons is:

$$\text{HSD} = q_{k,v} \sqrt{\frac{s_W^2}{n}} \qquad (10.12)$$

The value of q is found from table M in Appendix B using the within-groups degrees of freedom ($N_T - k$). The value of k differs depending upon whether the comparison is between individual cell means or subgroup means. The value of k when comparing a pair of means from different levels (subgroups) of an independent variable is equal to the number of levels (subgroups) within that independent variable (experimental factor). The value of k for comparing pairs of individual cell means is equal to the total number of cells or experimental factor combinations. For example, using the data from table 11.1, if the researcher wants to compare the differences between discussion and lecture method means, the value of k is equal to 2. If the comparison is between Lecture–Instructor A and Discussion–Instructor A, the value of k is 6. In cases where the n's are not quite equal, but are not radically different from each other, the harmonic mean of the n's is substituted for n. From Chapter 10, Formula 10.13 gives the harmonic mean of the n's.

$$n_h = k/[(1/n_1) + (1/n_2) + \ldots + (1/n_k)] \qquad (10.13)$$

Example   Using the data from table 11.1, M_{R_1} (mean of row 1 = mean of discussion method) = 12.17 is compared to M_{R_2} (mean of row 2 = mean of lecture method) = 10.13. The difference is $12.17 - 10.13 = 2.04$.

$$N_T = 60, k = 2, s_W^2 = 11.90, n = 10,$$

$$q_{.05,54} = 2.84, q_{.01,54} = 3.79.$$

$$\text{HSD}_{.05} = 2.84 \sqrt{\frac{11.90}{10}} = 2.84 \sqrt{1.19} = (2.84)(1.091) = 3.098$$

The obtained value of 3.098 is more than the tabled value of 2.04, so the hypothesis of equal means is not rejected.

In comparing $M_{12} = 13.7$ with $M_{23} = 8.6$, the difference is 5.1. The HSD is

$$q_{.05,54}, k = 6 = 4.18$$

$$\text{HSD} = (4.18)(1.091) = 4.560$$

Since 5.1 is greater than 4.56, the hypothesis of equal means is rejected. Students score higher with Instructor B using discussion than with Instructor C using lecture.

SUMMARY OF STEPS

Tukey's HSD Method of Multiple Comparisons in a Two-Way ANOVA

Step 1   Compute the cell means and the factor means.

Step 2   Perform the two-way ANOVA calculations.

Step 3   Depending upon the choice of factor mean comparisons or cell mean comparisons, form the differences, then compute the critical value HSD given in Formula 10.12.

Step 4   If the absolute value of the difference is larger than HSD, reject the hypothesis of equal means. Otherwise do not reject the hypothesis.

References

Box, G. E. P., Hunter, W. G., & Hunter, J. S. (1978). *Statistics for experimenters.* New York: Wiley.

Daniel, C. (1976). *Applications of statistics to industrial experimentation.* New York: Wiley.

Dunnette, M. D., Fads, fashions and folderol in psychology, *American Psychologist,* 1966, 21, 343–52.

Gallo, P. S., Jamieson, K., and Christian, K. *The strength of experimental effects in the psychological literature.* Annual Convention, Western Psychological Association, Seattle, Washington, April 20–23, 1977.

Kirk, R. E. (1982). *Experimental design: Procedures for the behavioral sciences* (2d ed.). Belmont, CA: Brooks/Cole Publishing Co.

Linton, M. S. and Gallo, P. S. (1975). *The Practical Statistician: Simplified Handbook of Statistics.* Monterey, CA: Brooks/Cole.

Simon, C. W. *Analysis of human factors engineering experiments: Characteristics, results and applications.* Westlake Village, CA: Canyon Research Group, Inc., Technical Report No. CWS–02–76, August 1976, 104 pages.

Simon, C. W. *Design, analysis and interpretation of screening studies for human factors engineering research.* Westlake Village, CA: Canyon Research Group, Inc., Technical Report No. CWS–03–77. September 1977, 220 pages.

Simon, C. W. *Economical multifactor designs for human factors engineering experiments.* Hughes Aircraft Company, Technical Report No. P73–326. June 1973. 171 pages. (AD 767–739).

Winer, B. J. (1971). *Statistical principles in experimental design* (2d ed.). New York: McGraw-Hill.

Exercises

11-1.   An investigator wishes to test the effect of three weight reducing programs and also to determine if sex is a factor to be considered. Three random samples of women and three random samples of men are drawn from the population being studied. The three weight reducing programs are diet (D), exercise (E), and diet plus exercise (DE). Each sample is also either male (M) or female (F). The number of pounds lost per subject in the trial period is shown below for each of the six samples. Using the two-way ANOVA, test the main effects (row and column differences) and the interaction effect.

Individual	MD	ME	MDE	FD	FE	FDE
1	13	11	20	8	6	6
2	8	14	25	18	7	10
3	10	11	16	17	12	10
4	13	12	15	22	9	13
5	13	10	12	16	6	18
6	14	10	9	14	7	11
7	12	5	17	12	8	12
8	13	8	14	16	4	14
9	18	13	15	17	6	13
10	12	11	16	16	2	11

    a. State the null and alternative hypotheses for each factor and interaction between factors.

    b. Compute ΣX (sum of all scores), ΣX^2 (sum of all square scores), and Σx^2 (total sum of squares).

    c. Compute N, the total number of observations.

    d. Compute ΣX and ΣX^2 for each cell. Note: there are $k = R \times C$ cells where R = number of rows and C = number of columns.

    e. Find Σx^2 (sum of squared deviation scores) for each cell using the values computed in d.

    f. Add all of the Σx^2 in e to get the within-groups sum of squares.

    g. Find the within-groups degrees of freedom and calculate the within-groups variance estimate, s_W^2.

    h. Compute the cell means, each row mean, each column mean, and the grand mean.

    i. Compute the row sum of squares.

    j. Find the degrees of freedom for rows and compute the between-rows variance estimate, s_R^2.

    k. Compute the column sum of squares.

    l. Find the degrees of freedom for columns and compute the between-column variance estimate, s_C^2.

    m. Find the interaction sum of squares.

    n. Find the degrees of freedom for interaction and compute the interaction variance estimate.

    o. Calculate the F ratios for the row effect, the column effect, and interaction.

p. Find the critical F values, $F_{.05}$ and $F_{.01}$.

q. Determine the significance of each effect by comparing the critical value of F with the obtained value of F.

r. Compute the ANOVA summary table below:

Source	Sum of squares	df	Mean square	F value
Between-rows				
Between-columns				
Interaction				
Within-groups				
Total				

11–2.   In a certain coed liberal arts college, a controversy has arisen about the relative scholarly capacities of males and females and sorority and fraternity members as opposed to nonorganization students. To test these hypotheses, four random samples of 12 cases each are taken from senior fraternity men (MO), senior nonfraternity men (MN), senior sorority women (FS), and senior nonsorority women (FN). Grade point averages for these students are computed and rounded off to one decimal place and multiplied by 10 to give the scores below. Using the two-way ANOVA test the main effects for rows (male vs. female), columns (org. vs. nonorg.), and interaction.

Individual	MO	MN	FS	FN
1	24	29	31	29
2	25	28	27	35
3	20	32	30	33
4	27	23	26	31
5	37	29	29	39
6	30	27	27	27
7	23	30	33	32
8	27	26	25	30
9	26	36	28	21
10	31	30	30	32
11	28	29	35	31
12	25	33	28	30

a. State the null and alternative hypotheses for each factor and interaction between factors.

b. Compute ΣX, ΣX^2, and Σx^2.

c. Compute N, the total number of observations.

d. Compute ΣX and ΣX^2 for each cell. Note: there are $k = R \times C$ cells where R = number of rows and C = number of columns.

e. Find Σx^2 for each cell using the values computed in (d).

f. Add all of the Σx^2 in (e) to get the within-groups sum of squares.

g. Find the within-groups degrees of freedom and calculate the within-groups variance estimate, s_W^2.

h. Compute the cell means, each row mean, each column mean, and the grand mean.

i. Compute the row sum of squares.

j. Find the degrees of freedom for rows and compute the between-rows variance estimate, s_R^2.

k. Compute the column sum of squares.

l. Find the degrees of freedom for columns and compute the between-column variance estimate, s_C^2.

m. Find the interaction sum of squares.

n. Find the degrees of freedom for interaction and compute the interaction variance estimate.

o. Calculate the F ratios for the row effect, the column effect, and interaction.

p. Find the critical F values, $F_{.05}$ and $F_{.01}$.

q. Determine the significance of each effect by comparing the critical value of F with the obtained value of F.

r. Compute the ANOVA summary table below:

Source	Sum of squares	df	Mean square	F value
Between-rows				
Between-columns				
Interaction				
Within-groups				
Total				

11–3. If justified, make t-tests for the data in question 11–1, computing the Scheffé critical values $t'_{.05}$ and $t'_{.01}$. Interpret the results for the entire experiment.

a. Determine if the F-test in the ANOVA is significant for at least one effect. If one is significant, then proceed with (b), otherwise stop at this point and indicate that the F-test is not significant.

b. Using Formulas 11.5, 11.6 and 11.9, compute the t_{ij} values for all possible pairs of subgroup means.

c. Using Formulas 11.7 and 11.8, find the Scheffé $t'_{.05}$ and $t'_{.01}$.

d. Compare the computed t_{ij} value with the Scheffé t' values and determine which pair or pairs of means are significantly different.

11–4. Perform the Tukey HSD test for the six group means of the data in question 11–1.

a. Determine if the F-test is significant in the ANOVA for at least one effect. If one F-test for the overall difference between means is significant, proceed with (b), otherwise stop at this point and indicate that multiple comparisons cannot be performed on a nonsignificant F.

b. Form D_{ij}, the absolute differences between the means of groups i and j, subgroups i and j, or cell means i and j.

c. Compute the Tukey HSD critical value for 0.05 and 0.01 using Formula 10.12 and table M of Appendix B.

d. Compare the D_{ij} values in (b) with the critical values found in (c) and determine which pairs of group means are statistically different.

11–5.   For the data in question 11–1, perform the Tukey HSD test for the means between the reducing programs, holding sex constant.

    a. Determine if the F-test is significant in the ANOVA for at least one effect. If one F-test for the overall difference between means is significant, proceed with (b), otherwise stop at this point and indicate that multiple comparisons cannot be performed on a nonsignificant F.

    b. Form D_{ij}, the absolute differences between the means of groups i and j, subgroups i and j, or cell means i and j.

    c. Compute the Tukey HSD critical value for 0.05 and 0.01 using Formula 10.12 and table M of Appendix B.

    d. Compare the D_{ij} values in (b) with the critical values found in (c) and determine which pairs of group means are statistically different.

11–6.   If justified, make t-tests for the data in question 11–2, computing the Scheffé critical value $t'_{.05}$ and $t'_{.01}$. Interpret the results for the entire experiment.

    a. Determine if the F-test in the ANOVA is significant for at least one effect. If one is significant, then proceed with (b), otherwise stop at this point and indicate that the F-test is not significant.

    b. Using Formulas 11.5, 11.6, and 11.9, compute the t_{ij} values for all possible pairs of subgroup means.

    c. Using Formulas 11.7 and 11.8, find the Scheffé $t'_{.05}$ and $t'_{.01}$.

    d. Compare the computed t_{ij} value with the Scheffé t' values and determine which pair or pairs of means are significantly different.

11–7.   Four different instructional methods are used to teach seventh-grade students at an all-girls junior high school. Twelve seventh-grade students in each of three different English classes are randomly selected and assigned to the four methods of instruction in each classroom. At the end of the semester, an achievement test is administered to all seventh-grade students. The test scores for those students that participated in the study are isolated and recorded below. Develop the appropriate hypothesis test to test a difference between classrooms, a test for the difference among instructional methods, and the existence of an interaction between classroom and instructional method.

Classroom	Instructional Methods			
	T_1	T_2	T_3	T_4
C_1	64	67	64	76
	66	70	72	81
	62	74	69	74
C_2	79	72	83	79
	86	79	72	85
	82	90	82	71
C_3	80	75	76	66
	85	77	72	75
	82	73	74	65

a. State the null and alternative hypotheses for each factor and interaction between factors.
b. Compute ΣX, ΣX^2, and Σx^2.
c. Compute N, the total number of observations.
d. Compute ΣX and ΣX^2 for each cell. Note: there are $k = R \times C$ cells where R = number of rows and C = number of columns.
e. Find Σx^2 for each cell using the values computed in (d).
f. Add up all of the Σx^2 in (e) to get the within-groups sum of squares.
g. Find the within-groups degrees of freedom and calculate the within-groups variance estimate, s_W^2.
h. Compute the cell means, each row mean, each column mean, and the grand mean.
i. Compute the row sum of squares.
j. Find the degrees of freedom for rows and compute the between-rows variance estimate, s_R^2.
k. Compute the column sum of squares.
l. Find the degrees of freedom for columns and compute the between-column variance estimate, s_C^2.
m. Find the interaction sum of squares.
n. Find the degrees of freedom for interaction and compute the interaction variance estimate.
o. Calculate the F ratios for the row effect, the column effect, and interaction.
p. Find the critical F values, $F_{.05}$ and $F_{.01}$.
q. Determine the significance of each effect by comparing the critical value of F with the obtained value of F.
r. Compute the ANOVA summary table below:

Source	Sum of squares	df	Mean square	F value
Between-rows				
Between-columns				
Interaction				
Within-groups				
Total				

Chapter 12

Introduction to
Nonparametric Methods

I n the previous chapters (except Chapter 9), the inferential statistical methods described involve testing hypotheses about population parameters. The sampling distribution is usually assumed to have a particular form, normal or binomial. In research, many variables under study warrant such an assumption. However, there are situations where the underlying distribution of the data are not known and assumptions about the distribution are questionable. Nonparametric and distribution-free tests have been developed that are less restrictive both about what is measured and the form of the sampling distribution. Kirk (1984) and Ury (1967) more accurately call these tests "assumption-freer tests." These assumption-freer tests usually use either ranks or frequency counts. Parametric tests are concerned with hypotheses about the population parameters. The nonparametric tests do not involve hypotheses about the population parameters, but instead are concerned with the *form* of the population frequency distribution. Distribution-free tests are not concerned with the form of the population frequency distribution. For example, the tests concerning μ (mu), σ (sigma), and ρ (rho for correlation) are parametric. A hypothesis that the distributions of two populations are the same is nonparametric. Likewise, a hypothesis test that a population possesses a normal distribution without any specification of parameter values is nonparametric.

nonparametric statistics Statistical procedures that involve no hypothesis about population parameters.

The calculation and measuring methods used in **nonparametric** test **statistics** are much simpler than those used in parametric tests. There is, however, a trade-off in using nonparametric methods over parametric ones. The nonparametric methods do not make use of the magnitude of information that parametric methods commonly use. With this loss of information, nonparametric methods tend to be less efficient than parametric methods. That is, the use of nonparametric methods tend to lead to rejection of the null hypothesis less often than they should when compared with the equivalent parametric methods. However, if a nonparametric method is used on parametric data and the result is a rejection of the null hypothesis, then the use of the equivalent parametric method on the same set of data also leads to the rejection of the null hypothesis. What this means is that nonparametric methods are almost as efficient as their equivalent parametric methods when the assumptions underlying the parametric methods are true.

Many of these tests use their own set of tables to determine an appropriate critical value for the hypothesis test. In many of these nonparametric tests, as the sample size gets large a Z statistic is computed as the test statistic and evaluated using the standard normal curve table. Entire books and many research papers have been written about nonparametric methods. This chapter only introduces a few of the methods and is far from being comprehensive.

sign test A nonparametric test based on replacing all observed values by either $+$ or $-$, since the values observed are larger or smaller than some specified value, and then testing the resulting binomial distribution.

The first nonparametric method focused on in this chapter is the sign test. The sign test is considered one of the simplest of nonparametric tests. The **sign test** is used to compare two populations using data that are either paired or unpaired. The form of the null hypothesis for a sign test is an implication that the two population distributions are identical. This is another way of saying that the median of a population of differences scores is zero.

PROBLEM **45**

Comparing Two Populations with Either Paired or Unpaired Data Using the Sign Test

The first step of the sign test procedure is to pair the scores from the two groups. If the two groups are not matched groups, for example not twins, pair them randomly. After completing this operation, the signs of the differences are noted. For instance, if a researcher wants to compare some characteristics of group A with those of group B, then values in group A are first paired with the values in group B. Those pairs that have equal score values are dropped. The differences between the paired values are determined and the *signs* of the differences are noted. Whether the group A value is subtracted from the group B value, or *vice versa*, is not important, but it must be done the same way for all pairs.

The number of plus signs and the number of minus signs are recorded (hence the name sign test). Depending on which is smaller (the number of pluses or the number of minuses), the value of the test statistic V is set equal to that value. To obtain the critical value necessary to evaluate the test statistic, table G in Appendix B is used. To enter this table, the sample size, *n*, and the level of significance is required. The null hypothesis, in this case, states that the median of the differences between group A and group B is zero. If the test statistic is less than or equal to the critical value found from the table, the null hypothesis is rejected. Otherwise, the null hypothesis is not rejected.

Example An efficiency expert collects data concerning the effectiveness of the use of two types of typewriters in a business environment. Eleven typists are selected to participate in the study. The data are reported in the form of performance percentage, the ratio of the number of words typed correctly to the total number of words typed (see table 12.1).

TABLE 12.1

Typewriter effectiveness data

Typist	Manual %	Electric %	Manual–Electric %	Sign
1	65	86	−21	−
2	95	94	+ 1	+
3	98	97	+ 1	+
4	85	98	−13	−
5	99	99	0	none
6	95	98	− 3	−
7	73	88	−15	−
8	79	93	−14	−
9	97	98	− 1	−
10	80	96	−16	−
11	90	92	− 2	−

The number of pluses is 2. The number of minuses is 8. One typist performed equally well on both types of typewriters. The test statistic is equal to:

$$V = \min (\text{number "+"}, \text{number "−"}) \qquad (12.1)$$

Whichever is smaller, the number of "pluses" or number of "minuses," is set equal to V. For this example $V = 2$ and $n = 10$ (the typist with no differences is dropped), and the level of significance equals 0.05. The critical value found by entering table G in Appendix B is 1.0. Therefore, if V is less than or equal to 1.0, the null hypothesis of no difference in performance between the two types of typewriters is rejected. V, in this example however, is equal to 2 and 2 is greater than 1, so the null hypothesis is not rejected. There is not sufficient evidence that there are any real differences in performance between the two types of typewriters. For large samples, where table G in Appendix B is no longer useful, a Z statistic is computed using the following formulas:

If $V < .5n$, then

$$Z = \frac{(V + .5) - (.5n)}{.5\sqrt{n}} \qquad (12.2)$$

If $V > .5n$, then

$$Z = \frac{(V - .5) - (.5n)}{.5\sqrt{n}} \qquad (12.3)$$

The significance of the Z value is determined by using table B in Appendix B. Depending on whether V is larger or smaller than one-half the sample size, Formulas 12.2 or 12.3 are used to transform the V value to a Z value.

Example   One hundred people are asked to perform a simple problem-solving task under two conditions, normal and stressful. Blood pressure data are gathered on each person under each condition. Differences are formed between the pair of blood pressure readings, normal − stressful. There were 9 pluses, 88 minuses, and 3 no differences. Dropping the 3 no differences, $n = 100 − 3 = 97$. V, the test statistic, is min $(9, 88) = 9$. Using the 0.05 level of significance, there is no entry in table G of Appendix B for these values. Hence, the large sample estimate, Z, must be used.

$$.5n = .5 \times 97 = 48.5$$

$$V = 9, \text{ and is less than 48.5, so use Formula 12.2.}$$

$$Z = \frac{(V + .5) - (.5n)}{.5\sqrt{n}}$$

$$Z = \frac{(9 + .5) - (.5)(97)}{.5\sqrt{97}}$$

$$Z = \frac{(9.5 - 48.5)}{(.5)(9.8488578)}$$

$$Z = \frac{-39.0}{4.9244} = -7.9197 = -7.92$$

Referring to table B of Appendix B, the probability of observing a Z value of this magnitude is 0.0001. Since 0.0001 is smaller than 0.05, the hypothesis of no difference between different levels of stress is rejected. That is, there *is* evidence in this case that difference in stress levels causes differences in blood pressure. Using the alternative approach, the critical Z value for $\alpha = 0.01$, two-tailed, is ± 2.58. Since -7.92 is less than -2.58, the hypothesis of no difference is rejected.

SUMMARY OF STEPS

Comparing Two Populations Using the Sign Test

Step 1   Form the pairs of data.

Step 2   Determine the sign of the difference between the two values of each pair.

Step 3   Compute the value of n by counting the total number of pairs having either a plus or minus sign. Those with no signs are dropped.

Step 4   Count the number of plusses and the number of minuses.

Step 5   Set the test statistic V equal to the smaller count number.

Step 6   To find the critical value for evaluating V, use table G in Appendix B if n is less than or equal to 90, otherwise the Z value is computed for V and evaluated against the appropriate critical value obtained from the standard normal curve table (table B of Appendix B).

Step 7   If n is less than or equal to 90, then reject the null hypothesis if V is less than or equal to the critical value obtained in table G of Appendix B for n and either 0.05 or 0.01 levels of significance. If n is greater than 90, the V is converted to a Z value using Formula 12.2. The probability of that Z value occurring by chance is computed by using table B of Appendix B. If the probability is less than 0.05 or 0.01 (this choice is up to the researcher), reject the null hypothesis. Otherwise do not reject the null hypothesis.

Table H in Appendix B can handle sample sizes of 20 or less. For larger sample sizes, a Z value is generated and evaluated using the standard normal curve table (table B, Appendix B). This is the same procedure used to evaluate the sign test for large samples. In the Mann-Whitney U test, the large sample test statistic is

$$Z = \frac{U - \frac{nm}{2}}{\sqrt{\frac{nm(n + m + 1)}{12}}} \qquad (12.6)$$

The decision rule is to reject the null hypothesis, h_0, if the probability of observing this Z value by chance is less than or equal to 0.05 or 0.01. This formula is based on the expectation that when two population distributions are identical, $\mu_U = \frac{nm}{2}$ and $\sigma_U^2 = \frac{nm (n + m + 1)}{12}$.

Example   A real estate developer wants to know whether the property values of two census tracts differ. The property values of 25 homes in tract X are assessed, and 23 homes in tract Y are assessed. For the sake of brevity, only a summary of the data is presented here. The assessed values of the homes are ranked from 1 to 48. The value of s from Formula 12.4 is computed to be 741. The U statistic is

$$U = 741 - \frac{(25)(26)}{2} = 741 - 325 = 416$$

Using Formula 12.6 because the sample is too large for table H, gives

$$Z = \frac{416 - \frac{(25)(23)}{2}}{\sqrt{\frac{(25)(23)(25 + 23 + 1)}{12}}} = \frac{416 - 287.5}{\sqrt{2347.92}}$$

$$= \frac{128.5}{48.46} = 2.65$$

The probability associated with observing this Z value by chance is less than 0.001, which is less than either 0.05 or 0.01, so the null hypothesis is rejected. There is evidence that the relative frequency distributions of the two samples are from different populations.

Ties in the observations are handled by averaging the ranks that would have been assigned to the untied observations and assigning this average to each. Suppose two observations are tied. If they are not tied they receive ranks 7 and 8. However, since they *are* tied, the mean of the two ranks (7.5) is assigned to each observation. If there are many ties, do not use the Mann-Whitney U test.

The Mann-Whitney U test is designed to determine whether two random samples are from the same population. If the difference between the two population distributions is assumed to be due to a shift in location of the two distributions, then the Mann-Whitney U test is equivalent to testing whether or not the two population means are equal.

SUMMARY OF STEPS

Comparing Two Population Distributions Using the Mann-Whitney U Test

Step 1   Combine the two samples of size n and m, and then assign ranks 1 to $(n + m)$ to each observation. For observations that are tied (having the same value), compute the average rank and assign this average to each of the tied observations.

Step 2   For the first sample (or group), add the appropriate ranks (see Formula 12.4).

Step 3   Compute the test statistic U by using Formula 12.5.

Step 4   If the sample sizes n and m are both less than or equal to 20, then the critical value is found by entering table H, Appendix B, using n, m, and α or $\alpha/2$. If the sample size for either n or m is larger than 20, then the test statistic U is transformed into a Z value using Formula 12.6. Like the Z values in Chapters 5 and 7, this Z value is evaluated by finding the probability of the value occurring by chance.

Step 5   Entering table H, the null hypothesis is rejected if the value of U is less than $U_{(\alpha/2)}$ found in the table, or U is greater than $U_{(1-\alpha/2)}$. Otherwise the null hypothesis is not rejected. $U_{(1-\alpha/2)}$ is found by subtracting $U_{(\alpha/2)}$ from $n \times m$. To evaluate the Z value for large samples ($n > 20$), the probability of the Z value is found in table B, Appendix B. If the tabled value is less than 0.05 or 0.01, the null hypothesis is rejected.

PROBLEM 47

Comparing Two Population Distributions That Are Correlated Using the Wilcoxon Rank-Sum Test

Wilcoxon rank-sum test   A nonparametric test used to determine whether data (paired) collected from a within-subjects design is statistically different.

The **Wilcoxon rank-sum test** is used to analyze data from two samples when the samples are correlated. This test is the nonparametric equivalent to the parametric test of differences between two correlated means (Problem 31, Chapter 8). The test assumes that the data consist of n matched pairs. To represent the difference between the ith pair, d_i is used. The d_i's are assumed to be continuous random variables and the distribution of each d_i is symmetrical. Like the Mann-Whitney U test, the null hypothesis states that the relative frequency distributions for the two populations are the same. The test statistic is found by executing the following procedure.

For each pair, determine the difference score. If X_i is a score for the first group on the ith pair and Y_i is a value for the second group on the ith pair, then d_i is $X_i - Y_i$. If d_i is equal to zero, it is dropped from the test and the number of pairs, n, is decreased by 1. The absolute values of the d_i are ranked. If any ties occur, the ranks of the items involved in the tie are averaged and the average is used as the rank for each tied item. After the ranks are assigned, the sign associated with each difference score is noted. For those with a positive sign $(+)$, add the associated ranks. Let this sum be set equal to r^+. For those differences with a negative sign $(-)$, add the associated ranks. Let this sum be set equal to r^-. The test statistic is either r^+ or r^-, depending on which is smaller. That is

$$W = \text{minimum} \ (r^+, r^-) \tag{12.7}$$

The decision rule for this test is to reject the null hypothesis, h_0, at the α level of significance if W exceeds $W_{(1 - \alpha/2)}$ or W is less than $W_{(\alpha/2)}$, where $W_{(\alpha/2)}$ is found by entering table I in Appendix B. $W_{(1 - \alpha/2)}$ is found by computing: $\dfrac{n(n + 1)}{2} - W_{(\alpha/2)}$. Otherwise, the null hypothesis is not rejected. The decision rule is for a two-tailed test. For a one-tailed test, regardless of right or left, reject h_0 (the null hypothesis) if W is less than or equal to W_α.

Example   In a product recognition study, the amount of time (in seconds) is recorded for each of two differently colored advertising layouts. Individually, 8 people are subjected to both layouts in random order. The data for this study are in table 12.3.

The sum of the ranks for all those with positive differences is

$$r^+ = 2 + 4.5 + 7 + 8 = 21.5.$$

The sum of the ranks for all those with negative differences is

$$r^- = 2 + 6 + 2 + 4.5 = 14.5.$$

The test statistic is the minimum or smaller of r^+ and r^-,

$$W = \text{min} \ (21.5, 14.5) = 14.5.$$

Entering table I in Appendix B with n, the number of pairs and the level of significance ($.05/2 = .025$ for the two-tailed test), the critical value for $W_{.025}$ is 4.

$$W_{(1 - \alpha/2)} = \frac{n(n + 1)}{2} - W_{(\alpha/2)} = \frac{(8)(9)}{2} - 4 = 36 - 4 = 32.$$

STATISTICS IN THE WORLD AROUND YOU

Energy experts have estimated the consumption of petroleum in automobiles and the amount of raw resources left in the world. If an alternative form of energy is not found, the world's supply of oil crude will soon run out. Researchers have examined the possibility of using alcohol, natural gas, and butane gas as alternatives. The country of Brazil has used a combination of alcohol and gasoline, called gasohol, for a number of years. Tests have been conducted to compare the performance of gasohol with conventional gasoline. Gasohol, if effective, will allow petroleum companies to stretch the supply of petroleum.

In a study to examine the difference in performance between gasoline and gasohol, 20 police cars in a large metropolitan city were used. Ten cars were run for a period of time with ordinary gasoline and then later with gasohol. The other 10 cars were run for a period of time with gasohol and then switched to ordinary gasoline. An index of performance was computed using average miles per gallon, the wear on engine parts, the smoothness of engine performance, ease of cold start-up, and quickness in acceleration. This index is, at best, an indication of rank in performance so parametric methods of statistical analysis are not the most appropriate. Since each police car was measured twice, once using gasoline and again using gasohol, the data for each police car are paired. The most appropriate method of statistical analysis for this problem is the Wilcoxon rank-sum test. The results of this test will indicate whether a difference exists between gasoline and gasohol in terms of preference.

TABLE 12.3

Product recognition data

Person	Layout 1	Layout 2	d	|d|	Rank of |d|
1	4	3	1	1	2
2	3	1	2	2	4.5
3	1	2	−1	1	2
4	5	1	4	4	7
5	2	5	−3	3	6
6	6	1	5	5	8
7	4	5	−1	1	2
8	1	3	−2	2	4.5

Reject h_0 if the test statistic W is greater than 32 or less than 4. The calculated W value is 14.5, which is neither greater than 32 nor less than 4. Therefore, the null hypothesis is not rejected. There is insufficient evidence that the time of recognition for each type of layout is different. For a one-tail test (regardless of right or left), the critical value is W_α found in Appendix B, table I for n pairs.

Table I in Appendix B is only usable if the study consists of 20 or fewer pairs. If there are more than 20 pairs, the test statistic, W, is transformed into a Z value. This Z value is evaluated using the standard normal curve table (table B, Appendix B). If the probability is less than 0.05 or 0.01, the null hypothesis is rejected.

The test statistic for the Wilcoxon rank-sum test with large samples (if $n > 20$), is

$$Z = \frac{W - \dfrac{n(n + 1)}{4}}{\sqrt{\dfrac{n(n + 1)(2n + 1)}{24}}} \tag{12.8}$$

This formula is based on the assumption that for large samples, the W statistic is approximately normally distributed with mean and standard deviation of $\mu_W = \dfrac{n(n + 1)}{4}$, $\sigma_W \sqrt{\dfrac{n(n + 1)(2n + 1)}{24}}$.

Example   Two methods of appraising property values are compared at each of 52 single family homes in a major area of the city. Differences between the estimated property values are formed and ranked according to the Wilcoxon procedure. The value of W, the smaller rank sum, is 449. The test statistic is:

$$Z = \frac{449 - \dfrac{(52)(53)}{4}}{\sqrt{\dfrac{(52)(53)(105)}{24}}} = \frac{449 - 689}{\sqrt{12057.50}} = \frac{-240}{109.81} = -2.19$$

Using the standard normal curve table given in table B of Appendix B, the probability of observing this Z value by chance is 0.0143 for a one-tailed test, and 0.0286 (2 × 0.0143) for a two-tailed test. Since these probabilities are less than 0.05, the null hypothesis is rejected at the 0.05 level of significance.

SUMMARY OF STEPS

Comparing Two Population Distributions That Are Correlated Using the Wilcoxon Rank-Sum Test

Step 1   For each pair, compute the difference between the pairs.

Step 2   Ignoring the signs, find the ranks for each difference score. For tied scores, assign the average of the tied ranks.

Step 3   Add up the ranks that correspond to all the positive difference scores found in Step 1.

Step 4   Add up the ranks that correspond to all the negative difference scores found in Step 1.

Step 5   Determine W, which equals the smaller of the sums found in Steps 3 and 4.

Step 6   If n is less than or equal to 20, enter table I, Appendix B with n and α, the level of significance, to find the critical value. If W is less than the critical value, $(W_{\alpha/2})$ or greater than $W_{1-\alpha/2}$, reject h_0.

Step 7   If n is larger than 20, the test statistic is computed using Formula 12.8. This Z value is evaluated by finding the probability that it occurred by chance. If the probability is less than 0.05, or 0.01, then reject h_0. If the test is a one-tailed test, the probability of the Z value is used without modification. On a two-tailed test, the probability is doubled before being compared with 0.05 or 0.01.

PROBLEM 48

Compute the Spearman Rank-Order Correlation and Significance Test

Spearman correlation   A correlation coefficient computed on two variables measured on an ordinal (rank order) scale.

The **Spearman rank-order correlation,** or the Spearman rho, is a nonparametric equivalent to the Pearson product-moment correlation coefficient described in Chapter 6. As is the case with the parametric correlation method, the Spearman rho's values range from -1.00 to $+1.00$. A coefficient of zero indicates no linear relationship or association between the two paired sets of data. Similarly, the Spearman rho can be subjected to significance tests to see if the obtained rho value is statistically different from zero.

For this measure of association, both variables' values are ranked separately from 1 to n, where n is the number of pairs. For variable Y, a set of ranks are found, $r(y_i)$. The same thing is done for variable X to get the ranks $r(x_i)$. Ties are handled in the usual way previously described. Once the two sets of ranks are found, the value of rho is found with the following formula:

$$\text{rho} = 1 - \frac{6\Sigma d_i^2}{n(n^2 - 1)} \tag{12.9}$$

where $d_i = r(y_i) - r(x_i)$, or the difference between the ranks, and

$n =$ the number of pairs.

If there are no tied ranks, the Formula 12.9 is a simplification of using ranks $[r(y_i), r(x_i)]$ in the Pearson correlation coefficient formula given in Chapter 6. Instead of scores for X and Y, the ranks are used. Formula 12.9 is used with tied ranks as long as there are not very many ties. The shortcut Formula of 12.9 yields misleading results if it is used for data containing too many tied ranks. The mathematical equivalence between Formula 12.9 and the Pearson coefficient using ranks is shown in Siegel (1956).

To test whether rho is significantly different from zero, a table has been prepared to give the appropriate critical value for a given number of pairs of data points and the level of significance. Table J in Appendix B contains the critical values for $\alpha = 0.05, 0.025, 0.01,$ and 0.005 for one-tailed tests. By doubling these α values, the table gives the critical values for two-tailed tests. For example, the critical value of 0.829 for 6 pairs is used to test a one-tailed hypothesis at the 0.05 level or a two-tailed hypothesis at the 0.10 level of significance.

If the absolute value of the computed rho is greater than or equal to the critical value, the null hypothesis of no linear correlation between the two sets of scores is rejected. This means that the observed correlation or association is large enough to be considered different from zero.

Example   A school psychologist suspects that those students who have a higher grade point average also have higher scores on a national aptitude test. To show this, the data from 12 randomly chosen students are gathered from school records (table 12.4).

In this example the data are ranked; the GPA values are ranked from 1 to 12 and the aptitude test scores are ranked from 1 to 12. The differences between the ranks (d_i) are computed, and the differences are squared (d_i^2). Using Formula 12.9, the value of the Spearman rho is computed. The sum of the differences squared is 88.

$$\text{rho} = 1 - \frac{(6)(88)}{(12)(144 - 1)} = 1 - \frac{528}{1716} = 1 - 0.3077 = 0.6923$$

TABLE 12.4

GPA/National Aptitude Test correlation data

Student	GPA (Max = 4.00)	Rank	Score (Max = 800)	Rank	d_i	d_i^2
1	2.95	(6)	550	(9)	−3	9
2	3.07	(5)	605	(6)	−1	1
3	2.20	(11)	470	(11)	0	0
4	3.36	(3)	610	(5)	−2	4
5	1.90	(12)	430	(12)	0	0
6	3.75	(2)	620	(4)	−2	4
7	2.76	(8)	510	(10)	−2	4
8	3.88	(1)	750	(1)	0	0
9	2.58	(10)	690	(2)	8	64
10	2.61	(9)	555	(8)	1	1
11	2.84	(7)	590	(7)	0	0
12	3.13	(4)	660	(3)	1	1
					Sum =	88

To determine whether rho is significantly different from zero, enter table J in Appendix B with $n = 12$ pairs and 0.05 two-tailed (equivalent to 0.025 one-tailed). From the table, the critical value is 0.591 (0.025 column, $n = 12$ row). The absolute value of rho is 0.6923. If rho is larger than 0.591, then the null hypothesis of no linear relationship is rejected. In this example, rho = 0.6923, and rho is larger than the critical value, 0.591. Hence, the null hypothesis is rejected. There appears to be a real relationship between these two variables.

Since the data for this example did not have any tied ranks, the use of the ranks in the Pearson correlation formula given in Chapter 6 (Formula 6.14) gave the same value as the rho Formula 12.9.

$$\text{rho} = \frac{n\Sigma r(X_i Y_i) - \Sigma r(X_i)\, \Sigma r(Y_i)}{\sqrt{n\Sigma r(X_i)^2 - [\Sigma r(X_i)]^2}\, \sqrt{n\Sigma r(Y_i)^2 - [\Sigma r(Y_i)]^2}} \qquad (12.10)$$

where $\Sigma r(X_i) = \Sigma r(Y_i) = 78,$

$\Sigma r(X_i)^2 = \Sigma r(Y_i)^2 = 650,$

$\Sigma r(X_i Y_i) = 606,$ and

$n = 12.$

$$\text{rho} = \frac{[12(606) - 78(78)]}{[(12)(650) - (78)^2]} = \frac{1188}{1716} = 0.6923$$

SUMMARY OF STEPS

Computing the Spearman Rank-Order Correlation and Significance Test

Step 1   Find the ranks for the observations on the X variable.

Step 2   Find the ranks for the observations on the Y variable. If there are any tied ranks, the ranks are averaged and assigned to those observations that have the same value. Tied ranks for the X variable are handled in the same way.

Step 3   Compute the square of the difference for each pair of values.

Step 4   Compute the square of each paired difference score.

Step 5   Add up the squared differences and enter this sum into Formula 12.9 along with n = the number of pairs. Use Formula 12.9 only if the number of tied ranks is small, otherwise use Formula 12.10.

Step 6   To determine whether the value of rho is statistically different from zero, enter table J, Appendix B to find the critical value. The critical value is based on *n,* the number of pairs, α, the level of significance, and whether the hypothesis is one- or two-tailed. If the absolute value of rho is larger than the critical value, reject the null hypothesis of no linear relationship between the two variables.

The efficiency of the Spearman rho when compared with the more powerful Pearson correlation of Chapter 6 is about 91%. This means that if a correlation exists between variables X and Y in the population (assuming data are parametric), with 100 observations, rho reveals that true correlation at the same level of significance that the Pearson correlation reveals with 91 observations.

PROBLEM 49

Determining the Randomness of a Set of Observations Using the Runs Test

In some research studies, the researcher wishes to measure the characteristics of a population. For economical and practical reasons, when using inferential statistics, a sample from the population is evaluated rather than trying to study the whole population. By choosing an appropriate sample, the researcher has a reasonable level of confidence that the conclusions that are drawn from the sample tell something about the population. In many research studies, the samples are chosen randomly from the population. There are several reasons for doing this, the most important among them are the lack of bias and representativeness. In quality control studies, a test for randomness is used to examine whether the occurrences of defective manufactured items are random or due to the manufacturing process. If some trend exists, such as every tenth item being defective, the occurrence of defects is not random. Several methods have been developed to test the hypothesis that the occurrence is random. These methods involve the sequence or order in which the individual observations were obtained. One method for determining randomness is the runs test. A **run** is defined as a maximal succession of identical occurrences. A run is usually followed by and preceded by another run. For example, to determine whether a coin is fair or not, the total numbers of heads and tails are examined along with the sequence in which they occurred. It is then determined whether the sequence of heads and tails is random. Suppose a coin is tossed 24 times and the following sequence of heads (H) and tails (T) occurs:

run   A succession of identical letters or labels that is followed by and preceded by a different letter or label or no letter or label at all.

$$H\ H\ H\ T\ T\ T\ T\ T\ T\ H\ H\ H\ H\ H\ H\ H\ H\ H\ T\ T\ T\ T\ T\ T$$

The number of heads equals the number of tails (12). There are, however, four (4) runs. The first run consists of 3 heads (H), the next run 6 tails (T), followed by runs of 9 heads and 6 tails. The number of runs for this sequence is 4. At this point, it is a bit difficult to determine whether the sequence of heads and tails is random. The order of the events is important in the runs test, while the frequency of occurrence of individual observations is not important.

runs test   A nonparametric test used to determine if an order of individual items (e.g. runs) are randomly arranged.

The **runs test** simply involves counting the number of runs in a sequence of events and then finding the probability of observing that number of runs. Under conditions of the null hypothesis that the sequence is random, if the number of runs observed is not within a certain range, the null hypothesis is rejected. The runs test is a two-tailed test since the focus is on whether there is an unusually small or an unusually large number of runs. To find the two critical values, enter table K, Appendix B with n_1, the number of occurrences of the first type; n_2, the number of occurrences of the second type; α, the level of significance; and $1 - \alpha$. If R_t is the number of runs, then reject h_0 if R_t is greater than $C_{(1 - \alpha/2)}$ or R_t is less than $C_{(\alpha/2)}$. Otherwise, do not reject h_0.

$C_{(1 - \alpha/2)}$ and $C_{(\alpha/2)}$ are the critical values found from consulting table K.

Example   At a nonpartisan benefit, members from both major political parties are present. The number of Democrats (D) and Republicans (R) and the sequence of seating are recorded for one large table. A political researcher wants to determine whether or not the Republicans and Democrats are seated randomly. The sequence of seating is:

$$D\ D\ D\ R\ R\ D\ R\ R\ R\ D\ D\ D\ R\ R\ D\ D\ D\ D$$

From this sequence, there are 11 Democrats and 8 Republicans. There are 7 runs:

$$(D\ D\ D),\ (R\ R),\ (D),\ (R\ R\ R),\ (D\ D\ D),\ (R\ R),\ (D\ D\ D\ D)$$

$$\quad 1 \qquad\quad 2 \quad\ 3 \qquad 4 \qquad\quad 5 \qquad\quad 6 \qquad\quad 7$$

where $R_t = 7,$

n_1 = number of Republicans = 8, and

n_2 = number of Democrats = 11.

Using $\alpha = 0.05$, from table K, Appendix B, $C_{(1-\alpha/2)} = C_{.975} = 14$; $C_{(\alpha/2)} = C_{.025} = 6$. Reject the null hypothesis (h_0 the sequence is random) if $R_t > 14$ or $R_t < 6$. Since the computed R_t for this problem is 7, which is smaller than 14 but larger than 6, the null hypothesis is not rejected. This means that there is insufficient evidence that the seating sequence of Republicans and Democrats is not random.

The runs test is occasionally used to check the assumption in analysis of variance and regression that the residuals are independent (random).

For samples where n_1 and n_2 are greater than 20, a large sample Z value is computed as the test statistic. The evaluation of the Z value involves finding the probability that the Z value occurs by chance. If this probability is less than 0.05 or 0.01 then the null hypothesis is rejected. The formula for the large sample runs test is

$$Z = \frac{R_t - \left(\dfrac{2n_1n_2}{n_1 + n_2} + 1 \right)}{\sqrt{\dfrac{2n_1n_2(2n_1n_2 - n_1 - n_2)}{(n_1 + n_2)^2 (n_1 + n_2 - 1)}}} \tag{12.11}$$

where R_t = the number of runs, and

n_1 and n_2 = the number of occurrences.

Example   A true-false test is constructed by a teacher for use in an auto mechanics class. The test consists of 50 questions; 30 questions are true while the remaining 20 questions are false. The teacher has arranged the sequence of true-false answers in the following pattern:

T T F T F T T T T F T T F T T F T F T F T F T F F F T T F T F T

F T T T T F T F T F T F F T T T F T F T

The teacher wants to know if the sequence is random in order to minimize the probability of a student getting the right answer strictly by guessing the pattern of true and false answers.

The number of runs is 35, $R_t = 35$. The number true, n_1, equals 30. The number of false, n_2, is equal to 20. The number of n_1 is too large to use the tabled critical values, so the large sample Z (Formula 12.11) value is used.

$$Z = \frac{35 - \left(\frac{(2)(30)(20)}{30 + 20} + 1\right)}{\sqrt{\frac{[(2)(30)(20)][2(30)(20) - 30 - 20]}{(30 + 20)^2 (30 + 20 - 1)}}} = \frac{35 - \left(\frac{1200}{50} + 1\right)}{\sqrt{\frac{(1200)(1150)}{(2500)(49)}}}$$

$$= \frac{35 - 24 + 1}{\sqrt{\frac{1,380,000}{122,500}}} = \frac{10}{\sqrt{11.2653}} = \frac{10}{3.3564} = 2.9793 = 2.98$$

The probability of this Z value occurring under conditions of the null hypothesis is 0.0014 for one-tailed or 2(0.0014) = 0.0028 for two-tailed. Since 0.0028 is less than 0.05, the null hypothesis is rejected. The test result does not support the contention that the sequence of true and false answers is random.

SUMMARY OF STEPS

Determining the Randomness of a Set of Observations Using the Runs Test

Step 1   Place the n_1 and n_2 observations in their actual sequence of occurrences.

Step 2   Count the number of runs, R_t.

Step 3   If n_1 and n_2 are less than or equal to 20, use table K in Appendix B to find the appropriate critical values. To enter the table, the level of significance, n_1 and n_2, are necessary to find $C_{(1 - \alpha/2)}$ and $C_{(\alpha/2)}$, the critical values. Reject h_0 if R_t is greater than $C_{(1 - \alpha/2)}$ or R_t is less than $C_{(\alpha/2)}$. Otherwise, do not reject h_0.

Step 4   If n_1 and n_2 are greater than 20, determine the value of Z from Formula 12.11. The probability of occurrence for this Z value is computed. If the probability is less than 0.05 or 0.01, the null hypothesis is rejected.

PROBLEM 50

Test Several Independent Samples Using the Kruskal-Wallis Test

Kruskal-Wallis test   A direct extension of the Mann-Whitney U test for more than two independent samples (groups). A nonparametric equivalent to the one-way ANOVA.

The **Kruskal-Wallis test** is considered a nonparametric analogue to the parametric, one-way analysis of variance method presented in Chapter 10. Just as the one-way ANOVA is an extension of the two independent group t-test, the Kruskal-Wallis test is an extension of the Mann-Whitney U test. The Kruskal-Wallis test handles k independent groups or samples. Like the Mann-Whitney U test, the test uses ranks.

The first step for conducting a Kruskal-Wallis test is to arrange the raw data into appropriate groups. Next, ranks are assigned to all N observations without consideration of group membership. Let r_i be the sum of the ranks assigned to the ith sample. Let n_i be the sample size for the ith sample. If x_{ij} is the raw data value for the jth observation in the ith sample or group, then $r(x_{ij})$ is the rank assigned to the jth observation in the ith sample. From this, $r_i = \Sigma r(x_{ij})$ and $\bar{r} = r_i/n_i$ (this is the mean of the ranks for group i). If the $\bar{r}_i$ values are almost equal, it is reasonable to suppose that the null hypothesis (that the k samples are from the same population) is true: that is, the k population distributions are statistically equal. If the $\bar{r}_i$'s are not equal, then at least one population tends to yield a different distribution of observations than the remainder. Ties are handled using the average of the ranks.

The test statistic for the Kruskal-Wallis test is

$$KW = \left[\frac{12}{N(N+1)}\right]\left[\Sigma\frac{[r_i - .5n_i(N+1)]^2}{n_i}\right] \tag{12.12}$$

where $N =$ the total N in all groups used,

$r_i =$ the sum of the ranks for a given group, i, and

$n_i =$ the number of subjects in a given group, i.

Table L in Appendix B gives the critical values for exact levels of α. This table, however, is very limited in that it is used only for situations where the number of groups or samples, k, is equal to 3 and the sample sizes of the groups are less than or equal to 5. If the number of samples exceeds 3 and the sample sizes are larger than 5, then a χ^2 (chi square) distribution with $k - 1$ degrees of freedom is used to find the appropriate critical value. It can be proved mathematically that if the k samples are from the same or identical populations, then the KW statistic in Formula 12.12 is distributed theoretically as a chi square with $k - 1$ degrees of freedom. The sample sizes, n_i, have to be large in order for the KW statistic to approximate the chi square. When $k = 3$ and the number of cases in each of the 3 samples is 5 or less, the chi square approximation of the sampling distribution of KW is not very close. The null hypothesis is rejected if KW, the test statistic, is greater than the critical value found in table L or the critical value found in the χ^2 distribution in table D, Appendix B. Otherwise, the null hypothesis is not rejected.

Example   Three different diet plans make the same claim, their plan is the safest and most effective method for losing weight. To determine which plan, A, B, or C is the most effective, 15 overweight men are randomly chosen from a large group of overweight people. The 15 participants are randomly assigned to one of the three plans so that each plan has 5 participants. Each participant is instructed to follow the plan for a 6 month period. At the end, the percentage of body weight in pounds lost is recorded for each person. Table 12.5 presents the data.

TABLE 12.5

Percentage weight loss by diet plan

Person	1	2	3	4	5
Plan A	23 (11)	41 (3)	42 (1.5)	36 (5)	30 (7)
Plan B	20 (12)	24 (10)	25 (9)	26 (8)	*
Plan C	40 (4)	42 (1.5)	37 (6)	*	*

The "*" indicates unavailable data for this person because the person dropped out. To test the hypothesis that each plan is equally effective in reducing weight, the Kruskal-Wallis test is used. The level of significance is set at 0.05.

From Formula 12.12, the test statistic for this example is computed in the following manner. The persons in the diet study are rank ordered according to the percentage of body weight in pounds lost. The ranks are enclosed in parentheses next to the weight loss figures.

There are three samples, so $k = 3$. The sample sizes are $n_1 = 5$, $n_2 = 4$, and $n_3 = 3$. The sum of the ranks by sample group are

$$r_1 = 11 + 3 + 1.5 + 5 + 7 = 27.5,$$

$$r_2 = 12 + 10 + 9 + 8 \quad\quad = 39.0,$$

$$r_3 = 4 + 1.5 + 6 \quad\quad\quad = 11.5$$

$$N = 5 + 4 + 3 \quad\quad\quad\quad = 12$$

Using Formula 12.12,

$$KW = \left[\frac{12}{12(12 + 1)}\right]\left[\frac{[(27.5) - (.5)(5)(13)]^2}{5}\right.$$

$$\left. + \frac{[(39.0) - (.5)(4)(13)]^2}{4} + \frac{[(11.5) - (.5)(3)(13)]^2}{3}\right]$$

$$= \left[\frac{12}{156}\right]\left[\frac{25}{5} + \frac{169}{4} + \frac{64}{3}\right]$$

$$= (.0769)(5 + 42.25 + 21.33)$$

$$= (.0769)(68.58)$$

$$= 5.2738, \text{ or } 5.27$$

From table L in Appendix B, the critical value obtained is 5.6308. This value is found using $n_1 = 5$, $n_2 = 4$, $n_3 = 3$, and $\alpha = 0.05$. Since 5.27 is not larger than 5.6308, the null hypothesis is not rejected. If the χ^2 (chi square) distribution is used instead of table L, the critical value for $k - 1$ degrees of freedom for $\alpha = 0.05$ is 5.99. Since 5.27 is less than 5.99, the null hypothesis is not rejected.

SUMMARY OF STEPS

Test Several Independent Samples Using the Kruskal-Wallis Test

Step 1   Ignoring sample membership, rank all of the observations and then assign ranks to the observations. For tied values, the average rank is assigned.

Step 2   For each sample, sum the ranks.

Step 3   Compute the test statistic using Formula 12.12.

Step 4   If $k = 3$ and n_1, n_2, and n_3 is less than or equal to 5, use table L in Appendix B to determine the critical value for a specified level of significance. If the test statistic, KW, is larger than the critical value, the null hypothesis is rejected. If $k > 3$ or either n_1, n_2, n_3 is greater than 5, the critical value is found by entering the chi square distribution (table D, Appendix B) with $k - 1$ degrees of freedom. If the test statistic, KW, is larger than the chi square value, then the null hypothesis is rejected. Otherwise, the null hypothesis is not rejected.

MATHEMATICAL NOTES

The test statistic for the Kruskal-Wallis is Formula 12.12. In some textbooks such as Siegel (1956), Hollander and Wolfe (1973), Guilford and Fruchter (1977), and Kushner and DeMaio (1980), the test statistic is:

$$H = \left[\frac{12}{N(N + 1)} \right] \left[\sum \frac{r_i^2}{n_i} \right] - 3(N + 1) \tag{12.13}$$

Formulas 12.12 and 12.13 are equivalent even though they appear quite different. Either Formula 12.12 or 12.13 can be used for the Kruskal-Wallis test statistic. A set of algebraic steps to show that the two formulas are equivalent is given below. Formula 12.13 is derived from Formula 12.12.

$$KW = \frac{12}{N(N + 1)} \left[\sum \frac{[r_i - .5n_i(N + 1)]^2}{n_i} \right] \tag{12.12}$$

$$= \frac{12}{N(N + 1)} \left[\sum \frac{[r_i^2 - r_i n_i(N + 1) + .25n_i^2(N + 1)^2]}{n_i} \right]$$

$$= \frac{12}{N(N + 1)} \left[\sum \frac{r_i^2}{n_i} - \frac{r_i n_i(N + 1)}{n_i} + \frac{.25n_i^2(N + 1)^2}{n_i} \right]$$

$$= \frac{12}{N(N + 1)} \left[\sum \frac{r_i^2}{n_i} - \Sigma r_i(N + 1) + .25(N + 1)^2 \Sigma n_i \right]$$

$$= \frac{12}{N(N + 1)} \left(\Sigma\frac{r_i^2}{n_i} \right) - \frac{12}{N(N + 1)}[(N + 1) \, \Sigma r_i] \qquad (12.14)$$

$$+ \frac{12}{N(N + 1)} \, [.25(N + 1)^2 \Sigma n_i]$$

Note: $\Sigma n_i = N$, $\Sigma r_i = \dfrac{N(N + 1)}{2}$.

Using the information in the note above, Formula 12.14 is rewritten as:

$$\frac{12}{N(N + 1)} \, \Sigma\frac{r_i^2}{n_i} - \frac{12}{N(N + 1)} \left(\frac{(N + 1)(N + 1)(N)}{2} \right)$$

$$+ \frac{12}{N(N + 1)} \, (.25(N + 1)^2 N)$$

Further simplifications yield:

$$\frac{12}{N(N + 1)} \left(\Sigma\frac{r_i^2}{n_i} \right) - 6(N + 1) + 3(N + 1)$$

$$= \frac{12}{N(N + 1)} \left(\Sigma\frac{r_i^2}{n_i} \right) - 3 \, (N + 1) \qquad (12.13)$$

TABLE 12.6

When to use which statistical test

Type of data	One sample case	Two sample case		K sample case
		Dependent groups	Independent groups	
Parametric	t-test	t-test	t-test	ANOVA
Nonparametric	χ^2	Wilcoxon	Mann-Whitney	Kruskal-Wallis
		Sign Test	Sign Test	

References

Guilford, J. P., & Fruchter, B. (1977). *Fundamental statistics in psychology and education* (6th ed.). New York: McGraw-Hill.

Hollander, M., & Wolfe, D. A. (1973). *Nonparametric statistical methods.* New York: Wiley.

Kirk, R. E. (1984). *Elementary statistics* (2d ed.). Belmont, CA: Brooks/Cole.

Kushner, H. W., & DeMaio, G. (1980). *Understanding basic statistics.* San Francisco: Holden Day.

Siegel, S. (1956). *Nonparametric statistics for the behavioral sciences.* New York: McGraw-Hill.

Ury, H. (1967). In response to Noether's letter, "Needed—a New Name." *American Statistician, 21*(4), 53.

Exercises

12-1. Two real estate appraisers, John and Bobbi, are asked to independently appraise 11 properties. Their appraisals are given below in terms of $1000 units.

Property	1	2	3	4	5	6	7	8	9	10	11
John	56	92	87	67	70	59	100	120	66	112	134
Bobbi	62	99	80	67	75	62	95	110	68	99	135

Do the two appraisers differ in their assessed valuation of different properties? Use the sign test to test the hypothesis that the two appraisers are different.

a. State the null and alternative hypotheses.
b. Compute the difference between the two appraisers for each property.
c. Retain only the signs for the differences in (b) and add up the number with "−" sign and "+" sign.
d. Using Formula 12.1, find V.
e. Using V and n (number of properties), enter table G in Appendix B to find the critical value for $\alpha = 0.05$.
f. Reject h_0 if V is less than or equal to the critical value.

12-2. Use the Mann-Whitney U test to test the same hypothesis in question 12-1.

a. State the null and alternative hypotheses.
b. Find $n + m$ where n is the number of properties appraised by John and m is the number appraised by Bobbi.

c. For each $n + m$ value, assign a rank where the smallest number receives a rank of 1.

Property	1	2	3	4	5	6	7	8	9	10	11
John											
Bobbi											

d. Find s using Formula 12.4.
e. Find U using Formula 12.5.
f. Find the critical value $U_{(\alpha/2)}$ and $U_{(1 - \alpha/2)}$. Use table H.
g. Reject h_0 if $U < U_{.025}$ or $U > U_{(1 - .025)}$.

12–3. Using the data from question 12–1, compute the Spearman rank correlation and conduct a significance test on the coefficient.

a. State the null and alternative hypotheses.
b. Assign a rank from 1 to n to the first group and rank from 1 to m to the second group.

	1	2	3	4	5	6	7	8	9	10	11
John											
Bobbi											

c. Compute the difference between the ranks for each property and sum the difference scores.
d. Square each difference score and then sum the squares.
e. Compute rho using Formula 12.9.
f. Find the critical value using table J of Appendix B.
g. Reject h_0 if the absolute value of rho is greater than the critical value.

12–4. The following set of data represent odd and even digits from a table that is considered random. Test at the 0.05 level of significance to see if they are random.

E E E O O O E O O O O O E E E E E O E O E E O O O E E E E O

a. State the null and alternative hypotheses.
b. Compare the number of runs.
c. Find the two actual values (see page 375).
d. Reject h_0 if R_t is greater than or equal to $C_{.975}$ or less than $C_{.025}$.

12–5. An education researcher has collected data on academic performance of freshmen students from three community colleges. The data collected are given below:

College A	College B	College C
2.3	2.7	1.1
2.6	2.1	3.0
2.0	2.9	1.5
2.9	2.8	2.6
2.5	1.9	2.2
2.4		1.8
3.1		

Test the hypothesis that there is no difference in the distribution of academic performance between the students from the three colleges. Use $\alpha = 0.05$.

    a. State the null and alternative hypotheses.

    b. Rank the raw data without consideration of group membership, and find the sum of the ranks for each group.

    c. Compute KW, the test statistic using Formula 12.12.

    d. Find the critical value.

    e. Reject h_0 if KW is greater than the critical value.

12-6. Two samples of 5-year-old children are randomly chosen to test the educational effectiveness of two math-learning programs, X and Y. At the end of the course, each child is evaluated with an unstandardized test. Due to the questionable metric properties of the data, the test scores are given ranks. The data are given below:

Rank of child	Child's program	Rank of child	Child's program
1	X	9	X
2	X	10	Y
3	Y	11	X
4	X	12	Y
5	Y	13	Y
6	Y	14	Y
7	Y	15	X
8	X	16	X

Develop the appropriate test to test whether program Y is better than program X.

    a. State the null and alternative hypotheses.

    b. Add up the ranks for the children receiving program X and program Y.

        1. Program X: $\Sigma r_i =$

            $n_i =$

        2. Program Y: $\Sigma r_i =$

            $m_i =$

    c. Set s equal to the sum of the rank for program X.

    d. Compute U using Formula 12.5.

    e. Find the critical values $U_{(\alpha/2)}$ and $U_{(1-\alpha/2)}$.

    f. Reject h_0 if $U < U_{.025}$ or $U > U_{.975}$.

12-7. In an electronic component manufacturing company, the number of defective components is recorded daily for each of two production lines, A and B. The results are given below:

Day	1	2	3	4	5	6	7	8	9	10
Line A	100	150	172	123	145	124	167	112	109	187
Line B	105	156	190	104	178	108	188	132	120	152

Compare the number of defective components produced by the two lines. Let x denote the number of times when Line B exceeds Line A. Conduct an appropriate test to see if Line B is producing more defective components, on the average, than Line A. Use the sign test.

    a. State the null and alternative hypotheses.

    b. Compare the difference score per day between lines A and B.

    c. Determine the number of differences that received a minus sign and plus sign respectively.

    d. Find V using Formula 12.1.

    e. Using V and n, enter table G in Appendix B to find the critical value for $\alpha = 0.05$.

    f. Reject h_0 if V is less than or equal to the critical value.

12–8. Use the Wilcoxon rank-sum test to test the hypothesis in question 12–7.

    a. State the null and alternative hypotheses.

    b. Compute the difference scores per day between lines A and B.

    c. Take the absolute value of each difference and assign a rank to each.

    d. For those with negative differences in (b) add up the corresponding ranks found in (c) and assign the sum to r^-.

    e. Repeat (d) for those with positive differences and assign the sum to r^+.

    f. Compute W using Formula 12.7.

    g. Find the critical values $W_{(1 - \alpha/2)}$, $W_{\alpha/2}$.

    h. Reject h_0 if $W < W_{.025}$ or $W > W_{.975}$.

12–9. In question 12–7, how is it determined whether a rank-sum test is used instead of a Mann-Whitney U test?

12–10. Fifteen volunteers viewed and rated 2 potential 30 second television commercials. Each viewer rated the commercials on a 10 point scale where $10 =$ "definitely like" the commercial to $1 =$ "definitely do not like" the commercial. The rating data are given below:

Person	Commercial 1	Commercial 2
1	5	8
2	6	2
3	7	6
4	5	5
5	10	9
6	4	5
7	9	1
8	2	4
9	8	8
10	7	9
11	1	2
12	8	7
13	9	8
14	7	8
15	6	6

Compute the Spearman rank order correlation. Is the coefficient statistically different from zero?

    a. State the null and alternative hypotheses.

    b. Assign a rank to each person's rating of commercial 1 and 2.

Person	1	2	3	4	5	6	7	8	9	10	11	12	13	14	15
Com #1															
Com #2															

    c. Compute rho.

       1. $r =$, $N =$, $\Sigma r(x_i y_i) =$.

       2. $\Sigma r(x_i) =$.

       3. $\Sigma r(y_i) =$.

       4. rho $=$.

    d. Find the critical value in table J of Appendix B.

    e. Reject h_0 if rho $\geq$ rho (critical).

12–11. Using data from question 12–10, do the distributions of the ratings for the two commercials differ significantly from each other? Develop the appropriate nonparametric test.

    a. State the null and alternative hypotheses.

    b. Compute the difference scores for each person between commercial 1 and 2.

    c. Take the absolute value of each difference and assign a rank to each. Drop the differences that are equal to zero.

    d. For those with negative differences in (b), add up the corresponding ranks found in (c). Assign the sum to r^-.

    e. Repeat (d) for those with positive differences. Assign sum to r^+.

    f. Compute W using Formula 12.7.

    g. Find the critical values $W_{(\alpha/2)}$, $W_{(1-\alpha/2)}$, with $\alpha = 0.05$.

    h. Reject h_0 if $W \leq W_{.025}$ or $\geq W_{.975}$.

12–12. Two professors in a large university are debating over which of them is the best professor in the university in terms of overall teaching evaluations from students in their courses. The students are asked to rate the professor in terms of overall teaching effectiveness. The optional ratings are 1 through 5, where a 5 = excellent, 4 = good, 3 = average, 2 = marginal, and 1 = poor. Professor T only taught 6 courses during the year, while Professor S taught 8 courses. The data are given below where the numbers are the averages of the student ratings per course.

Professor T:	4.1, 4.5, 3.8, 3.9, 4.1, 3.6
Professor S:	3.8, 4.4, 4.3, 4.0, 3.7, 3.5, 4.6, 4.2

Are the ratings for the two professors independent? Explain.

12–13. Using data from problem 12–12, compute the Mann-Whitney U test to test the hypothesis that the professors have different ratings.
   a. State the null and alternative hypotheses.
   b. Assign ranks to the ratings regardless of which professor they belong to.
   c. Sum the ranks for Professor T.
   d. Compute U using Formula 12.5.
   e. Find the critical values $U_{(\alpha/2)}$, $U_{(1-\alpha/2)}$.
   f. Reject h_0 if $U \leq U_{.025}$ or $U \geq U_{.975}$.

12–14. Two judges at the Olympic Games gave the following ratings to 8 female participants in a gymnastic event. A 10 is a perfect score.

Participant	Judge S	Judge R
Olga	9.5	9.9
Nadia	9.9	10.0
Mary	8.9	8.5
Netta	9.1	9.5
Sweta	9.6	9.5
Linda	9.0	8.0
Judy	9.1	8.5
Tasha	9.3	9.7

   a. Use the sign test to see if the two judges are giving statistically similar ratings.
      1. State the null and alternative hypotheses.
      2. Compute the difference scores for each participant.
      3. Find the number of difference scores that are minus and the number that are plus.
      4. Compute V using Formula 12.1.
      5. Using V and n, find the critical value for $\alpha = 0.05$.
      6. Reject h_0 if V $\leq$ critical value.
   b. Use the Wilcoxon test to see if the two judges are giving statistically similar ratings.
      1. State the null and alternative hypotheses.
      2. Compute the difference scores for each participant.
      3. Take the absolute value of each difference and assign a rank to each value.
      4. For those with negative differences in 2, add up the corresponding ranks found in 3. Assign the sum to r^-.
      5. Repeat 4 for those with positive differences. Assign the sum of ranks to r^+.
      6. Find W using Formula 12.7.
      7. Find the critical values $W_{(\alpha/2)}$ and $W_{(1-\alpha/2)}$ for $\alpha = 0.05$.
      8. Reject h_0 if W $\leq$ $W_{.025}$ or W $\geq$ $W_{.975}$.

12–15. In a test between two leading brands of dry cell batteries, the order in which each battery failed is recorded. These data are given below. "A" represents battery manufacturer A and "B" represents manufacturer B.

B B B A A A A A B A A A B B B B B A B B A A A B B B B A A A

Using the runs test, determine if this ordered sequence of failure is random.
   a. State the null and alternative hypotheses.
   b. Compute the number of runs.
   c. Find the critical values $C_{(\alpha/2)}$ and $C_{(1-\alpha/2)}$.
   d. Reject h_0 if R_t is greater than or equal to $C_{.975}$ or less than or equal to $C_{.025}$.

12–16. If the sequence in question 12–15 is thought of as rankings, then two groups of rankings can be created. For example, the first "B" receives a rank of 1. Using the Mann-Whitney U test, test the hypothesis that the rankings received by battery manufacturer A are different from the rankings received by battery manufacturer B.
   a. State the null and alternative hypotheses.
   b. Assign ranks to A and B.
   c. Sum the ranks for B.
   d. Compute U using Formula 12.5.
   e. Find the critical values $U_{(\alpha/2)}$, $U_{(1-\alpha/2)}$ for $\alpha = 0.05$.
   f. Reject h_0 if $U \le U_{.025}$ or $U \ge U_{.975}$.

12–17. Thirty overweight men decide to participate in a weight-loss program. Each participant is weighed before and after the program. If the person loses weight at the end of the program he receives a "$-$." If he gains weight as a result of the program, he receives a "$+$." If there is no weight loss or gain, that person's data are dropped from the analysis. Using the "$-$" and "$+$" in a sign test, test the hypothesis that the overweight men are as likely to gain weight as they are to lose weight using this weight reduction program. The test results are found below.

Participant	Results	Participant	Results
1	$-$	17	$+$
2	$-$	18	$-$
3	$-$	19	$+$
4	$-$	20	$-$
5	$+$	21	$-$
6	$-$	22	$-$
7	$+$	23	$-$
8	$+$	24	$-$
9	$+$	25	$+$
10	$-$	26	0
11	$-$	27	$-$
12	$+$	28	$-$
13	$-$	29	$-$
14	$-$	30	$-$
15	$-$		
16	$-$		

a. State the null and alternative hypotheses.
b. Count the number of pluses and minuses.
c. Find V using Formula 12.1.
d. Using V and n, find the critical value for $\alpha = 0.05$.
e. Reject h_0 if $V <$ critical value.

12–18. A teacher has developed a true–false examination and wants to know if the answers are in random order. Given below is the sequence of true-false answers to the 31 questions. Develop a hypothesis test to determine if the answers are arranged in random order. Use $\alpha = 0.05$.

T T F F T F F F F T T T T T T T T F F T T T F F F F F F T T T T T

a. State the null and alternative hypotheses.
b. Compute the number of runs, R_t.
c. Find the critical values $C_{(\alpha/2)}$ and $C_{(1 - \alpha/2)}$ for $\alpha = 0.05$.
d. Reject h_0 if R_t is greater than or equal to $C_{.975}$ or less than or equal to $C_{.025}$.

12–19. For some reason or another a researcher is interested in the length of the small right fingers of Caucasian men. He hypothesizes that those in northern California are smaller than those in southern California. He measured 6 Caucasian men in northern California and 7 in southern California. The data are given below (in inches).

N. Cal	2.8	2.6	3.0	3.1	2.7	2.9	
S. Cal	3.3	2.2	3.6	2.6	3.5	3.4	3.2

Develop a hypothesis test for the researcher. Use $\alpha = 0.10$.
a. State the null and alternative hypotheses.
b. Assign ranks to the measurements, disregarding the location of their residence.

N. Cal.						
S. Cal.						

c. Sum the ranks for northern California.
d. Compute U using Formula 12.5.
e. Find the critical values $U_{(\alpha/2)}$, and $U_{(1 - \alpha/2)}$ for $\alpha = 0.10$.
f. Reject h_0 if $U \leq U_{.05}$ or $U \geq U_{.95}$.

12–20. The same offbeat researcher hypothesized that in left-hemispheric Caucasian men, the right index finger is longer than the left index finger. To prove his point, he measured the left and right index fingers of 10 Caucasian men. The data are in inches.

	Men									
	1	2	3	4	5	6	7	8	9	10
L. i.f.	2.81	2.75	3.34	2.96	3.26	3.61	3.08	3.16	3.72	3.54
R. i.f.	2.92	2.82	3.51	2.91	3.32	3.64	3.01	3.25	3.80	3.67

Develop the appropriate nonparametric test for this researcher's hypothesis. Use $\alpha = 0.05$.

    a. State the null and alternative hypotheses.

    b. Form the difference score per participant.

    c. Take the absolute values of each difference and assign a rank to each.

    d. For those with negative differences in (b), add up the corresponding ranks found in (c). Assign sum to r^-.

    e. Repeat (d) for positive differences. Assign sum to r^+.

    f. Find W using Formula 12.7.

    g. Find critical values $W_{(\alpha/2)}$, $W_{(1 - \alpha/2)}$ for $\alpha = 0.05$.

    h. Reject h_0 if $W \leq W_{.025}$ or $\geq W_{.975}$.

12-21. For the purpose of predicting results, a poll researcher wants to determine whether or not the percentage of women are different in 3 separate areas within a city. The data available consist of 1980 census information for each census tract. The researcher randomly selected 6 census tracts from each area and retrieved the percentage of women data. The information is given below.

Area A	42.6	39.7	53.1	55.6	43.7	47.8
Area B	36.4	52.6	58.5	39.5	45.8	50.3
Area C	47.5	40.8	56.8	52.3	48.1	41.8

Develop a hypothesis test to determine whether or not there is a difference in the percentage of women across the three areas. Use $\alpha = 0.05$.

    a. State the null and alternative hypotheses.

    b. Rank the raw data without consideration of group membership. Find the sum of the ranks for each group.

Area A						
Area B						
Area C						

    c. Compute KW, the test statistic, using Formula 12.12 or Formula 12.13.

    d. Find the critical value using table L of Appendix B.

    f. Reject h_0 if KW is greater than or equal to the critical value.

Appendix A
Summary of Formulas

Formula for the Median

$$Mdn = X_{LL} + \left(\frac{N/2 - \text{Cum } f_{LL}}{f}\right)i \qquad \text{(3.1) page 57}$$

Raw-Score Formula for the Mean

$$M = \frac{\Sigma X}{N} \qquad \text{(3.2) page 61}$$

Formula for the Semi-Interquartile Range (Q)

$$Q = \frac{Q_3 - Q_1}{2} \qquad \text{(4.1) page 70}$$

Formula for the Third Quartile

$$Q_3 = X_{LL} + \left(\frac{3N/4 - \text{Cum } f_{LL}}{f}\right)i \qquad \text{(4.2a) page 71}$$

Formula for the First Quartile

$$Q_1 = X_{LL} + \left(\frac{N/4 - \text{Cum } f_{LL}}{f}\right)i \qquad \text{(4.2b) page 71}$$

Deviation-Score Formula for the Standard Deviation (S.D.)

$$\text{S.D.} = \sqrt{\frac{\Sigma x^2}{N}} \qquad \text{(4.3) page 79}$$

Raw-Score Formula for the Standard Deviation (S.D.)

$$\text{S.D.} = \frac{1}{N}\sqrt{N\Sigma X^2 - (\Sigma X)^2} \qquad \text{(4.4) page 80}$$

Independent Samples:

95% confidence interval: $(M_1 - M_2) \pm t_{.05} \times s_{D_M}$ (8.8b) page 241

99% confidence interval: $(M_1 - M_2) \pm t_{.01} \times s_{D_M}$ (8.9b) page 241

Dependent Samples

95% confidence interval: $M_d \pm t_{.05} \times s_{d_M}$ (8.8c) page 249

99% confidence interval: $M_d \pm t_{.01} \times s_{d_M}$ (8.9c) page 249

Formula for the Conversion of Sample Correlation to a t-Score Value

$$t = \frac{r\sqrt{N-2}}{\sqrt{1-r^2}}$$ (8.10) page 224

z' Formula

$$z' = 1/2 \left(\log_e (1+r) - \log_e (1-r)\right)$$ (8.11) page 226

Formula for Comparing a Sample Correlation to a True
Population Correlation Value

$$Z = \frac{z' - z'_0}{\sqrt{1/(N-3)}}$$ (8.12) page 226

Formula for the F Ratio between Two Variance Estimates

$$F = \frac{s_x^2}{s_y^2}$$ (8.14) page 229

Chi Square Test Formula

$$\chi^2 = \frac{(N)\text{ S.D.}^2}{\sigma_0^2}$$ (8.15) page 232

Formula for the Standard Deviation of the Sampling Distribution (Difference)

$$s_{D_M} = \sqrt{\frac{\Sigma x_1^2 + \Sigma x_2^2}{N_1 + N_2 - 2}\left(\frac{N_1 + N_2}{N_1 \cdot N_2}\right)}$$ (8.16) page 236

Formula for Testing the Difference between Two Means (Independent Samples)

$$t = \frac{(M_1 - M_2) - 0}{s_{D_M}}$$ (8.18) page 236

Formula for the Standard Error of the Mean for Difference Scores

$$s_{d_M} = \frac{s_d}{\sqrt{N}}$$

(8.20) page 245

Formula for Testing the Null Hypothesis with Matched Samples

$$t = \frac{M_d - 0}{s_{d_M}}$$

(8.21) page 246

Formula for Chi Square

$$\chi^2 = \Sigma \frac{(f_o - f_t)^2}{f_t}$$

(9.1) page 266

Formula for Determining Between-Groups Variance (ANOVA)

$$s^2 = \frac{N \sum_{i=1}^{k} (M_i - M)^2}{k - 1}$$

(10.4) page 300

Formula for Multiple Comparisons t-Test

$$t_{ij} = \frac{M_i - M_j - 0}{\sqrt{s_W^2 \left[(n_i + n_j)/n_i n_j\right]}}$$

(10.9) page 314

Formula for the Scheffé Test

$$t'_{.01} = \sqrt{(k - 1)F_{.01}}$$

(10.11) page 316

Formula for the Tukey HSD Test

$$HSD = q_{k,v} \sqrt{s_W^2/n}$$

(10.12) page 318

Formula to Compare any Two Subgroup Means (ANOVA)

$$t = \frac{M_{L_i} - M_{L_j} - 0}{\sqrt{s_W^2 \dfrac{N_{L_i} + N_{L_j}}{N_{L_i} \times N_{L_j}}}}$$

(11.6) page 348

Formula for Comparing Two Population Distributions
Using the Mann-Whitney U Test

$$s = \Sigma r(x_i) \qquad\qquad (12.4)\ \text{page } 364$$

$$U = s - \frac{n(n+1)}{2} \qquad\qquad (12.5)\ \text{page } 364$$

Formula for the Large Sample Statistic Using the Mann-Whitney U Test

$$Z = \frac{U - \dfrac{nm}{2}}{\sqrt{\dfrac{nm(n+m+1)}{12}}} \qquad\qquad (12.6)\ \text{page } 366$$

Formula for the Wilcoxon Rank-Sum Test

$$W = \text{minimum } (r^+, r^-) \qquad\qquad (12.7)\ \text{page } 368$$

Formula for the Wilcoxon Rank-Sum Test with Large Samples

$$Z = \frac{W - \dfrac{n(n+1)}{4}}{\sqrt{\dfrac{n(n+1)(2n+1)}{24}}} \qquad\qquad (12.8)\ \text{page } 370$$

Formula for Computing the Spearman Rank-Order Correlation

$$\text{rho} = 1 - \frac{6\Sigma d_i^2}{n(n^2-1)} \qquad\qquad (12.9)\ \text{page } 372$$

Formula for the Large Sample Runs Test

$$Z = \frac{R_t - \left(\dfrac{2n_1 n_2}{n_1 + n_2} + 1\right)}{\sqrt{\dfrac{2n_1 n_2(2n_1 n_2 - n_1 - n_2)}{(n_1 + n_2)^2\,(n_1 + n_2 - 1)}}} \qquad\qquad (12.11)\ \text{page } 376$$

Formula for the Kruskal-Wallis Test

$$KW = \left[\frac{12}{N(N+1)}\right]\left[\Sigma \frac{[r_i - .5n_i(N+1)]^2}{n_i}\right] \qquad\qquad (12.12)\ \text{page } 378$$

Appendix B Tables

TABLE A

List of symbols used in the text

Symbol	Meaning	Alternate symbols
α	Alpha, probability level of Type I error	
a	Constant	
β	Beta, probability level of Type II error	
C	Critical values for the Runs test	
cf	Cumulative frequency	
$\text{Cum } f_{LL}$	Cumulative frequency at lower true limit of class interval	
D	Index of Dispersion	
df	Degrees of freedom	
d_i	Difference score for the ith pair	
η^2	Eta squared	
f	Frequency, count	
f_i	Frequency count in category i	
f_o	Frequency observed	
f_t	Theoretical, or expected frequency (chi square)	f_e
F	F ratio	
F_α	Critical value for F	
h_0	Null hypothesis	H_0
h_1	Alternative hypothesis	H_1, H_A
HSD_α	Tukey Honestly Significant Difference Value	
i	Class interval size	
k	Number of categories—Index of Dispersion (Chapter 4) Number of groups—ANOVA (Chapter 10)	
KW	Test statistic for Kruskal-Wallis test	
M	Mean (ungrouped) of a sample, or Grand Mean (ANOVA)	$\overline{X}, \overline{Y}$
Mdn	Median	
M_i	Subgroup means	
μ	Population mean (mu)	
N	Total number of scores (observations), numbers of pairs (correlation)	
n_i	Subgroup sample sizes	
N_p	Population sample size	
ω^2	Omega squared	
p	Probability of a specified outcome, population proportion	
PR	Percentile or centile rank	P_s, P
Q	Semi-interquartile range	
Q_1	First quartile	
Q_2	Second quartile (median)	
Q_3	Third quartile	
$q_{k,v}$	Studentized range statistic	
ρ	Rho, population parameter for correlation	
r	Pearson correlation coefficient	

TABLE A
Continued

Symbol	Meaning	Alternate symbols
r^-	Sum of the ranks for negative scores	
r^+	Sum of ranks for positive scores	
$r(x_i)$	Rank values assigned to raw data	
R_t	Test statistic for runs test	
rho	Spearman rank correlation	
S	Scaled score	
s	Sample standard deviation (unbiased), sum of ranks in group 1 (Mann-Whitney U test)	
s_M	Standard error of the mean (single sample)	
s_{D_M}	Standard error of the difference between two means (independent samples)	
s_{d_M}	Standard error of the difference between two means (dependent samples)	
$s_W{}^2$	Within-groups mean variance	
σ	Sigma, population standard deviation	
σ^2	Sigma squared, population variance	
S.D.	Sample standard deviation (biased)	s'
S.D.$_i$	Standard deviation for i-type scores	
Σ	Summation	
$\Sigma\Sigma$	Double summation (sum all scores in a two-dimensional table)	
SS_B	Sum of squares between-groups (ANOVA)	
SS_W	Sum of squares within-groups (ANOVA)	
t	Symbol for t distribution, t-test, and t value	
t'_α	Scheffé t value	
T	T-score (McCall)	
θ	Theta, population parameter	
$\hat{\theta}$	Estimate (sample) of population parameter	
U	Mann-Whitney test statistic	
V	Sign test test statistic	
W	Wilcoxon test statistic	
X	Measurement, score, or interval of scores (raw score)	Y
x	Deviation score in variable X ($X_i - M$)	$X - \overline{X}$
X_i	ith measurement or observation (raw score)	X
X_j	jth measurement or observation (raw score)	
X_{LL}	Score at lower true limit of class interval containing X	
χ^2	Chi square	
y	Deviation score in variable Y ($Y_i - M$)	$Y - \overline{Y}$
Y'	Predicted raw-score in regression	$\hat{Y}$
Z	Standard score	z
z'	Fisher transformation correlation	

TABLE B

Normal curve table

Z	Area −∞ to Z	Area between 0.00 and Z	Area beyond Z	y	Z	Area −∞ to Z	Area between 0.00 and Z	Area beyond Z	y
A	B	C	D	E	A	B	C	D	E
0.00	.5000	.0000	.5000	.3989	0.40	.6554	.1554	.3446	.3683
0.01	.5040	.0040	.4960	.3989	0.41	.6591	.1591	.3409	.3668
0.02	.5080	.0080	.4920	.3989	0.42	.6628	.1628	.3372	.3653
0.03	.5120	.0120	.4880	.3988	0.43	.6664	.1664	.3336	.3637
0.04	.5160	.0160	.4840	.3986	0.44	.6700	.1700	.3300	.3621
0.05	.5199	.0199	.4801	.3984	0.45	.6736	.1736	.3264	.3605
0.06	.5239	.0239	.4761	.3982	0.46	.6772	.1772	.3228	.3589
0.07	.5279	.0279	.4721	.3980	0.47	.6808	.1808	.3192	.3572
0.08	.5319	.0319	.4681	.3977	0.48	.6844	.1844	.3156	.3555
0.09	.5359	.0359	.4641	.3973	0.49	.6879	.1879	.3121	.3538
0.10	.5398	.0398	.4602	.3970	0.50	.6915	.1915	.3085	.3521
0.11	.5438	.0438	.4562	.3965	0.51	.6950	.1950	.3050	.3503
0.12	.5478	.0478	.4522	.3961	0.52	.6985	.1985	.3015	.3485
0.13	.5517	.0517	.4483	.3956	0.53	.7019	.2019	.2981	.3467
0.14	.5557	.0557	.4443	.3951	0.54	.7054	.2054	.2946	.3448
0.15	.5596	.0596	.4404	.3945	0.55	.7088	.2088	.2912	.3429
0.16	.5636	.0636	.4364	.3939	0.56	.7123	.2123	.2877	.3410
0.17	.5675	.0675	.4325	.3932	0.57	.7157	.2157	.2843	.3391
0.18	.5714	.0714	.4286	.3925	0.58	.7190	.2190	.2810	.3372
0.19	.5753	.0753	.4247	.3918	0.59	.7224	.2224	.2776	.3352
0.20	.5793	.0793	.4207	.3910	0.60	.7257	.2257	.2743	.3332
0.21	.5832	.0832	.4168	.3902	0.61	.7291	.2291	.2709	.3312
0.22	.5871	.0871	.4129	.3894	0.62	.7324	.2324	.2676	.3292
0.23	.5910	.0910	.4090	.3885	0.63	.7357	.2357	.2643	.3271
0.24	.5948	.0948	.4052	.3876	0.64	.7389	.2389	.2611	.3251
0.25	.5987	.0987	.4013	.3867	0.65	.7422	.2422	.2578	.3230
0.26	.6026	.1026	.3974	.3857	0.66	.7454	.2454	.2546	.3209
0.27	.6064	.1064	.3936	.3847	0.67	.7486	.2486	.2514	.3187
0.28	.6103	.1103	.3897	.3836	0.68	.7517	.2517	.2483	.3166
0.29	.6141	.1141	.3859	.3825	0.69	.7549	.2549	.2451	.3144
0.30	.6179	.1179	.3821	.3814	0.70	.7580	.2580	.2420	.3123
0.31	.6217	.1217	.3783	.3802	0.71	.7611	.2611	.2389	.3101
0.32	.6255	.1255	.3745	.3790	0.72	.7642	.2642	.2358	.3079
0.33	.6293	.1293	.3707	.3778	0.73	.7673	.2673	.2327	.3056
0.34	.6331	.1331	.3669	.3765	0.74	.7704	.2704	.2296	.3034
0.35	.6368	.1368	.3632	.3752	0.75	.7734	.2734	.2266	.3011
0.36	.6406	.1406	.3594	.3739	0.76	.7764	.2764	.2236	.2989
0.37	.6443	.1443	.3557	.3726	0.77	.7794	.2794	.2206	.2966
0.38	.6480	.1480	.3520	.3712	0.78	.7823	.2823	.2177	.2943
0.39	.6517	.1517	.3483	.3697	0.79	.7852	.2852	.2148	.2920

TABLE B

Normal curve table (continued)

Z	Area −∞ to Z	Area between 0.00 and Z	Area beyond Z	y	Z	Area −∞ to Z	Area between 0.00 and Z	Area beyond Z	y
A	B	C	D	E	A	B	C	D	E
0.80	.7881	.2881	.2119	.2897	1.20	.8849	.3849	.1151	.1942
0.81	.7910	.2910	.2090	.2874	1.21	.8869	.3869	.1131	.1919
0.82	.7939	.2939	.2061	.2850	1.22	.8888	.3888	.1112	.1895
0.83	.7967	.2967	.2033	.2827	1.23	.8907	.3907	.1093	.1872
0.84	.7995	.2995	.2005	.2803	1.24	.8925	.3925	.1075	.1849
0.85	.8023	.3023	.1977	.2780	1.25	.8944	.3944	.1056	.1826
0.86	.8051	.3051	.1949	.2756	1.26	.8962	.3962	.1038	.1804
0.87	.8078	.3078	.1922	.2732	1.27	.8980	.3980	.1020	.1781
0.88	.8106	.3106	.1894	.2709	1.28	.8997	.3997	.1003	.1758
0.89	.8133	.3133	.1867	.2685	1.29	.9015	.4015	.0985	.1736
0.90	.8159	.3159	.1841	.2661	1.30	.9032	.4032	.0968	.1714
0.91	.8186	.3186	.1814	.2637	1.31	.9049	.4049	.0951	.1691
0.92	.8212	.3212	.1788	.2613	1.32	.9066	.4066	.0934	.1669
0.93	.8238	.3238	.1762	.2589	1.33	.9082	.4082	.0918	.1647
0.94	.8264	.3264	.1736	.2565	1.34	.9099	.4099	.0901	.1626
0.95	.8289	.3289	.1711	.2541	1.35	.9115	.4115	.0885	.1604
0.96	.8315	.3315	.1685	.2516	1.36	.9131	.4131	.0869	.1582
0.97	.8340	.3340	.1660	.2492	1.37	.9147	.4147	.0853	.1561
0.98	.8365	.3365	.1635	.2468	1.38	.9162	.4162	.0838	.1539
0.99	.8389	.3389	.1611	.2444	1.39	.9177	.4177	.0823	.1518
1.00	.8413	.3413	.1587	.2420	1.40	.9192	.4192	.0808	.1497
1.01	.8438	.3438	.1562	.2396	1.41	.9207	.4207	.0793	.1476
1.02	.8461	.3461	.1539	.2371	1.42	.9222	.4222	.0778	.1456
1.03	.8485	.3485	.1515	.2347	1.43	.9236	.4236	.0764	.1435
1.04	.8508	.3508	.1492	.2323	1.44	.9251	.4251	.0749	.1415
1.05	.8531	.3531	.1469	.2299	1.45	.9265	.4265	.0735	.1394
1.06	.8554	.3554	.1446	.2275	1.46	.9279	.4279	.0721	.1374
1.07	.8577	.3577	.1423	.2251	1.47	.9292	.4292	.0708	.1354
1.08	.8599	.3599	.1401	.2227	1.48	.9306	.4306	.0694	.1334
1.09	.8621	.3621	.1379	.2203	1.49	.9319	.4319	.0681	.1315
1.10	.8643	.3643	.1357	.2179	1.50	.9332	.4332	.0668	.1295
1.11	.8665	.3665	.1335	.2155	1.51	.9345	.4345	.0655	.1276
1.12	.8686	.3686	.1314	.2131	1.52	.9357	.4357	.0643	.1257
1.13	.8708	.3708	.1292	.2107	1.53	.9370	.4370	.0630	.1238
1.14	.8729	.3729	.1271	.2083	1.54	.9382	.4382	.0618	.1219
1.15	.8749	.3749	.1251	.2059	1.55	.9394	.4394	.0606	.1200
1.16	.8770	.3770	.1230	.2036	1.56	.9406	.4406	.0594	.1182
1.17	.8790	.3790	.1210	.2012	1.57	.9418	.4418	.0582	.1163
1.18	.8810	.3810	.1190	.1989	1.58	.9429	.4429	.0571	.1145
1.19	.8830	.3830	.1170	.1965	1.59	.9441	.4441	.0559	.1127

TABLE B

Normal curve table (continued)

Z	Area −∞ to Z	Area between 0.00 and Z	Area beyond Z	y	Z	Area −∞ to Z	Area between 0.00 and Z	Area beyond Z	y
A	B	C	D	E	A	B	C	D	E
1.60	.9452	.4452	.0548	.1109	2.00	.9772	.4772	.0228	.0540
1.61	.9463	.4463	.0537	.1092	2.01	.9778	.4778	.0222	.0529
1.62	.9474	.4474	.0526	.1074	2.02	.9783	.4783	.0217	.0519
1.63	.9484	.4484	.0516	.1057	2.03	.9788	.4788	.0212	.0508
1.64	.9495	.4495	.0505	.1040	2.04	.9793	.4793	.0207	.0498
1.65	.9505	.4505	.0495	.1023	2.05	.9798	.4798	.0202	.0488
1.66	.9515	.4515	.0485	.1006	2.06	.9803	.4803	.0197	.0478
1.67	.9525	.4525	.0475	.0989	2.07	.9808	.4808	.0192	.0468
1.68	.9535	.4535	.0465	.0973	2.08	.9812	.4812	.0188	.0459
1.69	.9545	.4545	.0455	.0957	2.09	.9817	.4817	.0183	.0449
1.70	.9554	.4554	.0446	.0940	2.10	.9821	.4821	.0179	.0440
1.71	.9564	.4564	.0436	.0925	2.11	.9826	.4826	.0174	.0431
1.72	.9573	.4573	.0427	.0909	2.12	.9830	.4830	.0170	.0422
1.73	.9582	.4582	.0418	.0893	2.13	.9834	.4834	.0166	.0413
1.74	.9591	.4591	.0409	.0878	2.14	.9838	.4838	.0162	.0404
1.75	.9599	.4599	.0401	.0863	2.15	.9842	.4842	.0158	.0395
1.76	.9608	.4608	.0392	.0848	2.16	.9846	.4846	.0154	.0387
1.77	.9616	.4616	.0384	.0833	2.17	.9850	.4850	.0150	.0379
1.78	.9625	.4625	.0375	.0818	2.18	.9854	.4854	.0146	.0371
1.79	.9633	.4633	.0367	.0804	2.19	.9857	.4857	.0143	.0363
1.80	.9641	.4641	.0359	.0790	2.20	.9861	.4861	.0139	.0355
1.81	.9649	.4649	.0351	.0775	2.21	.9864	.4864	.0136	.0347
1.82	.9656	.4656	.0344	.0761	2.22	.9868	.4868	.0132	.0339
1.83	.9664	.4664	.0336	.0748	2.23	.9871	.4871	.0129	.0332
1.84	.9671	.4671	.0329	.0734	2.24	.9875	.4875	.0125	.0325
1.85	.9678	.4678	.0322	.0721	2.25	.9878	.4878	.0122	.0317
1.86	.9686	.4686	.0314	.0707	2.26	.9881	.4881	.0119	.0310
1.87	.9693	.4693	.0307	.0694	2.27	.9884	.4884	.0116	.0303
1.88	.9699	.4699	.0301	.0681	2.28	.9887	.4887	.0113	.0297
1.89	.9706	.4706	.0294	.0669	2.29	.9890	.4890	.0110	.0290
1.90	.9713	.4713	.0287	.0656	2.30	.9893	.4893	.0107	.0283
1.91	.9719	.4719	.0281	.0644	2.31	.9896	.4896	.0104	.0277
1.92	.9726	.4726	.0274	.0632	2.32	.9898	.4898	.0102	.0270
1.93	.9732	.4732	.0268	.0620	2.33	.9901	.4901	.0099	.0264
1.94	.9738	.4738	.0262	.0608	2.34	.9904	.4904	.0096	.0258
1.95	.9744	.4744	.0256	.0596	2.35	.9906	.4906	.0094	.0252
1.96	.9750	.4750	.0250	.0584	2.36	.9909	.4909	.0091	.0246
1.97	.9756	.4756	.0244	.0573	2.37	.9911	.4911	.0089	.0241
1.98	.9761	.4761	.0239	.0562	2.38	.9913	.4913	.0087	.0235
1.99	.9767	.4767	.0233	.0551	2.39	.9916	.4916	.0084	.0229

TABLE B

Normal curve table (continued)

Z	Area −∞ to Z	Area between 0.00 and Z	Area beyond Z	y	Z	Area −∞ to Z	Area between 0.00 and Z	Area beyond Z	y
A	B	C	D	E	A	B	C	D	E
2.40	.9918	.4918	.0082	.0224	2.80	.9974	.4974	.0026	.0079
2.41	.9920	.4920	.0080	.0219	2.81	.9975	.4975	.0025	.0077
2.42	.9922	.4922	.0078	.0213	2.82	.9976	.4976	.0024	.0075
2.43	.9925	.4925	.0075	.0208	2.83	.9977	.4977	.0023	.0073
2.44	.9927	.4927	.0073	.0203	2.84	.9977	.4977	.0023	.0071
2.45	.9929	.4929	.0071	.0198	2.85	.9978	.4978	.0022	.0069
2.46	.9931	.4931	.0069	.0194	2.86	.9979	.4979	.0021	.0067
2.47	.9932	.4932	.0068	.0189	2.87	.9979	.4979	.0021	.0065
2.48	.9934	.4934	.0066	.0184	2.88	.9980	.4980	.0020	.0063
2.49	.9936	.4936	.0064	.0180	2.89	.9981	.4981	.0019	.0061
2.50	.9938	.4938	.0062	.0175	2.90	.9981	.4981	.0019	.0060
2.51	.9940	.4940	.0060	.0171	2.91	.9982	.4982	.0018	.0058
2.52	.9941	.4941	.0059	.0167	2.92	.9982	.4982	.0018	.0056
2.53	.9943	.4943	.0057	.0163	2.93	.9983	.4983	.0017	.0055
2.54	.9945	.4945	.0055	.0158	2.94	.9984	.4984	.0016	.0053
2.55	.9946	.4946	.0054	.0154	2.95	.9984	.4984	.0016	.0051
2.56	.9948	.4948	.0052	.0151	2.96	.9985	.4985	.0015	.0050
2.57	.9949	.4949	.0051	.0147	2.97	.9985	.4985	.0015	.0048
2.58	.9951	.4951	.0049	.0143	2.98	.9986	.4986	.0014	.0047
2.59	.9952	.4952	.0048	.0139	2.99	.9986	.4986	.0014	.0046
2.60	.9953	.4953	.0047	.0136	3.00	.9987	.4987	.0013	.0044
2.61	.9955	.4955	.0045	.0132	3.01	.9987	.4987	.0013	.0043
2.62	.9956	.4956	.0044	.0129	3.02	.9987	.4987	.0013	.0042
2.63	.9957	.4957	.0043	.0126	3.03	.9988	.4988	.0012	.0040
2.64	.9959	.4959	.0041	.0122	3.04	.9988	.4988	.0012	.0039
2.65	.9960	.4960	.0040	.0119	3.05	.9989	.4989	.0011	.0038
2.66	.9961	.4961	.0039	.0116	3.06	.9989	.4989	.0011	.0037
2.67	.9962	.4962	.0038	.0113	3.07	.9989	.4989	.0011	.0036
2.68	.9963	.4963	.0037	.0110	3.08	.9990	.4990	.0010	.0035
2.69	.9964	.4964	.0036	.0107	3.09	.9990	.4990	.0010	.0034
2.70	.9965	.4965	.0035	.0104	3.10	.9990	.4990	.0010	.0033
2.71	.9966	.4966	.0034	.0101	3.11	.9991	.4991	.0009	.0032
2.72	.9967	.4967	.0033	.0099	3.12	.9991	.4991	.0009	.0031
2.73	.9968	.4968	.0032	.0096	3.13	.9991	.4991	.0009	.0030
2.74	.9969	.4969	.0031	.0093	3.14	.9992	.4992	.0008	.0029
2.75	.9970	.4970	.0030	.0091	3.15	.9992	.4992	.0008	.0028
2.76	.9971	.4971	.0029	.0088	3.16	.9992	.4992	.0008	.0027
2.77	.9972	.4972	.0028	.0086	3.17	.9992	.4992	.0008	.0026
2.78	.9973	.4973	.0027	.0084	3.18	.9993	.4993	.0007	.0025
2.79	.9974	.4974	.0026	.0081	3.19	.9993	.4993	.0007	.0025

TABLE B

Normal curve table (continued)

Z	Area −∞ to Z	Area between 0.00 and Z	Area beyond Z	y
A	B	C	D	E
3.20	.9993	.4993	.0007	.0024
3.21	.9993	.4993	.0007	.0023
3.22	.9994	.4994	.0006	.0022
3.23	.9994	.4994	.0006	.0022
3.24	.9994	.4994	.0006	.0021
3.25	.9994	.4994	.0006	.0020
3.30	.9995	.4995	.0005	.0017
3.35	.9996	.4996	.0004	.0015
3.40	.9997	.4997	.0003	.0012
3.45	.9997	.4997	.0003	.0010
3.50	.9998	.4998	.0002	.0009
3.60	.9998	.4998	.0002	.0006
3.70	.9999	.4999	.0001	.0004
3.80	.9999	.4999	.0001	.0003
3.90		.49995	.00005	.0002
4.00		.49997	.00003	.0001

TABLE C

t distribution

Degrees of freedom	Probability				
	0.50	0.10	0.05	0.02	0.01
1	1.000	6.34	12.71	31.82	63.66
2	0.816	2.92	4.30	6.96	9.92
3	.765	2.35	3.18	4.54	5.84
4	.741	2.13	2.78	3.75	4.60
5	.727	2.02	2.57	3.36	4.03
6	.718	1.94	2.45	3.14	3.71
7	.711	1.90	2.36	3.00	3.50
8	.706	1.86	2.31	2.90	3.36
9	.703	1.83	2.26	2.82	3.25
10	.700	1.81	2.23	2.76	3.17
11	.697	1.80	2.20	2.72	3.11
12	.695	1.78	2.18	2.68	3.06
13	.694	1.77	2.16	2.65	3.01
14	.692	1.76	2.14	2.62	2.98
15	.691	1.75	2.13	2.60	2.95
16	.690	1.75	2.12	2.58	2.92
17	.689	1.74	2.11	2.57	2.90
18	.688	1.73	2.10	2.55	2.88
19	.688	1.73	2.09	2.54	2.86
20	.687	1.72	2.09	2.53	2.84
21	.686	1.72	2.08	2.52	2.83
22	.686	1.72	2.07	2.51	2.82
23	.685	1.71	2.07	2.50	2.81
24	.685	1.71	2.06	2.49	2.80
25	.684	1.71	2.06	2.48	2.79
26	.684	1.71	2.06	2.48	2.78
27	.684	1.70	2.05	2.47	2.77
28	.683	1.70	2.05	2.47	2.76
29	.683	1.70	2.04	2.46	2.76
30	.683	1.70	2.04	2.46	2.75
35	.682	1.69	2.03	2.44	2.72
40	.681	1.68	2.02	2.42	2.71
45	.680	1.68	2.02	2.41	2.69
50	.679	1.68	2.01	2.40	2.68
60	.678	1.67	2.00	2.39	2.66
70	.678	1.67	2.00	2.38	2.65
80	.677	1.66	1.99	2.38	2.64
90	.677	1.66	1.99	2.37	2.63
100	.677	1.66	1.98	2.36	2.63
125	.676	1.66	1.98	2.36	2.62
150	.676	1.66	1.98	2.35	2.61
200	.675	1.65	1.97	2.35	2.60
300	.675	1.65	1.97	2.34	2.59
400	.675	1.65	1.97	2.34	2.59
500	.674	1.65	1.96	2.33	2.59
1000	.674	1.65	1.96	2.33	2.58
∞	.674	1.64	1.96	2.33	2.58

'Table C is taken from Table III of Fisher & Yates': *Statistical Tables for Biological, Agricultural and Medical Research* published by Longman Group UK Ltd. London (previously published by Oliver and Boyd Ltd, Edinburgh) and by permission of the authors and publishers.

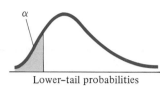

Lower–tail probabilities

Chi square distribution (lower-tail probabilities)

$df \backslash \alpha$	.001	.005	.010	.025	.050	.100
1	.000	.000	.000	.001	.004	.016
2	.002	.010	.020	.051	.103	.211
3	.024	.072	.115	.216	.352	.584
4	.091	.207	.297	.484	.711	1.06
5	.210	.412	.554	.831	1.15	1.61
6	.381	.676	.872	1.24	1.64	2.20
7	.598	.989	1.24	1.69	2.17	2.83
8	.857	1.34	1.65	2.18	2.73	3.49
9	1.15	1.73	2.09	2.70	3.33	4.17
10	1.48	2.16	2.56	3.25	3.94	4.87
11	1.83	2.60	3.05	3.82	4.57	5.58
12	2.21	3.07	3.57	4.40	5.23	6.30
13	2.62	3.57	4.11	5.01	5.89	7.04
14	3.04	4.07	4.66	5.63	6.57	7.79
15	3.48	4.60	5.23	6.26	7.26	8.55
16	3.94	5.14	5.81	6.91	7.96	9.31
17	4.42	5.70	6.41	7.56	8.67	10.1
18	4.90	6.26	7.01	8.23	9.39	10.9
19	5.41	6.84	7.63	8.91	10.1	11.7
20	5.92	7.43	8.26	9.59	10.9	12.4
21	6.45	8.03	8.90	10.3	11.6	13.2
22	6.98	8.64	9.54	11.0	12.3	14.0
23	7.53	9.26	10.2	11.7	13.1	14.8
24	8.08	9.89	10.9	12.4	13.8	15.7
25	8.65	10.5	11.5	13.1	14.6	16.5
26	9.22	11.2	12.2	13.8	15.4	17.3
27	9.80	11.8	12.9	14.6	16.2	18.1
28	10.4	12.5	13.6	15.3	16.9	18.9
29	11.0	13.1	14.3	16.0	17.7	19.8
30	11.6	13.8	15.0	16.8	18.5	20.6
35	14.7	17.2	18.5	20.6	22.5	24.8
40	17.9	20.7	22.2	24.4	26.5	29.1
45	21.3	24.3	25.9	28.4	30.6	33.4
50	24.7	28.0	29.7	32.4	34.8	37.7
55	28.2	31.7	33.6	36.4	39.0	42.1
60	31.7	35.5	37.5	40.5	43.2	46.5
65	35.4	39.4	41.4	44.6	47.4	50.9
70	39.0	43.3	45.4	48.8	51.7	55.3
75	42.8	47.2	49.5	52.9	56.1	59.8
80	46.5	51.2	53.5	57.2	60.4	64.3
85	50.3	55.2	57.6	61.4	64.7	68.8
90	54.2	59.2	61.8	65.6	69.1	73.3
95	58.0	63.2	65.9	69.9	73.5	77.8
100	61.9	67.3	70.1	74.2	77.9	82.4

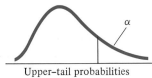

Upper–tail probabilities

TABLE D

Continued (upper-tail probabilities)

$df \backslash \alpha$	.100	.050	.025	.010	.005	.001
1	2.71	3.84	5.02	6.63	7.88	10.8
2	4.61	5.99	7.38	9.21	10.6	13.8
3	6.25	7.81	9.35	11.3	12.8	16.3
4	7.78	9.49	11.1	13.3	14.9	18.5
5	9.24	11.1	12.8	15.1	16.7	20.5
6	10.6	12.6	14.4	16.8	18.5	22.5
7	12.0	14.1	16.0	18.5	20.3	24.3
8	13.4	15.5	17.5	20.1	22.0	26.1
9	14.7	16.9	19.0	21.7	23.6	27.9
10	16.0	18.3	20.5	23.2	25.2	29.6
11	17.3	19.7	21.9	24.7	26.8	31.3
12	18.5	21.0	23.3	26.2	28.3	32.9
13	19.8	22.4	24.7	27.7	29.8	34.5
14	21.1	23.7	26.1	29.1	31.3	36.1
15	22.3	25.0	27.5	30.6	32.8	37.7
16	23.5	26.3	28.8	32.0	34.3	39.3
17	24.8	27.6	30.2	33.4	35.7	40.8
18	26.0	28.9	31.5	34.8	37.2	42.3
19	27.2	30.1	32.9	36.2	38.6	43.8
20	28.4	31.4	34.2	37.6	40.0	45.3
21	29.6	32.7	35.5	38.9	41.4	46.8
22	30.8	33.9	36.8	40.3	42.8	48.3
23	32.0	35.2	38.1	41.6	44.2	49.7
24	33.2	36.4	39.4	43.0	45.6	51.2
25	34.4	37.7	40.6	44.3	46.9	52.6
26	35.6	38.9	41.9	45.6	48.3	54.1
27	36.7	40.1	43.2	47.0	49.6	55.5
28	37.9	41.3	44.5	48.3	51.0	56.9
29	39.1	42.6	45.7	49.6	52.3	58.3
30	40.3	43.8	47.0	50.9	53.7	59.7
35	46.1	49.8	53.2	57.3	60.3	66.6
40	51.8	55.8	59.3	63.7	66.8	73.4
45	57.5	61.7	65.4	70.0	73.2	80.1
50	63.2	67.5	71.4	76.2	79.5	86.7
55	68.8	73.3	77.4	82.3	85.7	93.2
60	74.4	79.1	83.3	88.4	92.0	99.6
65	80.0	84.8	89.2	94.4	98.1	106.0
70	85.5	90.5	95.0	100.4	104.2	112.3
75	91.1	96.2	100.8	106.4	110.3	118.6
80	96.6	101.9	106.6	112.3	116.3	124.8
85	102.1	107.5	112.4	118.2	122.3	131.0
90	107.6	113.1	118.1	124.1	128.3	137.2
95	113.0	118.8	123.9	130.0	134.2	143.3
100	118.5	124.3	129.6	135.8	140.2	149.4

TABLE E

Critical values of F (0.05 level in medium type, 0.01 level in boldface type)

Degrees of freedom for greater mean square [numerator]

df (denom.)	1	2	3	4	5	6	7	8	9	10	11	12	14	16	20	24	30	40	50	75	100	200	500	∞
1	161 **4,052**	200 **4,999**	216 **5,403**	225 **5,625**	230 **5,764**	234 **5,859**	237 **5,928**	239 **5,981**	241 **6,022**	242 **6,056**	243 **6,082**	244 **6,106**	245 **6,142**	246 **6,169**	248 **6,208**	249 **6,234**	250 **6,261**	251 **6,286**	252 **6,302**	253 **6,323**	253 **6,334**	254 **6,352**	254 **6,361**	254 **6,366**
2	18.51 **98.49**	19.00 **99.00**	19.16 **99.17**	19.25 **99.25**	19.30 **99.30**	19.33 **99.33**	19.36 **99.36**	19.37 **99.37**	19.38 **99.39**	19.39 **99.40**	19.40 **99.41**	19.41 **99.42**	19.42 **99.43**	19.43 **99.44**	19.44 **99.45**	19.45 **99.46**	19.46 **99.47**	19.47 **99.48**	19.47 **99.48**	19.48 **99.49**	19.49 **99.49**	19.49 **99.49**	19.50 **99.50**	19.50 **99.50**
3	10.13 **34.12**	9.55 **30.82**	9.28 **29.46**	9.12 **28.71**	9.01 **28.24**	8.94 **27.91**	8.88 **27.67**	8.84 **27.49**	8.81 **27.34**	8.78 **27.23**	8.76 **27.13**	8.74 **27.05**	8.71 **26.92**	8.69 **26.83**	8.66 **26.69**	8.64 **26.60**	8.62 **26.50**	8.60 **26.41**	8.58 **26.35**	8.57 **26.27**	8.56 **26.23**	8.54 **26.18**	8.54 **26.14**	8.53 **26.12**
4	7.71 **21.20**	6.94 **18.00**	6.59 **16.69**	6.39 **15.98**	6.26 **15.52**	6.16 **15.21**	6.09 **14.98**	6.04 **14.80**	6.00 **14.66**	5.96 **14.54**	5.93 **14.45**	5.91 **14.37**	5.87 **14.24**	5.84 **14.15**	5.80 **14.02**	5.77 **13.93**	5.74 **13.83**	5.71 **13.74**	5.70 **13.69**	5.68 **13.61**	5.66 **13.57**	5.65 **13.52**	5.64 **13.48**	5.63 **13.46**
5	6.61 **16.26**	5.79 **13.27**	5.41 **12.06**	5.19 **11.39**	5.05 **10.97**	4.95 **10.67**	4.88 **10.45**	4.82 **10.29**	4.78 **10.15**	4.74 **10.05**	4.70 **9.96**	4.68 **9.89**	4.64 **9.77**	4.60 **9.68**	4.56 **9.55**	4.53 **9.47**	4.50 **9.38**	4.46 **9.29**	4.44 **9.24**	4.42 **9.17**	4.40 **9.13**	4.38 **9.07**	4.37 **9.04**	4.36 **9.02**
6	5.99 **13.74**	5.14 **10.92**	4.76 **9.78**	4.53 **9.15**	4.39 **8.75**	4.28 **8.47**	4.21 **8.26**	4.15 **8.10**	4.10 **7.98**	4.06 **7.87**	4.03 **7.79**	4.00 **7.72**	3.96 **7.60**	3.92 **7.52**	3.87 **7.39**	3.84 **7.31**	3.81 **7.23**	3.77 **7.14**	3.75 **7.09**	3.72 **7.02**	3.71 **6.99**	3.69 **6.94**	3.68 **6.90**	3.67 **6.88**
7	5.59 **12.25**	4.74 **9.55**	4.35 **8.45**	4.12 **7.85**	3.97 **7.46**	3.87 **7.19**	3.79 **7.00**	3.73 **6.84**	3.68 **6.71**	3.63 **6.62**	3.60 **6.54**	3.57 **6.47**	3.52 **6.35**	3.49 **6.27**	3.44 **6.15**	3.41 **6.07**	3.38 **5.98**	3.34 **5.90**	3.32 **5.85**	3.29 **5.78**	3.28 **5.75**	3.25 **5.70**	3.24 **5.67**	3.23 **5.65**
8	5.32 **11.26**	4.46 **8.65**	4.07 **7.59**	3.84 **7.01**	3.69 **6.63**	3.58 **6.37**	3.50 **6.19**	3.44 **6.03**	3.39 **5.91**	3.34 **5.82**	3.31 **5.74**	3.28 **5.67**	3.23 **5.56**	3.20 **5.48**	3.15 **5.36**	3.12 **5.28**	3.08 **5.20**	3.05 **5.11**	3.03 **5.06**	3.00 **5.00**	2.98 **4.96**	2.96 **4.91**	2.94 **4.88**	2.93 **4.86**
9	5.12 **10.56**	4.26 **8.02**	3.86 **6.99**	3.63 **6.42**	3.48 **6.06**	3.37 **5.80**	3.29 **5.62**	3.23 **5.47**	3.18 **5.35**	3.13 **5.26**	3.10 **5.18**	3.07 **5.11**	3.02 **5.00**	2.98 **4.92**	2.93 **4.80**	2.90 **4.73**	2.86 **4.64**	2.82 **4.56**	2.80 **4.51**	2.77 **4.45**	2.76 **4.41**	2.73 **4.36**	2.72 **4.33**	2.71 **4.31**
10	4.96 **10.04**	4.10 **7.56**	3.71 **6.55**	3.48 **5.99**	3.33 **5.64**	3.22 **5.39**	3.14 **5.21**	3.07 **5.06**	3.02 **4.95**	2.97 **4.85**	2.94 **4.78**	2.91 **4.71**	2.86 **4.60**	2.82 **4.52**	2.77 **4.41**	2.74 **4.33**	2.70 **4.25**	2.67 **4.17**	2.64 **4.12**	2.61 **4.05**	2.59 **4.01**	2.56 **3.96**	2.55 **3.93**	2.54 **3.91**
11	4.84 **9.65**	3.98 **7.20**	3.59 **6.22**	3.36 **5.67**	3.20 **5.32**	3.09 **5.07**	3.01 **4.88**	2.95 **4.74**	2.90 **4.63**	2.86 **4.54**	2.82 **4.46**	2.79 **4.40**	2.74 **4.29**	2.70 **4.21**	2.65 **4.10**	2.61 **4.02**	2.57 **3.94**	2.53 **3.86**	2.50 **3.80**	2.47 **3.74**	2.45 **3.70**	2.42 **3.66**	2.41 **3.62**	2.40 **3.60**
12	4.75 **9.33**	3.88 **6.93**	3.49 **5.95**	3.26 **5.41**	3.11 **5.06**	3.00 **4.82**	2.92 **4.65**	2.85 **4.50**	2.80 **4.39**	2.76 **4.30**	2.72 **4.22**	2.69 **4.16**	2.64 **4.05**	2.60 **3.98**	2.54 **3.86**	2.50 **3.78**	2.46 **3.70**	2.42 **3.61**	2.40 **3.56**	2.36 **3.49**	2.35 **3.46**	2.32 **3.41**	2.31 **3.38**	2.30 **3.36**
13	4.67 **9.07**	3.80 **6.70**	3.41 **5.74**	3.18 **5.20**	3.02 **4.86**	2.92 **4.62**	2.84 **4.44**	2.77 **4.30**	2.72 **4.19**	2.67 **4.10**	2.63 **4.02**	2.60 **3.96**	2.55 **3.85**	2.51 **3.78**	2.46 **3.67**	2.42 **3.59**	2.38 **3.51**	2.34 **3.42**	2.32 **3.37**	2.28 **3.30**	2.26 **3.27**	2.24 **3.21**	2.22 **3.18**	2.21 **3.16**
14	4.60 **8.86**	3.74 **6.51**	3.34 **5.56**	3.11 **5.03**	2.96 **4.69**	2.85 **4.46**	2.77 **4.28**	2.70 **4.14**	2.65 **4.03**	2.60 **3.94**	2.56 **3.86**	2.53 **3.80**	2.48 **3.70**	2.44 **3.62**	2.39 **3.51**	2.35 **3.43**	2.31 **3.34**	2.27 **3.26**	2.24 **3.21**	2.21 **3.14**	2.19 **3.11**	2.16 **3.06**	2.14 **3.02**	2.13 **3.00**
15	4.54 **8.68**	3.68 **6.36**	3.29 **5.42**	3.06 **4.89**	2.90 **4.56**	2.79 **4.32**	2.70 **4.14**	2.64 **4.00**	2.59 **3.89**	2.55 **3.80**	2.51 **3.73**	2.48 **3.67**	2.43 **3.56**	2.39 **3.48**	2.33 **3.36**	2.29 **3.29**	2.25 **3.20**	2.21 **3.12**	2.18 **3.07**	2.15 **3.00**	2.12 **2.97**	2.10 **2.92**	2.08 **2.89**	2.07 **2.87**
16	4.49 **8.53**	3.63 **6.23**	3.24 **5.29**	3.01 **4.77**	2.85 **4.44**	2.74 **4.20**	2.66 **4.03**	2.59 **3.89**	2.54 **3.78**	2.49 **3.69**	2.45 **3.61**	2.42 **3.55**	2.37 **3.45**	2.33 **3.37**	2.28 **3.25**	2.24 **3.18**	2.20 **3.10**	2.16 **3.01**	2.13 **2.96**	2.09 **2.89**	2.07 **2.86**	2.04 **2.80**	2.02 **2.77**	2.01 **2.75**
17	4.45 **8.40**	3.59 **6.11**	3.20 **5.18**	2.96 **4.67**	2.81 **4.34**	2.70 **4.10**	2.62 **3.93**	2.55 **3.79**	2.50 **3.68**	2.45 **3.59**	2.41 **3.52**	2.38 **3.45**	2.33 **3.35**	2.29 **3.27**	2.23 **3.16**	2.19 **3.08**	2.15 **3.00**	2.11 **2.92**	2.08 **2.86**	2.04 **2.79**	2.02 **2.76**	1.99 **2.70**	1.97 **2.67**	1.96 **2.65**
18	4.41 **8.28**	3.55 **6.01**	3.16 **5.09**	2.93 **4.58**	2.77 **4.25**	2.66 **4.01**	2.58 **3.85**	2.51 **3.71**	2.46 **3.60**	2.41 **3.51**	2.37 **3.44**	2.34 **3.37**	2.29 **3.27**	2.25 **3.19**	2.19 **3.07**	2.15 **3.00**	2.11 **2.91**	2.07 **2.83**	2.04 **2.78**	2.00 **2.71**	1.98 **2.68**	1.95 **2.62**	1.93 **2.59**	1.92 **2.57**
19	4.38 **8.18**	3.52 **5.93**	3.13 **5.01**	2.90 **4.50**	2.74 **4.17**	2.63 **3.94**	2.55 **3.77**	2.48 **3.63**	2.43 **3.52**	2.38 **3.43**	2.34 **3.36**	2.31 **3.30**	2.26 **3.19**	2.21 **3.12**	2.15 **3.00**	2.11 **2.92**	2.07 **2.84**	2.02 **2.76**	2.00 **2.70**	1.96 **2.63**	1.94 **2.60**	1.91 **2.54**	1.90 **2.51**	1.88 **2.49**

Degrees of freedom for lesser mean square [denominator]

Degrees of freedom for greater mean square [numerator]

(denominator)	1	2	3	4	5	6	7	8	9	10	11	12	14	16	20	24	30	40	50	75	100	200	500	∞
20	4.35 **8.10**	3.49 **5.85**	3.10 **4.94**	2.87 **4.43**	2.71 **4.10**	2.60 **3.87**	2.52 **3.71**	2.45 **3.56**	2.40 **3.45**	2.35 **3.37**	2.31 **3.30**	2.28 **3.23**	2.23 **3.13**	2.18 **3.05**	2.12 **2.94**	2.08 **2.86**	2.04 **2.77**	1.99 **2.69**	1.96 **2.63**	1.92 **2.56**	1.90 **2.53**	1.87 **2.47**	1.85 **2.44**	1.84 **2.42**
21	4.32 **8.02**	3.47 **5.78**	3.07 **4.87**	2.84 **4.37**	2.68 **4.04**	2.57 **3.81**	2.49 **3.65**	2.42 **3.51**	2.37 **3.40**	2.32 **3.31**	2.28 **3.24**	2.25 **3.17**	2.20 **3.07**	2.15 **2.99**	2.09 **2.88**	2.05 **2.80**	2.00 **2.72**	1.96 **2.63**	1.93 **2.58**	1.89 **2.51**	1.87 **2.47**	1.84 **2.42**	1.82 **2.38**	1.81 **2.36**
22	4.30 **7.94**	3.44 **5.72**	3.05 **4.82**	2.82 **4.31**	2.66 **3.99**	2.55 **3.76**	2.47 **3.59**	2.40 **3.45**	2.35 **3.35**	2.30 **3.26**	2.26 **3.18**	2.23 **3.12**	2.18 **3.02**	2.13 **2.94**	2.07 **2.83**	2.03 **2.75**	1.98 **2.67**	1.93 **2.58**	1.91 **2.53**	1.87 **2.46**	1.84 **2.42**	1.81 **2.37**	1.80 **2.33**	1.78 **2.31**
23	4.28 **7.88**	3.42 **5.66**	3.03 **4.76**	2.80 **4.26**	2.64 **3.94**	2.53 **3.71**	2.45 **3.54**	2.38 **3.41**	2.32 **3.30**	2.28 **3.21**	2.24 **3.14**	2.20 **3.07**	2.14 **2.97**	2.10 **2.89**	2.04 **2.78**	2.00 **2.70**	1.96 **2.62**	1.91 **2.53**	1.88 **2.48**	1.84 **2.41**	1.82 **2.37**	1.79 **2.32**	1.77 **2.28**	1.76 **2.26**
24	4.26 **7.82**	3.40 **5.61**	3.01 **4.72**	2.78 **4.22**	2.62 **3.90**	2.51 **3.67**	2.43 **3.50**	2.36 **3.36**	2.30 **3.25**	2.26 **3.17**	2.22 **3.09**	2.18 **3.03**	2.13 **2.93**	2.09 **2.85**	2.02 **2.74**	1.98 **2.66**	1.94 **2.58**	1.89 **2.49**	1.86 **2.44**	1.82 **2.36**	1.80 **2.33**	1.76 **2.27**	1.74 **2.23**	1.73 **2.21**
25	4.24 **7.77**	3.38 **5.57**	2.99 **4.68**	2.76 **4.18**	2.60 **3.86**	2.49 **3.63**	2.41 **3.46**	2.34 **3.32**	2.28 **3.21**	2.24 **3.13**	2.20 **3.05**	2.16 **2.99**	2.11 **2.89**	2.06 **2.81**	2.00 **2.70**	1.96 **2.62**	1.92 **2.54**	1.87 **2.45**	1.84 **2.40**	1.80 **2.32**	1.77 **2.29**	1.74 **2.23**	1.72 **2.19**	1.71 **2.17**
26	4.22 **7.72**	3.37 **5.53**	2.98 **4.64**	2.74 **4.14**	2.59 **3.82**	2.47 **3.59**	2.39 **3.42**	2.32 **3.29**	2.27 **3.17**	2.22 **3.09**	2.18 **3.02**	2.15 **2.96**	2.10 **2.86**	2.05 **2.77**	1.99 **2.66**	1.95 **2.58**	1.90 **2.50**	1.85 **2.41**	1.82 **2.36**	1.78 **2.28**	1.76 **2.25**	1.72 **2.19**	1.70 **2.15**	1.69 **2.13**
27	4.21 **7.68**	3.35 **5.49**	2.96 **4.60**	2.73 **4.11**	2.57 **3.79**	2.46 **3.56**	2.37 **3.39**	2.30 **3.26**	2.25 **3.14**	2.20 **3.06**	2.16 **2.98**	2.13 **2.93**	2.08 **2.83**	2.03 **2.74**	1.97 **2.63**	1.93 **2.55**	1.88 **2.47**	1.84 **2.38**	1.80 **2.33**	1.76 **2.25**	1.74 **2.21**	1.71 **2.16**	1.68 **2.12**	1.67 **2.10**
28	4.20 **7.64**	3.34 **5.45**	2.95 **4.57**	2.71 **4.07**	2.56 **3.76**	2.44 **3.53**	2.36 **3.36**	2.29 **3.23**	2.24 **3.11**	2.19 **3.03**	2.15 **2.95**	2.12 **2.90**	2.06 **2.80**	2.02 **2.71**	1.96 **2.60**	1.91 **2.52**	1.87 **2.44**	1.81 **2.35**	1.78 **2.30**	1.75 **2.22**	1.72 **2.18**	1.69 **2.13**	1.67 **2.09**	1.65 **2.06**
29	4.18 **7.60**	3.33 **5.42**	2.93 **4.54**	2.70 **4.04**	2.54 **3.73**	2.43 **3.50**	2.35 **3.33**	2.28 **3.20**	2.22 **3.08**	2.18 **3.00**	2.14 **2.92**	2.10 **2.87**	2.05 **2.77**	2.00 **2.68**	1.94 **2.57**	1.90 **2.49**	1.85 **2.41**	1.80 **2.32**	1.77 **2.27**	1.73 **2.19**	1.71 **2.15**	1.68 **2.10**	1.65 **2.06**	1.64 **2.03**
30	4.17 **7.56**	3.32 **5.39**	2.92 **4.51**	2.69 **4.02**	2.53 **3.70**	2.42 **3.47**	2.34 **3.30**	2.27 **3.17**	2.21 **3.06**	2.16 **2.98**	2.12 **2.90**	2.09 **2.84**	2.04 **2.74**	1.99 **2.66**	1.93 **2.55**	1.89 **2.47**	1.84 **2.38**	1.79 **2.29**	1.76 **2.24**	1.72 **2.16**	1.69 **2.13**	1.66 **2.07**	1.64 **2.03**	1.62 **2.01**
32	4.15 **7.50**	3.30 **5.34**	2.90 **4.46**	2.67 **3.97**	2.51 **3.66**	2.40 **3.42**	2.32 **3.25**	2.25 **3.12**	2.19 **3.01**	2.14 **2.94**	2.10 **2.86**	2.07 **2.80**	2.02 **2.70**	1.97 **2.62**	1.91 **2.51**	1.86 **2.42**	1.82 **2.34**	1.76 **2.25**	1.74 **2.20**	1.69 **2.12**	1.67 **2.08**	1.64 **2.02**	1.61 **1.98**	1.59 **1.96**
34	4.13 **7.44**	3.28 **5.29**	2.88 **4.42**	2.65 **3.93**	2.49 **3.61**	2.38 **3.38**	2.30 **3.21**	2.23 **3.08**	2.17 **2.97**	2.12 **2.89**	2.08 **2.82**	2.05 **2.76**	2.00 **2.66**	1.95 **2.58**	1.89 **2.47**	1.84 **2.38**	1.80 **2.30**	1.74 **2.21**	1.71 **2.15**	1.67 **2.08**	1.64 **2.04**	1.61 **1.98**	1.59 **1.94**	1.57 **1.91**
36	4.11 **7.39**	3.26 **5.25**	2.86 **4.38**	2.63 **3.89**	2.48 **3.58**	2.36 **3.35**	2.28 **3.18**	2.21 **3.04**	2.15 **2.94**	2.10 **2.86**	2.06 **2.78**	2.03 **2.72**	1.98 **2.62**	1.93 **2.54**	1.87 **2.43**	1.82 **2.35**	1.78 **2.26**	1.72 **2.17**	1.69 **2.12**	1.65 **2.04**	1.62 **2.00**	1.59 **1.94**	1.56 **1.90**	1.55 **1.87**
38	4.10 **7.35**	3.25 **5.21**	2.85 **4.34**	2.62 **3.86**	2.46 **3.54**	2.35 **3.32**	2.26 **3.15**	2.19 **3.02**	2.14 **2.91**	2.09 **2.82**	2.05 **2.75**	2.02 **2.69**	1.96 **2.59**	1.92 **2.51**	1.85 **2.40**	1.80 **2.32**	1.76 **2.22**	1.71 **2.14**	1.67 **2.08**	1.63 **2.00**	1.60 **1.97**	1.57 **1.90**	1.54 **1.86**	1.53 **1.84**
40	4.08 **7.31**	3.23 **5.18**	2.84 **4.31**	2.61 **3.83**	2.45 **3.51**	2.34 **3.29**	2.25 **3.12**	2.18 **2.99**	2.12 **2.88**	2.07 **2.80**	2.04 **2.73**	2.00 **2.66**	1.95 **2.56**	1.90 **2.49**	1.84 **2.37**	1.79 **2.29**	1.74 **2.20**	1.69 **2.11**	1.66 **2.05**	1.61 **1.97**	1.59 **1.94**	1.55 **1.88**	1.53 **1.84**	1.51 **1.81**
42	4.07 **7.27**	3.22 **5.15**	2.83 **4.29**	2.59 **3.80**	2.44 **3.49**	2.32 **3.26**	2.24 **3.10**	2.17 **2.96**	2.11 **2.86**	2.06 **2.77**	2.02 **2.70**	1.99 **2.64**	1.94 **2.54**	1.89 **2.46**	1.82 **2.35**	1.78 **2.26**	1.73 **2.17**	1.68 **2.08**	1.64 **2.02**	1.60 **1.94**	1.57 **1.91**	1.54 **1.85**	1.51 **1.80**	1.49 **1.78**
44	4.06 **7.24**	3.21 **5.12**	2.82 **4.26**	2.58 **3.78**	2.43 **3.46**	2.31 **3.24**	2.23 **3.07**	2.16 **2.94**	2.10 **2.84**	2.05 **2.75**	2.01 **2.68**	1.98 **2.62**	1.92 **2.52**	1.88 **2.44**	1.81 **2.32**	1.76 **2.24**	1.72 **2.15**	1.66 **2.06**	1.63 **2.00**	1.58 **1.92**	1.56 **1.88**	1.52 **1.82**	1.50 **1.78**	1.48 **1.75**

Degrees of freedom for lesser mean square [denominator]

Degrees of freedom for greater mean square [numerator]

	1	2	3	4	5	6	7	8	9	10	11	12	14	16	20	24	30	40	50	75	100	200	500	∞
46	4.05 **7.21**	3.20 **5.10**	2.81 **4.24**	2.57 **3.76**	2.42 **3.44**	2.30 **3.22**	2.22 **3.05**	2.14 **2.92**	2.09 **2.82**	2.04 **2.73**	2.00 **2.66**	1.97 **2.60**	1.91 **2.50**	1.87 **2.42**	1.80 **2.30**	1.75 **2.22**	1.71 **2.13**	1.65 **2.04**	1.62 **1.98**	1.57 **1.90**	1.54 **1.86**	1.51 **1.80**	1.48 **1.76**	1.46 **1.72**
48	4.04 **7.19**	3.19 **5.08**	2.80 **4.22**	2.56 **3.74**	2.41 **3.42**	2.30 **3.20**	2.21 **3.04**	2.14 **2.90**	2.08 **2.80**	2.03 **2.71**	1.99 **2.64**	1.96 **2.58**	1.90 **2.48**	1.86 **2.40**	1.79 **2.28**	1.74 **2.20**	1.70 **2.11**	1.64 **2.02**	1.61 **1.96**	1.56 **1.88**	1.53 **1.84**	1.50 **1.78**	1.47 **1.73**	1.45 **1.70**
50	4.03 **7.17**	3.18 **5.06**	2.79 **4.20**	2.56 **3.72**	2.40 **3.41**	2.29 **3.18**	2.20 **3.02**	2.13 **2.88**	2.07 **2.78**	2.02 **2.70**	1.98 **2.62**	1.95 **2.56**	1.90 **2.46**	1.85 **2.39**	1.78 **2.26**	1.74 **2.18**	1.69 **2.10**	1.63 **2.00**	1.60 **1.94**	1.55 **1.86**	1.52 **1.82**	1.48 **1.76**	1.46 **1.71**	1.44 **1.68**
55	4.02 **7.12**	3.17 **5.01**	2.78 **4.16**	2.54 **3.68**	2.38 **3.37**	2.27 **3.15**	2.18 **2.98**	2.11 **2.85**	2.05 **2.75**	2.00 **2.66**	1.97 **2.59**	1.93 **2.53**	1.88 **2.43**	1.83 **2.35**	1.76 **2.23**	1.72 **2.15**	1.67 **2.06**	1.61 **1.96**	1.58 **1.90**	1.52 **1.82**	1.50 **1.78**	1.46 **1.71**	1.43 **1.66**	1.41 **1.64**
60	4.00 **7.08**	3.15 **4.98**	2.76 **4.13**	2.52 **3.65**	2.37 **3.34**	2.25 **3.12**	2.17 **2.95**	2.10 **2.82**	2.04 **2.72**	1.99 **2.63**	1.95 **2.56**	1.92 **2.50**	1.86 **2.40**	1.81 **2.32**	1.75 **2.20**	1.70 **2.12**	1.65 **2.03**	1.59 **1.93**	1.56 **1.87**	1.50 **1.79**	1.48 **1.74**	1.44 **1.68**	1.41 **1.63**	1.39 **1.60**
65	3.99 **7.04**	3.14 **4.95**	2.75 **4.10**	2.51 **3.62**	2.36 **3.31**	2.24 **3.09**	2.15 **2.93**	2.08 **2.79**	2.02 **2.70**	1.98 **2.61**	1.94 **2.54**	1.90 **2.47**	1.85 **2.37**	1.80 **2.30**	1.73 **2.18**	1.68 **2.09**	1.63 **2.00**	1.57 **1.90**	1.54 **1.84**	1.49 **1.76**	1.46 **1.71**	1.42 **1.64**	1.39 **1.60**	1.37 **1.56**
70	3.98 **7.01**	3.13 **4.92**	2.74 **4.08**	2.50 **3.60**	2.35 **3.29**	2.23 **3.07**	2.14 **2.91**	2.07 **2.77**	2.01 **2.67**	1.97 **2.59**	1.93 **2.51**	1.89 **2.45**	1.84 **2.35**	1.79 **2.28**	1.72 **2.15**	1.67 **2.07**	1.62 **1.98**	1.56 **1.88**	1.53 **1.82**	1.47 **1.74**	1.45 **1.69**	1.40 **1.62**	1.37 **1.56**	1.35 **1.53**
80	3.96 **6.96**	3.11 **4.88**	2.72 **4.04**	2.48 **3.56**	2.33 **3.25**	2.21 **3.04**	2.12 **2.87**	2.05 **2.74**	1.99 **2.64**	1.95 **2.55**	1.91 **2.48**	1.88 **2.41**	1.82 **2.32**	1.77 **2.24**	1.70 **2.11**	1.65 **2.03**	1.60 **1.94**	1.54 **1.84**	1.51 **1.78**	1.45 **1.70**	1.42 **1.65**	1.38 **1.57**	1.35 **1.52**	1.32 **1.49**
100	3.94 **6.90**	3.09 **4.82**	2.70 **3.98**	2.46 **3.51**	2.30 **3.20**	2.19 **2.99**	2.10 **2.82**	2.03 **2.69**	1.97 **2.59**	1.92 **2.51**	1.88 **2.43**	1.85 **2.36**	1.79 **2.26**	1.75 **2.19**	1.68 **2.06**	1.63 **1.98**	1.57 **1.89**	1.51 **1.79**	1.48 **1.73**	1.42 **1.64**	1.39 **1.59**	1.34 **1.51**	1.30 **1.46**	1.28 **1.43**
125	3.92 **6.84**	3.07 **4.78**	2.68 **3.94**	2.44 **3.47**	2.29 **3.17**	2.17 **2.95**	2.08 **2.79**	2.01 **2.65**	1.95 **2.56**	1.90 **2.47**	1.86 **2.40**	1.83 **2.33**	1.77 **2.23**	1.72 **2.15**	1.65 **2.03**	1.60 **1.94**	1.55 **1.85**	1.49 **1.75**	1.45 **1.68**	1.39 **1.59**	1.36 **1.54**	1.31 **1.46**	1.27 **1.40**	1.25 **1.37**
150	3.91 **6.81**	3.06 **4.75**	2.67 **3.91**	2.43 **3.44**	2.27 **3.14**	2.16 **2.92**	2.07 **2.76**	2.00 **2.62**	1.94 **2.53**	1.89 **2.44**	1.85 **2.37**	1.82 **2.30**	1.76 **2.20**	1.71 **2.12**	1.64 **2.00**	1.59 **1.91**	1.54 **1.83**	1.47 **1.72**	1.44 **1.66**	1.37 **1.56**	1.34 **1.51**	1.29 **1.43**	1.25 **1.37**	1.22 **1.33**
200	3.89 **6.76**	3.04 **4.71**	2.65 **3.88**	2.41 **3.41**	2.26 **3.11**	2.14 **2.90**	2.05 **2.73**	1.98 **2.60**	1.92 **2.50**	1.87 **2.41**	1.83 **2.34**	1.80 **2.28**	1.74 **2.17**	1.69 **2.09**	1.62 **1.97**	1.57 **1.88**	1.52 **1.79**	1.45 **1.69**	1.42 **1.62**	1.35 **1.53**	1.32 **1.48**	1.26 **1.39**	1.22 **1.33**	1.19 **1.28**
400	3.86 **6.70**	3.02 **4.66**	2.62 **3.83**	2.39 **3.36**	2.23 **3.06**	2.12 **2.85**	2.03 **2.69**	1.96 **2.55**	1.90 **2.46**	1.85 **2.37**	1.81 **2.29**	1.78 **2.23**	1.72 **2.12**	1.67 **2.04**	1.60 **1.92**	1.54 **1.84**	1.49 **1.74**	1.42 **1.64**	1.38 **1.57**	1.32 **1.47**	1.28 **1.42**	1.22 **1.32**	1.16 **1.24**	1.13 **1.19**
1000	3.85 **6.66**	3.00 **4.62**	2.61 **3.80**	2.38 **3.34**	2.22 **3.04**	2.10 **2.82**	2.02 **2.66**	1.95 **2.53**	1.89 **2.43**	1.84 **2.34**	1.80 **2.26**	1.76 **2.20**	1.70 **2.09**	1.65 **2.01**	1.58 **1.89**	1.53 **1.81**	1.47 **1.71**	1.41 **1.61**	1.36 **1.54**	1.30 **1.44**	1.26 **1.38**	1.19 **1.28**	1.13 **1.19**	1.08 **1.11**
∞	3.84 **6.64**	2.99 **4.60**	2.60 **3.78**	2.37 **3.32**	2.21 **3.02**	2.09 **2.80**	2.01 **2.64**	1.94 **2.51**	1.88 **2.41**	1.83 **2.32**	1.79 **2.24**	1.75 **2.18**	1.69 **2.07**	1.64 **1.99**	1.57 **1.87**	1.52 **1.79**	1.46 **1.69**	1.40 **1.59**	1.35 **1.52**	1.28 **1.41**	1.24 **1.36**	1.17 **1.25**	1.11 **1.15**	1.00 **1.00**

Degrees of freedom for lesser mean square [denominator]

Reprinted by permission from *Statistical Methods, Seventh Edition* by G. W. Snedecor and W. G. Cochran © 1980 by the Iowa State University Press, Ames, Iowa 50010.

TABLE F

Conversion of a Pearson r into a Fisher z' coefficient

r	z'	r	z'	r	z'	r	z'	r	z'
.000	.000	.200	.203	.400	.424	.600	.693	.800	1.099
.005	.005	.205	.208	.405	.430	.605	.701	.805	1.113
.010	.010	.210	.213	.410	.436	.610	.709	.810	1.127
.015	.015	.215	.218	.415	.442	.615	.717	.815	1.142
.020	.020	.220	.224	.420	.448	.620	.725	.820	1.157
.025	.025	.225	.229	.425	.454	.625	.733	.825	1.172
.030	.030	.230	.234	.430	.460	.630	.741	.830	1.188
.035	.035	.235	.239	.435	.466	.635	.750	.835	1.204
.040	.040	.240	.245	.440	.472	.640	.758	.840	1.221
.045	.045	.245	.250	.445	.478	.645	.767	.845	1.238
.050	.050	.250	.255	.450	.485	.650	.775	.850	1.256
.055	.055	.255	.261	.455	.491	.655	.784	.855	1.274
.060	.060	.260	.266	.460	.497	.660	.793	.860	1.293
.065	.065	.265	.271	.465	.504	.665	.802	.865	1.313
.070	.070	.270	.277	.470	.510	.670	.811	.870	1.333
.075	.075	.275	.282	.475	.517	.675	.820	.875	1.354
.080	.080	.280	.288	.480	.523	.680	.829	.880	1.376
.085	.085	.285	.293	.485	.530	.685	.838	.885	1.398
.090	.090	.290	.299	.490	.536	.690	.848	.890	1.422
.095	.095	.295	.304	.495	.543	.695	.858	.895	1.447
.100	.100	.300	.310	.500	.549	.700	.867	.900	1.472
.105	.105	.305	.315	.505	.556	.705	.877	.905	1.499
.110	.110	.310	.321	.510	.563	.710	.887	.910	1.528
.115	.116	.315	.326	.515	.570	.715	.897	.915	1.557
.120	.121	.320	.332	.520	.576	.720	.908	.920	1.589
.125	.126	.325	.337	.525	.583	.725	.918	.925	1.623
.130	.131	.330	.343	.530	.590	.730	.929	.930	1.658
.135	.136	.335	.348	.535	.597	.735	.940	.935	1.697
.140	.141	.340	.354	.540	.604	.740	.950	.940	1.738
.145	.146	.345	.360	.545	.611	.745	.962	.945	1.783
.150	.151	.350	.365	.550	.618	.750	.973	.950	1.832
.155	.156	.355	.371	.555	.626	.755	.984	.955	1.886
.160	.161	.360	.377	.560	.633	.760	.996	.960	1.946
.165	.167	.365	.383	.565	.640	.765	1.008	.965	2.014
.170	.172	.370	.388	.570	.648	.770	1.020	.970	2.092
.175	.177	.375	.394	.575	.655	.775	1.033	.975	2.185
.180	.182	.380	.400	.580	.662	.780	1.045	.980	2.298
.185	.187	.385	.406	.585	.670	.785	1.058	.985	2.443
.190	.192	.390	.412	.590	.678	.790	1.071	.990	2.647
.195	.198	.395	.418	.595	.685	.795	1.085	.995	2.994

Reprinted with permission of Macmillan Publishing Company from *Statistical Methods for Research Workers*, 14th edition, by Ronald A. Fisher. Copyright © 1970 by University of Adelaide.

TABLE G

Critical values of V for the sign test

N	.01	.05	N	.01	.05
1			46	13	15
2			47	14	16
3			48	14	16
4			49	15	17
5			50	15	17
6		0	51	15	18
7		0	52	16	18
8	0	0	53	16	18
9	0	1	54	17	19
10	0	1	55	17	19
11	0	1	56	17	20
12	1	2	57	18	20
13	1	2	58	18	21
14	1	2	59	19	21
15	2	3	60	19	21
16	2	3	61	20	22
17	2	4	62	20	22
18	3	4	63	20	23
19	3	4	64	21	23
20	3	5	65	21	24
21	4	5	66	22	24
22	4	5	67	22	25
23	4	6	68	22	25
24	5	6	69	23	25
25	5	7	70	23	26
26	6	7	71	24	26
27	6	7	72	24	27
28	6	8	73	25	27
29	7	8	74	25	28
30	7	9	75	25	28
31	7	9	76	26	28
32	8	9	77	26	29
33	8	10	78	27	29
34	9	10	79	27	30
35	9	11	80	28	30
36	9	11	81	28	31
37	10	12	82	28	31
38	10	12	83	29	32
39	11	12	84	29	32
40	11	13	85	30	32
41	11	13	86	30	33
42	12	14	87	31	33
43	12	14	88	31	34
44	13	15	89	31	34
45	13	15	90	32	35

TABLE H

Critical values of the Mann-Whitney test statistic

n	α	m=2	3	4	5	6	7	8	9	10	11	12	13	14	15	16	17	18	19	20
	.001	0	0	0	0	0	0	0	0	0	0	0	0	0	0	0	0	0	0	0
	.005	0	0	0	0	0	0	0	0	0	0	0	0	0	0	0	0	0	1	1
2	.01	0	0	0	0	0	0	0	0	0	0	0	1	1	1	1	1	1	2	2
	.025	0	0	0	0	0	0	1	1	1	1	2	2	2	2	2	3	3	3	3
	.05	0	0	0	1	1	1	2	2	2	2	3	3	4	4	4	4	5	5	5
	.10	0	1	1	2	2	2	3	3	4	4	5	5	5	6	6	7	7	8	8
	.001	0	0	0	0	0	0	0	0	0	0	0	0	0	0	0	1	1	1	1
	.005	0	0	0	0	0	0	0	1	1	1	2	2	2	3	3	3	3	4	4
3	.01	0	0	0	0	0	1	1	2	2	2	3	3	3	4	4	5	5	5	6
	.025	0	0	0	1	2	2	3	3	4	4	5	5	6	6	7	7	8	8	9
	.05	0	1	1	2	3	3	4	5	5	6	6	7	8	8	9	10	10	11	12
	.10	1	2	2	3	4	5	6	6	7	8	9	10	11	11	12	13	14	15	16
	.001	0	0	0	0	0	0	0	0	1	1	1	2	2	2	3	3	4	4	4
	.005	0	0	0	0	1	1	2	2	3	3	4	4	5	6	6	7	7	8	9
4	.01	0	0	0	1	2	2	3	4	4	5	6	6	7	8	9	9	10	10	11
	.025	0	0	1	2	3	4	5	5	6	7	8	9	10	11	12	12	13	14	15
	.05	0	1	2	3	4	5	6	7	8	9	10	11	12	13	15	16	17	18	19
	.10	1	2	4	5	6	7	8	10	11	12	13	14	16	17	18	19	21	22	23
	.001	0	0	0	0	0	0	1	2	2	3	3	4	4	5	6	6	7	8	8
	.005	0	0	0	1	2	2	3	4	5	6	7	8	8	9	10	11	12	13	14
5	.01	0	0	1	2	3	4	5	6	7	8	9	10	11	12	13	14	15	16	17
	.025	0	1	2	3	4	6	7	8	9	10	12	13	14	15	16	18	19	20	21
	.05	1	2	3	5	6	7	9	10	12	13	14	16	17	19	20	21	23	24	26
	.10	2	3	5	6	8	9	11	13	14	16	18	19	21	23	24	26	28	29	31
	.001	0	0	0	0	0	0	2	3	4	5	5	6	7	8	9	10	11	12	13
	.005	0	0	1	2	3	4	5	6	7	8	10	11	12	13	14	16	17	18	19
6	.01	0	0	2	3	4	5	7	8	9	10	12	13	14	16	17	19	20	21	23
	.025	0	2	3	4	6	7	9	11	12	14	15	17	18	20	22	23	25	26	28
	.05	1	3	4	6	8	9	11	13	15	17	18	20	22	24	26	27	29	31	33
	.10	2	4	6	8	10	12	14	16	18	20	22	24	26	28	30	32	35	37	39
	.001	0	0	0	0	1	2	3	4	6	7	8	9	10	11	12	14	15	16	17
	.005	0	0	1	2	4	5	7	8	10	11	13	14	16	17	19	20	22	23	25
7	.01	0	1	2	4	5	7	8	10	12	13	15	17	18	20	22	24	25	27	29
	.025	0	2	4	6	7	9	11	13	15	17	19	21	23	25	27	29	31	33	35
	.05	1	3	5	7	9	12	14	16	18	20	22	25	27	29	31	34	36	38	40
	.10	2	5	7	9	12	14	17	19	22	24	27	29	32	34	37	39	42	44	47
	.001	0	0	0	1	2	3	5	6	7	9	10	12	13	15	16	18	19	21	22
	.005	0	0	2	3	5	7	8	10	12	14	16	18	19	21	23	25	27	29	31
8	.01	0	1	3	5	7	8	10	12	14	16	18	21	23	25	27	29	31	33	35
	.025	1	3	5	7	9	11	14	16	18	20	23	25	27	30	32	35	37	39	42
	.05	2	4	6	9	11	14	16	19	21	24	27	29	32	34	37	40	42	45	48
	.10	3	6	8	11	14	17	20	23	25	28	31	34	37	40	43	46	49	52	55

TABLE H
Continued

n	α	m=2	3	4	5	6	7	8	9	10	11	12	13	14	15	16	17	18	19	20
	.001	0	0	0	2	3	4	6	8	9	11	13	15	16	18	20	22	24	26	27
	.005	0	1	2	4	6	8	10	12	14	17	19	21	23	25	28	30	32	34	37
9	.01	0	2	4	6	8	10	12	15	17	19	22	24	27	29	32	34	37	39	41
	.025	1	3	5	8	11	13	16	18	21	24	27	29	32	35	38	40	43	46	49
	.05	2	5	7	10	13	16	19	22	25	28	31	34	37	40	43	46	49	52	55
	.10	3	6	10	13	16	19	23	26	29	32	36	39	42	46	49	53	56	59	63
	.001	0	0	1	2	4	6	7	9	11	13	15	18	20	22	24	26	28	30	33
	.005	0	1	3	5	7	10	12	14	17	19	22	25	27	30	32	35	38	40	43
10	.01	0	2	4	7	9	12	14	17	20	23	25	28	31	34	37	39	42	45	48
	.025	1	4	6	9	12	15	18	21	24	27	30	34	37	40	43	46	49	53	56
	.05	2	5	8	12	15	18	21	25	28	32	35	38	42	45	49	52	56	59	63
	.10	4	7	11	14	18	22	25	29	33	37	40	44	48	52	55	59	63	67	71
	.001	0	0	1	3	5	7	9	11	13	16	18	21	23	25	28	30	33	35	38
	.005	0	1	3	6	8	11	14	17	19	22	25	28	31	34	37	40	43	46	49
11	.01	0	2	5	8	10	13	16	19	23	26	29	32	35	38	42	45	48	51	54
	.025	1	4	7	10	14	17	20	24	27	31	34	38	41	45	48	52	56	59	63
	.05	2	6	9	13	17	20	24	28	32	35	39	43	47	51	55	58	62	66	70
	.10	4	8	12	16	20	24	28	32	37	41	45	49	53	58	62	66	70	74	79
	.001	0	0	1	3	5	8	10	13	15	18	21	24	26	29	32	35	38	41	43
	.005	0	2	4	7	10	13	16	19	22	25	28	32	35	38	42	45	48	52	55
12	.01	0	3	6	9	12	15	18	22	25	29	32	36	39	43	47	50	54	57	61
	.025	2	5	8	12	15	19	23	27	30	34	38	42	46	50	54	58	62	66	70
	.05	3	6	10	14	18	22	27	31	35	39	43	48	52	56	61	65	69	73	78
	.10	5	9	13	18	22	27	31	36	40	45	50	54	59	64	68	73	78	82	87
	.001	0	0	2	4	6	9	12	15	18	21	24	27	30	33	36	39	43	46	49
	.005	0	2	4	8	11	14	18	21	25	28	32	35	39	43	46	50	54	58	61
13	.01	1	3	6	10	13	17	21	24	28	32	36	40	44	48	52	56	60	64	68
	.025	2	5	9	13	17	21	25	29	34	38	42	46	51	55	60	64	68	73	77
	.05	3	7	11	16	20	25	29	34	38	43	48	52	57	62	66	71	76	81	85
	.10	5	10	14	19	24	29	34	39	44	49	54	59	64	69	75	80	85	90	95
	.001	0	0	2	4	7	10	13	16	20	23	26	30	33	37	40	44	47	51	55
	.005	0	2	5	8	12	16	19	23	27	31	35	39	43	47	51	55	59	64	68
14	.01	1	3	7	11	14	18	23	27	31	35	39	44	48	52	57	61	66	70	74
	.025	2	6	10	14	18	23	27	32	37	41	46	51	56	60	65	70	75	79	84
	.05	4	8	12	17	22	27	32	37	42	47	52	57	62	67	72	78	83	88	93
	.10	5	11	16	21	26	32	37	42	48	53	59	64	70	75	81	86	92	98	103
	.001	0	0	2	5	8	11	15	18	22	25	29	33	37	41	44	48	52	56	60
	.005	0	3	6	9	13	17	21	25	30	34	38	43	47	52	56	61	65	70	74

TABLE H

Continued

n	α	m=2	3	4	5	6	7	8	9	10	11	12	13	14	15	16	17	18	19	20
15	.01	1	4	8	12	16	20	25	29	34	38	43	48	52	57	62	67	71	76	81
	.025	2	6	11	15	20	25	30	35	40	45	50	55	60	65	71	76	81	86	91
	.05	4	8	13	19	24	29	34	40	45	51	56	62	67	73	78	84	89	95	101
	.10	6	11	17	23	28	34	40	46	52	58	64	69	75	81	87	93	99	105	111
	.001	0	0	3	6	9	12	16	20	24	28	32	36	40	44	49	53	57	61	66
	.005	0	3	6	10	14	19	23	28	32	37	42	46	51	56	61	66	71	75	80
16	.01	1	4	8	13	17	22	27	32	37	42	47	52	57	62	67	72	77	83	88
	.025	2	7	12	16	22	27	32	38	43	48	54	60	65	71	76	82	87	93	99
	.05	4	9	15	20	26	31	37	43	49	55	61	66	72	78	84	90	96	102	108
	.10	6	12	18	24	30	37	43	49	55	62	68	75	81	87	94	100	107	113	120
	.001	0	1	3	6	10	14	18	22	26	30	35	39	44	48	53	58	62	67	71
	.005	0	3	7	11	16	20	25	30	35	40	45	50	55	61	66	71	76	82	87
17	.01	1	5	9	14	19	24	29	34	39	45	50	56	61	67	72	78	83	89	94
	.025	3	7	12	18	23	29	35	40	46	52	58	64	70	76	82	88	94	100	106
	.05	4	10	16	21	27	34	40	46	52	58	65	71	78	84	90	97	103	110	116
	.10	7	13	19	26	32	39	46	53	59	66	73	80	86	93	100	107	114	121	128
	.001	0	1	4	7	11	15	19	24	28	33	38	43	47	52	57	62	67	72	77
	.005	0	3	7	12	17	22	27	32	38	43	48	54	59	65	71	76	82	88	93
18	.01	1	5	10	15	20	25	31	37	42	48	54	60	66	71	77	83	89	95	101
	.025	3	8	13	19	25	31	37	43	49	56	62	68	75	81	87	94	100	107	113
	.05	5	10	17	23	29	36	42	49	56	62	69	76	83	89	96	103	110	117	124
	.10	7	14	21	28	35	42	49	56	63	70	78	85	92	99	107	114	121	129	136
	.001	0	1	4	8	12	16	21	26	30	35	41	46	51	56	61	67	72	78	83
	.005	1	4	8	13	18	23	29	34	40	46	52	58	64	70	75	82	88	94	100
19	.01	2	5	10	16	21	27	33	39	45	51	57	64	70	76	83	89	95	102	108
	.025	3	8	14	20	26	33	39	46	53	59	66	73	79	86	93	100	107	114	120
	.05	5	11	18	24	31	38	45	52	59	66	73	81	88	95	102	110	117	124	131
	.10	8	15	22	29	37	44	52	59	67	74	82	90	98	105	113	121	129	136	144
	.001	0	1	4	8	13	17	22	27	33	38	43	49	55	60	66	71	77	83	89
	.005	1	4	9	14	19	25	31	37	43	49	55	61	68	74	80	87	93	100	106
20	.001	2	6	11	17	23	29	35	41	48	54	61	68	74	81	88	94	101	108	115
	.025	3	9	15	21	28	35	42	49	56	63	70	77	84	91	99	106	113	120	128
	.05	5	12	19	26	33	40	48	55	63	70	78	85	93	101	108	116	124	131	139
	.10	8	16	23	31	39	47	55	63	71	79	87	95	103	111	120	128	136	144	152

TABLE I

Critical values of the Wilcoxon rank-sum test statistic

Sample size n	$W_{.005}$	$W_{.01}$	$W_{.025}$	$W_{.05}$	$W_{.10}$	$W_{.20}$	$n(n+1)/2$
4	0	0	0	0	1	3	10
5	0	0	0	1	3	4	15
6	0	0	1	3	4	6	21
7	0	1	3	4	6	9	28
8	1	2	4	6	9	12	36
9	2	4	6	9	11	15	45
10	4	6	9	11	15	19	55
11	6	8	11	14	18	23	66
12	8	10	14	18	22	28	78
13	10	13	18	22	27	33	91
14	13	16	22	26	32	39	105
15	16	20	26	31	37	45	120
16	20	24	30	36	43	51	136
17	24	28	35	42	49	58	153
18	28	33	41	48	56	66	171
19	33	38	47	54	63	74	190
20	38	44	53	61	70	82	210

Table entries are lower-tail critical values: W_α such that $P(W \le W_\alpha) = \alpha$. The upper-tail critical values may be determined by using the relationship:

$$W_{1-\alpha} = [n(n+1)/2] - W_\alpha, \alpha > 0.50$$

For example, if $n = 10$, $\alpha = 0.05$, and the test is two-tailed, $W_{\alpha/2} = W_{0.025} = 9$ from the table:

$$W_{1-\alpha/2} = W_{0.075} = [10(10+1)/2] - W_{0.025} = 55 - 9 = 46$$

For $n > 20$, the critical value W_α may be approximated from:

$$W_\alpha = [n(n+1)/4] + Z_\alpha \sqrt{n(n+1)(2n+1)/24}$$

where Z_α is the standard normal variate such that a proportion α of the area is to the right of Z_α.

From R. L. McCornack, "Extended Tables of the Wilcoxon Matched Pairs Signed Rank Statistics," *Journal of the American Statistical Association,* vol. 60, pp. 864–71. Copyright © 1965 American Statistical Association, Washington, DC. Reprinted by permission.

TABLE J

Critical values of the Spearman test statistic

n	$\alpha = 0.05$	$\alpha = 0.025$	$\alpha = 0.01$	$\alpha = 0.005$
5	0.900	—	—	—
6	0.829	0.886	0.943	—
7	0.714	0.786	0.893	—
8	0.643	0.738	0.833	0.881
9	0.600	0.683	0.783	0.833
10	0.564	0.648	0.745	0.818
11	0.523	0.623	0.736	0.794
12	0.497	0.591	0.703	0.780
13	0.475	0.566	0.673	0.745
14	0.457	0.545	0.646	0.716
15	0.441	0.525	0.623	0.689
16	0.425	0.507	0.601	0.666
17	0.412	0.490	0.582	0.645
18	0.399	0.476	0.564	0.625
19	0.388	0.462	0.549	0.608
20	0.377	0.450	0.534	0.591
21	0.368	0.438	0.521	0.576
22	0.359	0.428	0.508	0.562
23	0.351	0.418	0.496	0.549
24	0.343	0.409	0.485	0.537
25	0.336	0.400	0.475	0.526
26	0.329	0.392	0.465	0.515
27	0.323	0.385	0.456	0.505
28	0.317	0.377	0.448	0.496
29	0.311	0.370	0.440	0.487
30	0.305	0.364	0.432	0.478

From E. G. Olds, "Distribution of Sums of Squares of Rank Differences for Small Samples," *Annals of Mathematical Statistics,* vol. 9. Copyright © 1938 Institute of Mathematical Statistics, Hayward, CA. Reprinted with the permission of the Institute of Mathematical Statistics.

TABLE K ∘

Critical values of the runs statistic

N_1	N_2	$C_{.005}$	$C_{.01}$	$C_{.025}$	$C_{.05}$	$C_{.10}$	$C_{.90}$	$C_{.95}$	$C_{.975}$	$C_{.99}$	$C_{.995}$
				C_α					$C_{1-\alpha}$		
2	5	—	—	—	—	3	—	—	—	—	—
	8	—	—	—	3	3	—	—	—	—	—
	11	—	—	—	3	3	—	—	—	—	—
	14	—	—	3	3	3	—	—	—	—	—
	17	—	—	3	3	3	—	—	—	—	—
	20	—	3	3	3	4	—	—	—	—	—
5	5	—	3	3	4	4	8	8	9	9	—
	8	3	3	4	4	5	9	10	10	—	—
	11	4	4	5	5	6	10	—	—	—	—
	14	4	4	5	6	6	—	—	—	—	—
	17	4	5	5	6	7	—	—	—	—	—
	20	5	5	6	6	7	—	—	—	—	—
8	8	4	5	5	6	6	12	12	13	13	14
	11	5	6	6	7	8	13	14	14	15	15
	14	6	6	7	8	8	14	15	15	16	16
	17	6	7	8	8	9	15	15	16	—	—
	20	7	7	8	9	10	15	16	16	—	—
11	11	6	7	8	8	9	15	16	16	17	18
	14	7	8	9	9	10	16	17	18	19	19
	17	8	9	10	10	11	17	18	19	20	21
	20	9	9	10	11	12	18	19	20	21	21
14	14	8	9	10	11	12	18	19	20	21	22
	17	9	10	11	12	13	20	21	22	23	23
	20	10	11	12	13	14	21	22	23	24	24
17	17	11	11	12	13	14	22	23	24	25	25
	20	12	12	14	14	16	23	24	25	26	27
20	20	13	14	15	16	17	25	26	27	28	29

Table entries are critical values C_p such that $P(R_t \leq C_p) = p \cdot p = \alpha$ if $\alpha \leq 0.10$ or $p = 1 - \alpha$ if $\alpha \geq 0.90$.

For n or m greater than 20, the critical value $C_p(p = \alpha$ or $p = 1 - \alpha)$ may be approximated by:

$$C_p = \frac{2nm}{n + m} + 1 + z_p \sqrt{\frac{2nm(2nm - n - m)}{(n + m)^2(n + m - 1)}}$$

where z_p is the standard normal variate such that a proportion p of the area is to the right of z_p.

To use the table: Let N_1 be the smaller sample size and N_2 the larger. If the exact values of N_1 and N_2 are not listed, use the nearest values given as an approximation. Reject h_0 if R_t is less than C_α (or greater than $C_{1-\alpha}$) for the one-tailed test at the significance level. For the two-tailed test, reject h_0 if either $R_t > C_{1-\alpha/2}$ or $R_t < C_{\alpha/2}$ at the α significance level.

From F. S. Swed and C. Eisenhart, "Tables for Testing Randomness of Grouping in a Sequence of Alternatives," *Annals of Mathematical Statistics,* vol. 14, pp. 66–87. Copyright © 1943 Institute of Mathematical Statistics, Hayward, CA. Reprinted with the permission of the Institute of Mathematical Statistics.

TABLE L

Critical values of the Kruskal-Wallis test statistic

Sample sizes					Sample sizes				
n_1	n_2	n_3	Critical value	α	n_1	n_2	n_3	Critical value	α
2	1	1	2.7000	.500	4	3	2	6.4444	.009
2	2	1	3.6000	.267				6.3000	.011
2	2	2	4.5714	.067				5.4444	.046
			3.7143	.200				5.4000	.051
3	1	1	3.2000	.300				4.5111	.098
3	2	1	4.2857	.100				4.4444	.102
			3.8571	.133	4	3	3	6.7455	.010
								6.7091	.013
3	2	2	5.3572	.029				5.7909	.046
			4.7143	.048				5.7273	.050
			4.5000	.067				4.7091	.092
			4.4643	.105				4.7000	.101
3	3	1	5.1429	.043	4	4	1	6.6667	.010
			4.5714	.100				6.1667	.022
			4.0000	.129				4.9667	.048
3	3	2	6.2500	.011				4.8667	.054
			5.3611	.032				4.1667	.082
			5.1389	.061				4.0667	.102
			4.5556	.100	4	4	2	7.0364	.006
			4.2500	.121				6.8727	.011
3	3	3	7.2000	.004				5.4545	.046
			6.4889	.001				5.2364	.052
			5.6889	.029				4.5545	.098
			5.6000	.050				4.4455	.103
			5.0667	.086	4	4	3	7.1439	.010
			4.6222	.100				7.1364	.011
4	1	1	3.5714	.200				5.5985	.049
4	2	1	4.8214	.057				5.5758	.051
			4.5000	.076				4.5455	.099
			4.0179	.114				4.4773	.102
4	2	2	6.0000	.014	4	4	4	7.6538	.008
			5.3333	.033				7.5385	.011
			5.1250	.052				5.6923	.049
			4.3750	.100				5.6538	.054
			4.1667	.105				4.6539	.097
4	3	1	5.8333	.021				4.5001	.104
			5.2083	.050	5	1	1	3.8571	.143
			5.0000	.057	5	2	1	5.2500	.036
			4.0556	.093				5.0000	.048
			3.8889	.129				4.4500	.071
								4.2000	.095
								4.0500	.119

From W. H. Kruskal and W. A. Wallis, "Use of Ranks on One-Criterion Variance Analysis," *Journal of the American Statistical Association,* vol. 47, pp. 583–621. Copyright © 1952 American Statistical Association, Washington, DC. Reprinted by permission.

TABLE L

Continued

Sample sizes			Critical value	α	Sample sizes			Critical value	α
n_1	n_2	n_3			n_1	n_2	n_3		
5	2	2	6.5333	.008	5	4	4	7.7604	.009
			6.1333	.013				7.7440	.011
			5.1600	.034				5.6571	.049
			5.0400	.056				5.6176	.050
			4.3733	.090				4.6187	.100
			4.2933	.112				4.5527	.102
5	3	1	6.4000	.012	5	5	1	7.3091	.009
			4.9600	.048				6.8364	.011
			4.8711	.052				5.1273	.046
			4.0178	.095				4.9091	.053
			3.8400	.123				4.1091	.086
5	3	2	6.9091	.009				4.0364	.105
			6.8281	.010	5	5	2	7.3385	.010
			5.2509	.049				7.2692	.010
			5.1055	.052				5.3385	.047
			4.6509	.091				5.2462	.051
			4.4945	.101				4.6231	.097
5	3	3	7.0788	.009				4.5077	.100
			6.9818	.011	5	5	3	7.5780	.010
			5.6485	.049				7.5429	.010
			5.5152	.051				5.7055	.046
			4.5333	.097				5.6264	.051
			4.4121	.109				4.5451	.100
5	4	1	6.9545	.008				4.5363	.102
			6.8400	.011	5	5	4	7.8229	.010
			4.9855	.044				7.7914	.010
			4.8600	.056				5.6657	.049
			3.9873	.098				5.6429	.050
			3.9600	.102				4.5229	.100
5	4	2	7.2045	.009				4.5200	.101
			7.1182	.010	5	5	5	8.0000	.009
			5.2727	.049				7.9800	.010
			5.2682	.050				5.7800	.049
			4.5409	.098				5.6600	.051
			4.5182	.101				4.5600	.100
5	4	3	7.4449	.010				4.5000	.102
			7.3949	.011					
			5.6564	.049					
			5.6308	.050					
			4.5487	.099					
			4.5231	.103					

TABLE M

Distribution of the studentized range statistic

		d = Number of ordered means spanned by the comparison or k = number of means being compared													
df for MS_{Error}	α	2	3	4	5	6	7	8	9	10	11	12	13	14	15
1	.05	18.0	27.0	32.8	37.1	40.4	43.1	45.4	47.4	49.1	50.6	52.0	53.2	54.3	55.4
	.01	90.0	135	164	186	202	216	227	237	246	253	260	266	272	277
2	.05	6.09	8.3	9.8	10.9	11.7	12.4	13.0	13.5	14.0	14.4	14.7	15.1	15.4	15.7
	.01	14.0	19.0	22.3	24.7	26.6	28.2	29.5	30.7	31.7	32.6	33.4	34.1	34.8	35.4
3	.05	4.50	5.91	6.82	7.50	8.04	8.48	8.85	9.18	9.46	9.72	9.95	10.2	10.4	10.5
	.01	8.26	10.6	12.2	13.3	14.2	15.0	15.6	16.2	16.7	17.1	17.5	17.9	18.2	18.5
4	.05	3.93	5.04	5.76	6.29	6.71	7.05	7.35	7.60	7.83	8.03	8.21	8.37	8.52	8.66
	.01	6.51	8.12	9.17	9.96	10.6	11.1	11.5	11.9	12.3	12.6	12.8	13.1	13.3	13.5
5	.05	3.64	4.60	5.22	5.67	6.03	6.33	6.58	6.80	6.99	7.17	7.32	7.47	7.60	7.72
	.01	5.70	6.97	7.80	8.42	8.91	9.32	9.67	9.97	10.2	10.5	10.7	10.9	11.1	11.2
6	.05	3.46	4.34	4.90	5.31	5.63	5.89	6.12	6.32	6.49	6.65	6.79	6.92	7.03	7.14
	.01	5.24	6.33	7.03	7.56	7.97	8.32	8.61	8.87	9.10	9.30	9.49	9.65	9.81	9.95
7	.05	3.34	4.16	4.69	5.06	5.36	5.61	5.82	6.00	6.16	6.30	6.43	6.55	6.66	6.76
	.01	4.95	5.92	6.54	7.01	7.37	7.68	7.94	8.17	8.37	8.55	8.71	8.86	9.00	9.12
8	.05	3.26	4.04	4.53	4.89	5.17	5.40	5.60	5.77	5.92	6.05	6.18	6.29	6.39	6.48
	.01	4.74	5.63	6.20	6.63	6.96	7.24	7.47	7.68	7.87	8.03	8.18	8.31	8.44	8.55
9	.05	3.20	3.95	4.42	4.76	5.02	5.24	5.43	5.60	5.74	5.87	5.98	6.09	6.19	6.28
	.01	4.60	5.43	5.96	6.35	6.66	6.91	7.13	7.32	7.49	7.65	7.78	7.91	8.03	8.13
10	.05	3.15	3.88	4.33	4.65	4.91	5.12	5.30	5.46	5.60	5.72	5.83	5.93	6.03	6.11
	.01	4.48	5.27	5.77	6.14	6.43	6.67	6.87	7.05	7.21	7.36	7.48	7.60	7.71	7.81
11	.05	3.11	3.82	4.26	4.57	4.82	5.03	5.20	5.35	5.49	5.61	5.71	5.81	5.90	5.99
	.01	4.39	5.14	5.62	5.97	6.25	6.48	6.67	6.84	6.99	7.13	7.26	7.36	7.46	7.56
12	.05	3.08	3.77	4.20	4.51	4.75	4.95	5.12	5.27	5.40	5.51	5.62	5.71	5.80	5.88
	.01	4.32	5.04	5.50	5.84	6.10	6.32	6.51	6.67	6.81	6.94	7.06	7.17	7.26	7.36
13	.05	3.06	3.73	4.15	4.45	4.69	4.88	5.05	5.19	5.32	5.43	5.53	5.63	5.71	5.79
	.01	4.26	4.96	5.40	5.73	5.98	6.19	6.37	6.53	6.67	6.79	6.90	7.01	7.10	7.19
14	.05	3.03	3.70	4.11	4.41	4.64	4.83	4.99	5.13	5.25	5.36	5.46	5.55	6.64	5.72
	.01	4.21	4.89	5.32	5.63	5.88	6.08	6.26	6.41	6.54	6.66	6.77	6.87	6.96	7.05
16	.05	3.00	3.65	4.05	4.33	4.56	4.74	4.90	5.03	5.15	5.26	5.35	5.44	5.52	5.59
	.01	4.13	4.78	5.19	5.49	5.72	5.92	6.08	6.22	6.35	6.46	6.56	6.66	6.74	6.82
18	.05	2.97	3.61	4.00	4.28	4.49	4.67	4.82	4.96	5.07	5.17	5.27	5.35	5.43	5.50
	.01	4.07	4.70	5.09	5.38	5.60	5.79	5.94	6.08	6.20	6.31	6.41	6.50	6.58	6.65
20	.05	2.95	3.58	3.96	4.23	4.45	4.62	4.77	4.90	5.01	5.11	5.20	5.28	5.36	5.43
	.01	4.02	4.64	5.02	5.29	5.51	5.69	5.84	5.97	6.09	6.19	6.29	6.37	6.45	6.52
24	.05	2.92	3.53	3.90	4.17	4.37	4.54	4.68	4.81	4.92	5.01	5.10	5.18	5.25	5.32
	.01	3.96	4.54	4.91	5.17	5.37	5.54	5.69	5.81	5.92	6.02	6.11	6.19	6.26	6.33

From B. J. Winer, *Statistical Principles in Experimental Design,* McGraw-Hill, 1962; abridged from H. L. Harter, D. S. Clemm, and E. H. Guthrie, "The probability integrals of the range and of the studentized range," WADC Tech. Rep. 58–484, vol. 2, 1959, Wright Air Development Center, Table II.2, pp. 243–281. Reprinted by permission.

TABLE M

Continued

		d = Number of ordered means spanned by the comparison or k = number of means being compared												
df for MS_{Error}	α	2	3	4	5	6	7	8	9	10	11	12	13	14	15
30	.05	2.89	3.49	3.84	4.10	4.30	4.46	4.60	4.72	4.83	4.92	5.00	5.08	5.15	5.21
	.01	3.89	4.45	4.80	5.05	5.24	5.40	5.54	5.56	5.76	5.85	5.93	6.01	6.08	6.14
40	.05	2.86	3.44	3.79	4.04	4.23	4.39	4.52	4.63	4.74	4.82	4.91	4.98	5.05	5.11
	.01	3.82	4.37	4.70	4.93	5.11	5.27	5.39	5.50	5.60	5.69	5.77	5.84	5.90	5.96
60	.05	2.83	3.40	3.74	3.98	4.16	4.31	4.44	4.55	4.65	4.73	4.81	4.88	4.94	5.00
	.01	3.76	4.28	4.60	4.82	4.99	5.13	5.25	5.36	5.45	5.53	5.60	5.67	5.73	5.79
120	.05	2.80	3.36	3.69	3.92	4.10	4.24	4.36	4.48	4.56	4.64	4.72	4.78	4.84	4.90
	.01	3.70	4.20	4.50	4.71	4.87	5.01	5.12	5.21	5.30	5.38	5.44	5.51	5.56	5.61
∞	.05	2.77	3.31	3.63	3.86	4.03	4.17	4.29	4.39	4.47	4.55	4.62	4.68	4.74	4.80
	.01	3.64	4.12	4.40	4.60	4.76	4.88	4.99	5.08	5.16	5.23	5.29	5.35	5.40	5.45

TABLE N

Greek alphabet

Greek letter	Greek name	English equivalent	Greek letter	Greek name	English equivalent
A α	Alpha	a	N ν	Nu	n
B β	Beta	b	Ξ ξ	Xi	x
Γ γ	Gamma	g	O o	Omicron	ŏ
Δ δ	Delta	d	Π π	Pi	p
E ϵ	Epsilon	ĕ	P ρ	Rho	r
Z ζ	Zeta	z	Σ σ	Sigma	s
H η	Eta	ē	T τ	Tau	t
Θ θ	Theta	th	Υ υ	Upsilon	u
I ι	Iota	i	Φ ϕ	Phi	ph
K κ	Kappa	k	X χ	Chi	ch
Λ λ	Lambda	l	Ψ ψ	Psi	ps
M μ	Mu	m	Ω ω	Omega	ō

Appendix C
Answers to
Exercises

Note: Depending on how, or whether, you "chain" calculations using a calculator or whether you use a computer, rounding differences may cause some answers to differ slightly from those you find in this section.

Chapter 2

2–1. **a.** Range $= 145 - 58 + 1 = 88$
 b. $i = 5$
 c. 55–59
 d. Frequency distribution

X	Tally	f
145–149	/	1
140–144		0
135–139	/	1
130–134	//	2
125–129	///	3
120–124		1
115–119		5
110–114		9
105–109		11
100–104		10
95–99		8
90–94		11
85–89		7
80–84		4
75–79		3
70–74	/	1
65–69	/	1
60–64	/	1
55–59	/	1
N		80

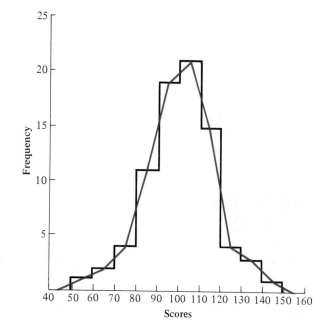

423

2–2.   Class examination scores
- **a.** Range $= 95 - 37 + 1 = 59$
- **b.** With $i = 5$,
- **c.** begin lowest class interval at 35. This gives 13 class intervals. With $i = 3$, begin lowest class interval at 36. This gives 20 class intervals, the maximum.
- **d. Frequency distribution**

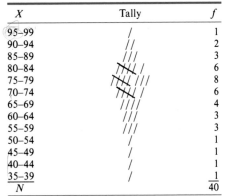

X	Tally	f
95–99		1
90–94		2
85–89		3
80–84		6
75–79		8
70–74		6
65–69		4
60–64		3
55–59		3
50–54		1
45–49		1
40–44		1
35–39		1
N		40

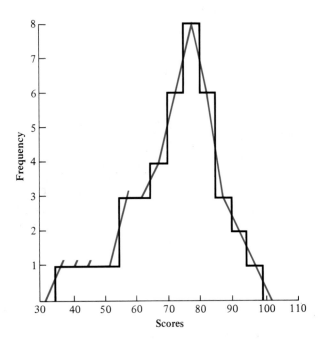

2–3.   Liquor store data
- **a.** Range $= 33 - 2 + 1 = 32$
- **b.** With $i = 3$,
- **c.** begin lowest class interval at zero. This gives 12 class intervals. With $i = 2$, begin lowest class interval at 2. This gives 16 class intervals.
- **d. Frequency distribution**

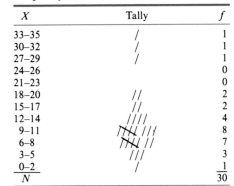

X	Tally	f
33–35		1
30–32		1
27–29		1
24–26		0
21–23		0
18–20		2
15–17		2
12–14		4
9–11		8
6–8		7
3–5		3
0–2		1
N		30

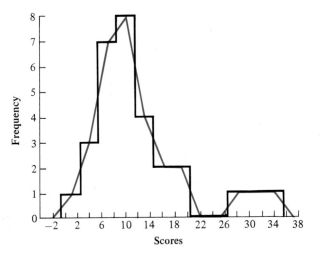

2-4. Hours of sleep
    a. Range $= 12 - 2 + 1 = 11$
    b. With $i = 1$,
    c. begin lowest class interval at 2. No other class interval size is acceptable.
    d. Frequency distribution

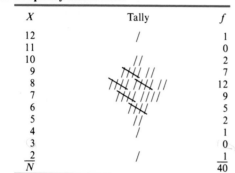

X	Tally	f
12	/	1
11		0
10	//	2
9		7
8		12
7		9
6		5
5	//	2
4	/	1
3		0
2	/	1
N		40

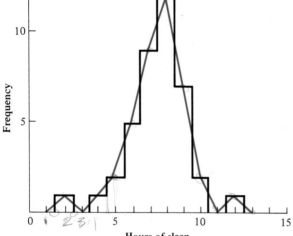

2-5. Typing errors
    a. Range $= 25 - 3 + 1 = 23$
    b. With $i = 2$,
    c. begin lowest class interval at 2. No other class interval is acceptable.
    d. Frequency distribution

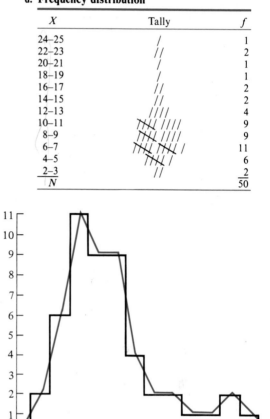

X	Tally	f
24–25	/	1
22–23	//	2
20–21	/	1
18–19	/	1
16–17	//	2
14–15	//	2
12–13	////	4
10–11		9
8–9		9
6–7		11
4–5		6
2–3	//	2
N		50

2–6. 100-yard dash times
    a. Range $= 10.2 - 9.2 + 0.1 = 1.1$ seconds
    b,c. Because scores are not whole numbers, the usual rules for selecting class intervals are not applicable here. In this exercise, $i = 0.1$ is the only suitable class interval size.
    d. Frequency distribution

X	Tally	f
10.2		1
10.1		1
10.0		2
9.9		5
9.8		6
9.7		9
9.6		7
9.5		4
9.4		3
9.3		1
9.2		1
N		40

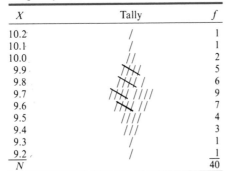

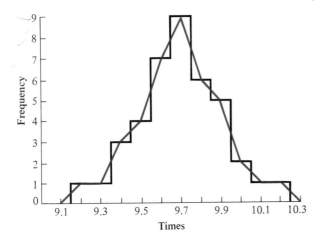

2–7. Grade point averages
    a. Range $= 3.7 - 1.9 + 0.1 = 1.9$
    b. Either $i = 0.1$ or $i = 0.2$ is acceptable as a class interval size.
    d. Frequency distribution

X	Tally	f
3.6–3.7		1
3.4–3.5		1
3.2–3.3		2
3.0–3.1		4
2.8–2.9		6
2.6–2.7		4
2.4–2.5		9
2.2–2.3		4
2.0–2.1		3
1.8–1.9		1
N		35

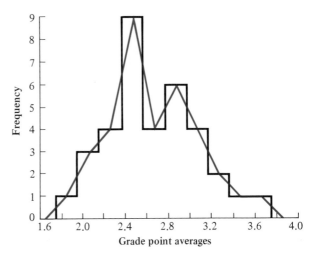

2–8. Frequency distribution

X	Tally	f
510–539	/	1
480–509	///	3
450–479	///	3
420–449	////	4
390–419		0
360–389	⫲⫶⫶ /	6
330–359	⫲⫶⫶ //	7
300–329	⫲⫶⫶ ⫲⫶⫶ /	11
270–299	⫲⫶⫶ ///	8
240–269	⫲⫶⫶	5
210–239	//	2
N		50

2–9. a. 155

   b. 240

   c. $240 - 155 + 1 = 86$

2–10. a. 4.5–5.5

   b. -2.35 to -2.25

   c. 100.5–101.5

   d. 76.315–76.325

   e. -61.5 to -60.5

2–11. a. 15

   b. 16.325

   c. 103.5

   d. -3.0

   e. 0.375

2–12. 1

2–13. 102

2–14. 35

2–15.

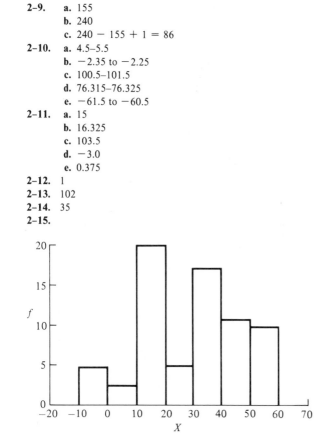

2–16. **a.** $50 - 0 + 1 = 51$

b. $i = 5$

c. 0–4

d.

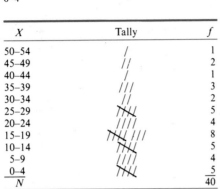

X	Tally	f
50–54	/	1
45–49	//	2
40–44	/	1
35–39	///	3
30–34	//	2
25–29	////	5
20–24	////	4
15–19	//// ///	8
10–14	////	5
5–9	////	4
0–4	////	5
N		40

2–17. **a.** $240 - 155 + 1 = 86$

b. $i = 5$

c. 155–159

d.

X	Tally	f
240–244	/	1
235–239		0
230–234	//	2
225–229	/	1
220–224	//	2
215–219	/	1
210–214	//	2
205–209	//	2
200–204	//// ///	8
195–199	////	4
190–194	////	4
185–189	////	4
180–184	///	3
175–179	///	3
170–174	/	1
165–169		0
160–164	/	1
155–159	/	1
N		40

alternative solution $i = 10$

X	Tally	f
240–249	/	1
230–239	//	2
220–229	///	3
210–219	///	3
200–209	//// ////	10
190–199	//// ///	8
180–189	//// //	7
170–179	////	4
160–169	/	1
150–159	/	1
N		40

2–18. **a.** Lowest $= 8$

b. Highest $= 49$

c. Range $= 49 - 8 + 1 = 42$

2–19. **a.** Range $= 49 - 8 + 1 = 42$

b. $42/1 = 42$

$42/2 = 21$

$42/3 = 14$

$42/5 = 8.4$

$42/10 = 4.2$

c. $42/3 = 14$, between 10 and 20, so $i = 3$

2–20. **a.** $i = 3$

b. Lowest score $= 8$

multiple of 3 is 6

2–21. See pages 19–20 of text.

2–22. See pages 20–21 of text.

Chapter 3

3–1. 74.5

3–2. 10.5

3–3. $Mdn = 59.5 + \left[\dfrac{(80.5 - 49)}{33}\right] \times 10 = 59.5 + \dfrac{315}{33}$

$= 59.5 + 9.545 = 69.045 = 69.0$

3–4. $Mdn = 10.5$

3–5. $Mdn = 24.5 + \left[\dfrac{(5 - 3)}{3}\right] \times 1 = 24.5 + \dfrac{2}{3}$

$= 24.5 + .67 = 25.17 = 25.2$

3–6. $Mdn = 24.5$

3–7. $Mdn = 34.5 + \left[\dfrac{(6 - 4)}{4}\right] \times 1 = 34.5 + \dfrac{2}{4}$

$= 34.5 + .5 = 35.0$

3–8. $Mdn = 0$

3–9. From the bottom up: $Mdn = 99.5 + \left[\dfrac{(40 - 37)}{21}\right] \times 10$

$= 99.5 + \dfrac{30}{21} = 99.5 + 1.43$

$= 100.93 = 100.9$

From the top down: $Mdn = 109.5 - \left[\dfrac{(40 - 22)}{21}\right] \times 10$

$= 109.5 - \dfrac{180}{21} = 109.5 - 8.57$

$= 100.93 = 100.9$

3–10. From the bottom up: $Mdn = 10$.

From the top down: $Mdn = 10$.

3–11. **a.** 89

b. 7.42

3–12. **a.** 1283

b. 98.69

3-13. a.

Midpoint	f	Midpoint $\times f$
134.5	2	269
124.5	1	124.5
114.5	4	458
104.5	0	0
94.5	6	567
84.5	8	676
74.5	11	819.5
64.5	7	451.5
54.5	5	272.5
44.5	3	133.5
34.5	2	69
24.5	0	0
14.5	1	14.5

b. Sum =   3855

c. $M = \dfrac{3855}{50} = 77.1$

3-14. $M = 33.19$

3-15. a.

Midpoint	f	Midpoint $\times f$
114.5	3	343.5
104.5	8	836
94.5	12	1134
84.5	20	1690
74.5	36	2682
64.5	33	2128.5
54.5	24	1308
44.5	16	712
34.5	7	241.5
24.5	2	49

b. Sum =   11124.5

c. $M = \dfrac{11124.5}{161} = 69.1$

3-16. $M = 10.18$

3-17. mode = 32 (midpoint of interval 30–34)

3-18. $Mdn = 38.25$

3-19. a.

Midpoint	f	Midpoint $\times f$
72	1	72
67	3	201
62	6	372
57	6	342
52	5	260
47	8	376
42	9	378
37	8	296
32	12	384
27	10	270
22	7	154
17	5	85

b. Sum =   3190

c. $M = \dfrac{3190}{80} = 39.875$

$= 39.9$

3-20. Weight: $M = 196$, $Mdn = 195.25$, mode = 200. Minutes: $M = 19.95$, $Mdn = 18.5$, mode = 0.

Chapter 4

4-1. a. $Q_1 = P_{25} = 15.00$; any number between 10.5 and 19.5 satisfies the definition, so 15.00, the middle point in this range is taken as Q_1

b. $Q_3 = P_{75} = 30.5 + (2/3) \times 1 = 30.5 + 0.67 = 31.17$

c. $Q = \dfrac{(31.17 - 15.00)}{2} = \dfrac{16.17}{2} = 8.08 = 8.1.$

4-2. $Q = 13.1$

4-3. a. $Q_1 = 3.5 + \left[\dfrac{2.5}{4}\right] \times 1 = 3.5 + 0.625 = 4.125;$

b. $Q_3 = 11.5 + \left[\dfrac{1.5}{3}\right] \times 1 = 11.5 + 0.5 = 12.00;$

c. $Q = \dfrac{(12.000 - 4.125)}{2} = \dfrac{7.875}{2} = 3.9375 = 3.9.$

4-4. $Q = 11.5$

4-5. **a.** $Q_1 = 89.5 + \left[\dfrac{(20 - 18)}{11}\right] \times 5 = 90.4$

b. $Q_3 = 109.5 + \left[\dfrac{(60 - 58)}{9}\right] \times 5 = 110.6$

c. $Q = \dfrac{110.6 - 90.4}{2} = 10.1$

$Q_2 = 99.5 + \left[\dfrac{(40 - 37)}{10}\right] \times 5 = 101$

$Q_3 - Q_2 = 9.6$

$Q_2 - Q_1 = 10.6$

The distribution is slightly negative skewed because $Q_3 - Q_2 < Q_2 - Q_1$

4-6. $Q = 3.57$

The distribution is positively skewed since $Q_3 - Q_2$ is greater than $Q_2 - Q_1$.

4-7. **a.** $Q_1 = 83.5 + \left[\dfrac{(52.25 - 38)}{16}\right] \times 1 = 83.5 + \dfrac{14.25}{16}$

$= 83.5 + 0.89 = 84.39;$

b. $Q_3 = 89.5 + \left[\dfrac{(156.75 - 156)}{15}\right] \times 1 = 89.5 + 0.05 = 89.55;$

c. $Q = \dfrac{(89.55 - 84.39)}{2} = \dfrac{5.16}{2} = 2.58 = 2.6.$

4-8. $Q = 12.8$

4-9. **a.** $Q_1 = 24.5 + \left[\dfrac{(20 - 12)}{10}\right] \times 5 = 24.5 + 4 = 28.5$

b. $Q_3 = 49.5 + \left[\dfrac{(60 - 59)}{5}\right] \times 5 = 49.5 + 1 = 50.5$

c. $Q = \dfrac{50.5 - 28.5}{2} = \dfrac{22}{2} = 11$

4-10. S.D. $= 3.256$

4-11. **a.** 80

b. 746

c. 10

d. 3.256

4-12. $M = 10$, S.D. $= 3.256$

4-13. $M = 10$

a. 100

b. 1106

c. 10

d. 3.256

4-14. $M = 24$, S.D. $= 9.77$

4-15. $M = 24$

a. 240

b. 6714

c. 10

d. S.D. $= 9.77$

4-16. Adding a constant to a set of scores changes the value of the mean [mean + constant] but leaves the S.D. unchanged. Multiplying each score by a constant changes the mean [mean × constant] and the S.D. [S.D. × constant].

4-17. Original

    a. 4.17

    b. $-3.17, -3.17, -1.17, -1.17, -1.17, -.17, .83, .83,$ 1.83, 1.83, 1.83, 2.83

    c. 10.05, 10.05, 1.37, 1.37, 1.37, .029, .69, .69, 3.35, 3.35, 3.35, 8.01

    d. 43.68

    e. 3.64

    f. 1.91

Divide 2 into each score

    a. 2.08

    b. $-1.59, -1.59, -.59, -.59, -.59, -.09, .41, .41, .92,$ .92, .92, 1.42

    c. 2.53, 2.53, .35, .35, .35, .01, .17, .17, .85, .85, .85, 2.02

    d. 11.03

    e. .92

    f. .96

Dividing the score by 2 yields a mean that is 1/2 of original mean and S.D. that is 1/2 of original S.D.

4–18. S.D. = 1.91, S.D. = .95.

Dividing each score by a constant results in a mean that is $\dfrac{\text{original mean}}{\text{constant}}$ and S.D. that is $\dfrac{\text{original S.D.}}{\text{constant}}$.

4–19. **a.** MTS 1
  b. 1. k = 4
  2. $N = 5 + 10 + 25 + 60 = 100$
  3. $f_1 = 5$ $f_2 = 10$ $f_3 = 25$ $f_4 = 60$
  4. $N^2 = 10000$
  5. $f_1^2 = 25$ $f_2^2 = 100$ $f_3^2 = 625$ $f_4^2 = 3600$ $\Sigma f_i^2 = 4350$
  6. $10000 - 4350 = 5650$
  7. $5650 \times 4 = 22600$
  8. $(10000(3) = 30000)$ D = $22600/30000 = .753$

4–20. **a.** Eye color = .85, **b.** Sex = .99, **c.** Soft drink preference = .99.

4–21. **a.** 99, $M = 19.8$
  b. 2291
  c. 5
  d. 8.13

4–22. $M = -1$, S.D. = 7.41

4–23. **a.** 149, $M = 29.8$
  b. 4771
  c. 5
  d. 8.13

4–24. $M = 9, 7.41$

4–25. When data are categorical (such as eye color) or ranks.

4–26. When there is no variability in the scores, i.e., when all values are the same (equal to one another).

4–27. Yes, especially if the frequency distribution is of a categorical variable.
  $k = 13, N = 160, f_i = 3, 8, 15, 25, 12, 10, 13, 18, 23, 16, 10, 5, 2$
  $N^2 = 25600, \Sigma f_i^2 = 2574,$ D = $299,338/(25600 \times 12) = .97$

4–28. 50%

4–29. **a.** 75%
  b. 25%

Chapter 5

5–1. **a.** $X = 89$
  b. ns = 140
  c. $n = 16$
  d. $LL = 88.5$
  e. $i = 1$
  f. $140 + \dfrac{(89 - 88.5)\,(16)}{1} = 140 + 8 = 148$
  g. Centile rank = $100(148)/209 = 70.8 = 71$

5–2. Centile rank = 67

5–3. **a.** $X = 40\text{–}49$
  b. ns = 9
  c. $n = 16$
  d. $LL = 39.5$
  e. $i = 10$
  f. $9 + \dfrac{(42 - 39.5)\,(16)}{10} = 9 + 4 = 13$
  g. Centile rank = $100\,(13)/161 = 8.1 = 8$

5–4. Centile rank = 28

5-24. Z-score $= 2.36$

5-25.   **a.** $Z_{89} = (89 - 100)/20 = -0.55$

$Z_{124} = (124 - 100)/20 = 1.20$

[For AGCT scores, $M = 100$, S.D. $= 20$]

b. Area of $-\infty$ to $Z = +0.55 = .7088$

c. $1.000 - .7088 = 0.2912$ (Area above $Z = +0.55$ and area below $Z = -0.55$)

d. Area of $-\infty$ to $Z = 1.20 = .8849$

e. Area between 89 and $124 = .8849 - .2912 = .5937$

f. Percentage of scores between 89 and $124 = .5937 \times 100 = 59.37\%$

5-26.   **a.** $Z_{53} = (53 - 50)/10 = 0.30$

$Z_{62} = (61 - 50)/10 = 1.10$

[Comrey Personality Scales have $M = 50$ and S.D. $= 10$]

b. Area from $-\infty$ to $Z_{53} = .6179$

c. Area from $-\infty$ to $Z_{61} = .8643$

d. Area between 53 and $61 = .8643 - .6179 = .2464$ Proportion $= .2464$

Chapter 6

6-1.   **a.**

X	Y
6	7
5	4
4	2
5	5
6	5
2	4
9	8
7	9
1	4
8	7
6	5
5	6

b. $\Sigma X = 64$

c. $X^2 = 36, 25, 16, 25, 36, 4, 81, 49, 1, 64, 36, 25$

d. $\Sigma X^2 = 398$

e. $\Sigma Y = 66$

f. $Y^2 = 49, 16, 4, 25, 25, 16, 64, 81, 16, 49, 25, 36$

g. $\Sigma Y^2 = 406$

h. $XY = 42, 20, 8, 25, 30, 8, 72, 63, 4, 56, 30, 30$

i. $\Sigma XY = 388$

j. $N = 12$

k. $r = \dfrac{12(388) - (64)(66)}{\sqrt{12(398) - (64)^2}\ \sqrt{12(406) - (66)^2}}$

$= \dfrac{4656 - 4224}{\sqrt{4776 - 4096}\ \sqrt{4872 - 4356}}$

$= \dfrac{432}{\sqrt{680}\ \sqrt{516}} = \dfrac{432}{592.35} = 0.729$

6-2. $r = -0.742$

6–3.   **a.**

Subject	X	Y
1	72	59
2	81	87
3	54	36
4	91	61
5	74	58
6	88	86
7	78	70
8	90	39
9	77	104
10	88	84
11	90	70
12	69	55
13	94	89
14	100	94
15	74	51
16	76	54
17	96	101
18	50	46
19	103	80
20	93	91

b. $\Sigma X = 1638$

c. $X^2 = 5184, 6561, 2916, 8281, 5476, 7744, 6084, 8100, 5929, 7744, 8100, 4761, 8836, 10000, 5476, 5776, 9216, 2500, 10609, 8649$

d. $\Sigma X^2 = 137942$

e. $\Sigma Y = 1415$

f. $Y^2 = 3481, 7569, 1296, 3721, 3364, 7396, 4900, 1521, 10816, 7056, 4900, 3025, 7921, 8836, 2601, 2916, 10201, 2116, 6400, 8281$

g. $\Sigma Y^2 = 108317$

h. $XY = 4248, 7047, 1944, 5551, 4292, 7568, 5460, 3510, 8008, 7392, 6300, 3795, 8366, 9400, 3774, 4104, 9696, 2300, 8240, 8463$

i. $\Sigma XY = 119458$

j. $N = 20$

k.
$$r = \frac{N\Sigma XY - \Sigma X \Sigma Y}{\sqrt{N\Sigma X^2 - (\Sigma X)^2}\ \sqrt{N\Sigma Y^2 - (\Sigma Y)^2}}$$

$$= \frac{(20)(119458) - (1638)(1415)}{\sqrt{20(137942) - (1638)^2}\ \sqrt{20(108317) - (1415)^2}}$$

$$= \frac{2389160 - 2317770}{\sqrt{2758840 - 2683044}\ \sqrt{2166340 - 2002225}}$$

$$= \frac{71390}{\sqrt{75796}\ \sqrt{164115}} = \frac{73190}{(275.3)(405.1)} = \frac{73190}{111524.0}$$

$$= 0.64$$

6–4. $r = 0.30$

6-5. **a.**

Subject	X	Y	Subject	X	Y	Subject	X	Y
1	72	59	21	93	87	41	77	69
2	81	87	22	111	83	42	56	53
3	54	36	23	70	53	43	97	83
4	91	61	24	103	81	44	71	50
5	74	58	25	77	82	45	71	84
6	88	86	26	97	93	46	88	97
7	78	70	27	91	58	47	103	84
8	90	39	28	53	95	48	81	92
9	77	104	29	81	67	49	56	56
10	88	84	30	73	101	50	53	76
11	90	70	31	93	72	51	77	77
12	69	55	32	70	49	52	104	73
13	94	89	33	82	66	53	83	60
14	100	94	34	86	92	54	81	48
15	74	51	35	66	57	55	70	44
16	76	54	36	81	90	56	71	66
17	96	101	37	101	95	57	75	105
18	50	46	38	94	68	58	76	62
19	103	80	39	108	86	59	94	73
20	93	91	40	69	51	60	94	85

b. $\Sigma X = 4915$

c. $X^2 =$ [Values for 1–40, in questions 6–3 and 6–4] plus values for 41–60: 5929, 3136, 9409, 5041, 5041, 7744, 10609, 6561, 3136, 2809, 5929, 10816, 6889, 6561, 4900, 5041, 5625, 5776, 8836, 8836

d. $\Sigma X^2 = 415411$

e. $\Sigma Y = 4378$

f. $Y^2 =$ [see questions 6–3 and 6–4 for 1–40]. Values for 41–60: 4761, 2809, 6889, 2500, 7056, 9409, 7056, 8464, 3136, 5776, 5929, 5329, 3600, 2304, 1936, 4356, 11025, 3844, 5329, 7225

g. $\Sigma Y^2 = 338730$

h. $XY =$ [see questions 6–3 and 6–4 for 1–40]. Values for 41–60: 5313, 2968, 8051, 3550, 5964, 8536, 8652, 7452, 3156, 4028, 5929, 7592, 4980, 3888, 3080, 4686, 7875, 4712, 6862, 7990

i. $\Sigma XY = 365803$

j. $N = 60$

k.
$$r = \frac{60(365803) - (4915)(4378)}{\sqrt{60(415411) - (4915)^2}\ \sqrt{60(338730) - (4378)^2}}$$
$$= \frac{21948180 - 21517870}{\sqrt{24924660 - 24157225}\ \sqrt{20323800 - 19166884}}$$
$$= \frac{430310}{\sqrt{767435}\ \sqrt{1156916}} = \frac{430310}{(876.0)(1075.6)} = \frac{430310}{942226} = 0.46$$

6-6. $r = 0.50$

6-7. **a.**

X	Y
9	10
7	2
1	5
12	14
3	0
6	8
9	7
10	11
10	5
2	1

b. $\Sigma X = 69$

c. $X^2 = 81, 49, 1, 144, 9, 36, 81, 100, 100, 4$

d. $\Sigma X^2 = 605$

e. $\Sigma Y = 63$

f. $Y^2 = 100, 4, 25, 196, 0, 64, 49, 121, 25, 1$

g. $\Sigma Y^2 = 585$

h. $XY = 90, 14, 5, 168, 0, 48, 63, 110, 50, 2$

i. $\Sigma XY = 550$

j. $N = 10$

k.
$$r = \frac{10(550) - (69)(63)}{\sqrt{10(605) - (69)^2}\ \sqrt{10(585) - (63)^2}}$$
$$= \frac{1153}{\sqrt{1289}\ \sqrt{1881}}$$
$$= 0.74$$

6–8. $r = -0.953$

6–9. **a.** $r = 0.50$

b. $S.D._y = 10$

c. $S.D._x = 15$

d. $x = (60 - 80) = -20$

e. $y' = r(S.D._y / S.D._x)(X - M_x)$
$y' = 0.5\,(10/15)\,(X - 80)$
$y' = 0.5\,(2/3)\,(60 - 80) = -6.667$

6–10. $Y_2 = 34.4$

6–11. **a.** $r = 0.50$

b. $Z'_y = r\,Z_x$ $Z'_y = 0.5Z_x$

c. $Y' = r\,(S.D._y/S.D._x)\,X + (M_y - r\,[S.D._y/S.D._x]M_x)$
$= 0.5\,(10/15)X + (50 - 0.5[10/15] \times [80])$
$= (1/3)X + 23.33$

d. $Y' = (1/3)\,(60) + 23.333 = 43.333$

6–12. $Z\text{-score}_2 = -1.1111$

$X_2 = 34.44$

6–13. See solution for question 6–5 to obtain details on (a–e).

a. $\Sigma O = 4915$

b. $\Sigma O^2 = 415411$

c. $\Sigma C = 4378$

d. $\Sigma C^2 = 338730$

e. $r = .457$

f. $S.D._O = 14.6$ $S.D._C = 17.93$

g. $M_O = 81.917$ $M_C = 72.967$

h. $O' = .457\,(14.6/17.93)\,C + (81.917 - .457\,[14.6/17.93]\,72.967)$
$= .372\,C + 54.8$
Line-of-best-fit predicting O scores for C scores.

6–14. Raw-score equation: $.366X - 51.82$
Deviation-score equation: $y' = .501(13.5/18.49)x = .366x$
Standard-score equation: $Z'_y = rZ_x = 0.501Z_x$

6–15. **a.**

X	Y
5	35
0	24
1	55
2	30
7	29
2	33
2	43
0	44
3	47
1	26
4	37
0	23

b. $\Sigma X = 27$

c. $X^2 = 25, 0, 1, 4, 49, 4, 4, 0, 9, 1, 16, 0$

d. $\Sigma X^2 = 113$

e. $\Sigma Y = 426$

f. $Y^2 = 1225, 576, 3025, 900, 841, 1089, 1849, 1936,$
$2209, 676, 1369, 529$

g. $\Sigma Y^2 = 16224$

h. $XY = 175, 0, 55, 60, 203, 66, 86, 0, 141, 26, 148, 0$

i. $\Sigma XY = 960$

j. $N = 12$

k. $r = \dfrac{12\,(960) - (27)\,(426)}{\sqrt{(12)\,113 - (27)^2}\;\sqrt{(12)\,16224 - (426)^2}}$

$= \dfrac{18}{\sqrt{627}\,\sqrt{13212}} = \dfrac{18}{2878.2}$

$= 0.0063$

$r^2 = 0.000039$

6–16. Salary$'$ = $-4760.83 + 7710.63$ (GPA)

6–17. **a.** $r = .804$

b. $r^2 = (.804)^2 = .646$

6–18. Salary$'$ = \$15,980.76

Salary$'$ = \$19,758.96

6–19. Let Space = X, Sales = Y. $\Sigma X = 72$ $\Sigma X^2 = 504$ $\Sigma Y = 180$ $\Sigma Y^2 = 2964$ $\Sigma XY = 1176$ $N = 12$

$$r = \frac{(12)(1176) - (72)(180)}{\sqrt{(12)(504) - (72)^2}\ \sqrt{(12)(2964) - (180)^2}}$$

$$= \frac{1152}{\sqrt{864}\ \sqrt{3168}} = 0.696$$

6–20. Sales$'$ = $7.006 + 1.332$ (Space), or Sales$'$ = $7 + 1\ 1/3$ Space

Chapter 7

7–1. **a.** $h_0 : p \leq 1/2$

b. One-tailed test, hypotheses emphasize direction, i.e., larger.

c. $n = 17$, $x = 13$ or 4. Use 4—easier and faster to compute.

d. $P(x=0) = \dfrac{17!}{0!(17-0)!}(0.5)(0.5)^{17-0} = 0.000$

$P(x=1) = P_1 = .0001$

$P(x=2) = P_2 = .0010$

$P(x=3) = P_3 = .0052$

$P(x=4) = P_4 = .0182$

e. Add the P_x's, $\Sigma P_x = .0245$

One-tailed test, $p = .0245$

f. $p < 0.05$, so reject h_0. Evidence supports Professor X's theory.

7–2. $h_0 : p = 1/2$, $p = 0.0308$, $0.0308 < 0.05$, so reject h_0. The vote sample indicates the bill will be defeated.

7–3. **a.** $h_0 : p = 0.10$ of failures or $p = 0.90$ of successes

b. One-tailed test

c. $n = 10$, $X = 3$ or 7

d. Use Formula 7.3

e. The first 3 terms of the binomial expansion for $[9/10 + 1/10]^{10}$ are 0.3487, 0.3874, 0.1937. Summing we get 0.9298

f. The probability of 3 successes under $h_0 : p = 0.10$, for 3 failures under $h_0 : p = 0.90$, the p value for 3 failures is $1 - 0.9298$

$= 0.0702$

$p = 0.0702$

$p > 0.05$, so do not reject h_0. Not sufficient evidence to reject h_0.

7–4. $h_0 : p = 1/36$, $p = 0.0045$, $0.0045 < 0.05$, so reject h_0. There is evidence the dice are not fair.

7–5. **a.** $h_0 : p = 1/2$

b. One-tailed test

c. $n = 17$, $X = 13$

d. Use Formula 7.8: $M_b = np + 17(1/2) = 8\ 1/2$

e. Use Formula 7.9: S.D.$_b + \sqrt{npq} = \sqrt{17(1/2)(1/2)} = \sqrt{4.25} = 2.0616$

f. $X - 1/2 = 12\ 1/2$

g. $Z = (12.5 - 8.5)/2.0616 = 1.940$

h. $P(Z > 1.94) = 1 - 0.9738 = 0.0262$

$p = 0.262$

i. $0.0262 < 0.05$, so reject h_0. Professor X's theory is supported.

Repeat (f–h), using $X + 1/2 = 13\ 1/2$

$Z = 13.5 - 8.5/2.0616 = 2.4253$

$P(Z > 2.4253) = 0.00765$

$P(X = 13) = 0.0262 - 0.00765 = 0.01855$, the binomial is 0.0182

7–6. $h_0 : p = 1/2$, $p = 0.0339$, $p < 0.05$, so reject h_0. There is evidence that the voting will be lopsided.

7–7.   **a.** Z for $12.5 = 1.94$
     b. Z for $13.5 = 2.43$
     c. Z for between 1.94 and 2.43 is $(1.94 + 2.43)/2 = 2.185$
     d. Interpolated ordinate $= 0.0367$
     e. Base $= 1/\text{S.D.}_b = 1/2.06$
     f. Area $=$ base $\times$ height $= (1/2.06)(0.0368) = 0.0178$

7–8.   Probability $= 0.0116$

7–9.   **a.**

	Yes I	No I
Yes II	72	9
No II	16	48

$a = 72 \quad c = 16$
$b = 48 \quad d = 9$

     b. $c + d = n \rightarrow 16 + 9 = 25$ h_0: $p = 1/2$
     c. larger of $c, d = 16, X = 16$
     d. $X - 0.5 = 16 - 0.5 = 15.5$
     e. $Md_b = np = 25\,(0.5) = 12.5$
        $\text{S.D.}_b = \sqrt{npq} = \sqrt{(25)(0.5)(0.5)} = \sqrt{6.25} = 2.5$
     f. $Z = 15.5 - 12.5/2.5 = 3/2.5 = 1.20$
     g. $p(Z > 1.20) = 0.1151$
     h. $0.1151 > 0.05$ Do not reject h_0. No significant difference in the switch.

7–10. $p = 0.0382 < 0.05$ Reject h_0.

7–11.   **a.** $n = 100, X = 2.33$
        $M_b = np = 100\,(0.5) = 50$
        $\text{S.D.}_b = \sqrt{npq} = \sqrt{(50)(0.5)} = \sqrt{25} = 5$
     b. $Z = (X - 1/2) - 50/5$
     c. $2.33 = (X - 0.5) - 50/5$
        $11.65 = X - 0.5 - 50$
        $11.65 + 50 = X - 0.5$
        $11.65 + 50 + 0.5 = X$
     d. $62.15 = X$. Since X must be an integer, in this case a dog, round up to 63.

7–12.   0.9994

7–13.   **a.** Use Formula 7.1
$$P = \frac{20!}{6!(20 - 6)!}\,(0.6)^6\,(0.4)^{20\,-\,6} = 0.00485 = 0.005$$
     b. Use Formula 7.1 five times for $X = 0,1,2,3,4$
$$P_0 = \frac{20!}{0!(20 - 0)!}\,(0.6)^0\,(0.4)^{20\,-\,0} = 0$$
        $P_1 = 0$
        $P_2 = 0$
        $P_3 = 0$
        $P_4 = 0$
        Add up the probabilities
        $P_0 + P_1 + P_2 + P_3 + P_4 = 0.003$
     c. Use Formula 7.1 for $X = 9$
$$P = \frac{20!}{9!(20 - 9)!}\,(0.6)^9\,(0.4)^{20\,-\,9} = .071$$

7–14.   See pages 174–75.

7–15.   A one-tailed test yields a greater number of rejections of h_0. For a fixed α in a one-tailed test, the entire α is in one tail instead of being spread into two tails. With more in one tail, the chances of reaching the rejection region are higher.

7–16. Use Formula 7.3

$$\left(\frac{1}{6} + \frac{5}{6}\right)^3 = \left(\frac{1}{6}\right)^3 + \frac{3}{1}\left(\frac{1}{6}\right)^2\left(\frac{5}{6}\right)^1 + \frac{3(2)}{2}\left(\frac{1}{6}\right)\left(\frac{5}{6}\right)^2$$

7–17. Use Formula 7.3

First 3 terms of $\left(\frac{1}{3} + \frac{2}{3}\right)^9 = \left(\frac{1}{3}\right)^9 + 9\left(\frac{1}{3}\right)^8\left(\frac{2}{3}\right) + \frac{(9)(8)}{2}\left(\frac{1}{3}\right)^7\left(\frac{2}{3}\right)^2$

Last 3 terms of $\left(\frac{1}{3} + \frac{2}{3}\right)^9 = \frac{9(8)}{2}\left(\frac{1}{3}\right)^2\left(\frac{2}{3}\right)^7 + 9\left(\frac{1}{3}\right)^1\left(\frac{2}{3}\right)^8 + \frac{9!}{9!}\left(\frac{1}{3}\right)^{9-9}\left(\frac{2}{3}\right)^9$

7–18. $P = 0.031, P = 0.7422$

7–19.   **a.** See page 167. $6! = 6 \cdot 5 \cdot 4 \cdot 3 \cdot 2 \cdot 1 = 720$

    b. See page 167. $11!/8! = 11 \cdot 10 \cdot 9 \cdot 8/8 = 990$

    c. See page 167.

$$\frac{20!}{17!3!} = \frac{20 \cdot 19 \cdot 18}{3!} = \frac{20 \cdot 19 \cdot 18}{3 \cdot 2 \cdot 1} = 20 \cdot 19 \cdot 3 = 1140$$

    d. $0!/0! = 1/1 = 1$

7–20. See page 175–77.

Chapter 8

8–1.   **a.** $h_0 : \mu \geq 0$ (good advice—break even or better)

      $h_1 : \mu < 0$ (bad advice—losses)

    b. $n = 12$

    c. Mean and standard deviation for the 12 values.

$$M = \frac{(-2) + (-10) + 5 - 7 - 6 - 8 - 3 + 1 + 2 - 12 - 8 - 4}{12}$$

$$= -4.33$$

$$s = \sqrt{\frac{\Sigma X^2 - (\Sigma X)^2/n}{n - 1}} = \sqrt{\frac{516 - (-52)^2/12}{11}} = 5.14$$

    d. Standard error

      $s_M = s/\sqrt{N} = 5.14/\sqrt{12} = 5.14/3.464 = 1.484$

    e. $t = (M - \mu_0)/s_M = (-4.33 - 0)/1.484 = -2.918$

    f. $df = n - 1 = 11$

    g. One-tailed (left tail) $t_{.01} = -2.72$

    h. $-2.918 < -2.72$, reject h_0

8–2. $h_0 : \mu \leq 33$

    $h_1 : \mu > 33$

    $2.33 \ngtr 2.62$, do not reject h_0.

8–3.   **a.** $M = 117$, S.D. $= 8$

    b. $s_M = \text{S.D.}/\sqrt{N - 1} = 8/\sqrt{50 - 1} = 8/\sqrt{49} = 8/7 = 1.14$

    c. $M \pm t_{.05} \cdot s_M = 117 \pm 2.01(1.14) = 117 \pm 2.3 = 114.7$ to 119.3

8–4. 18.86 to 41.34

8–5.   **a.** $t = \dfrac{r \sqrt{N - 2}}{\sqrt{1 - r^2}} = \dfrac{.30 \sqrt{100 - 2}}{\sqrt{1 - (.3)^2}} = \dfrac{2.97}{.954} = 3.11$

    b. $df = N - 2 = 98 \quad \alpha = 0.01 \quad t_{.01} = 2.63$

    c. $t = 3.11 > 2.63$, so reject h_0, where $h_0 : \rho = 0$

                                                   $h_1 : \rho \neq 0$

8–6. $t = 1.81 \ngtr 1.99$, do not reject h_0, where $h_0 : \rho = 0$

                                                          $h_1 : \rho \neq 0$

8-7. **a.** $h_0 : \sigma_1^2 = \sigma_2^2$ $h_1 : \sigma_1^2 \neq \sigma_2^2$

  b. $\text{S.D.}_1 = 1/N_1 \sqrt{N_1 \Sigma X_1^2 - (\Sigma X_1)^2} = 1/20 \sqrt{20(7128) - (226)^2}$

$$= 15.12$$

$$s_1 = \text{S.D.}_1 \sqrt{N_1/N_1 - 1} = 15.12 \sqrt{20/19} = 15.52$$

$$s_1^2 = 240.87$$

$$\text{S.D.}_2 = 1/N_2 \sqrt{N_2 \Sigma X_2^2 - (\Sigma X_2)^2} = 1/18 \sqrt{18 (8321) - (275)^2}$$

$$= 15.128$$

$$s_2 = \text{S.D.}_2 \sqrt{N_2/N_2 - 1} = 15.128 \sqrt{18/17} = 15.57$$

$$s_2^2 = 15.57^2 = 242.42$$

  c. $F = 242.42/240.87 = 1.006$

  d. $df_1 = 17$ $df_2 = 19$ $\alpha = 0.05$ $F_{.05} = 2.21$

  e. $1.007 \not> 2.21$, so do not reject h_0. Yes, it is appropriate to proceed with the t-test.

8-8. $h_0 : \mu_E = \mu_C$, $h_1 : \mu_E \neq \mu_C$, two-tailed test, $F = 2.86 > 2.81 > 2.06$, so reject h_0. No, it might not be appropriate to conduct a t-test.

8-9. **a.** $h_0 : \mu_1 = \mu_2$, $h_1 : \mu_1 \neq \mu_2$

  b. $\text{S.D.}_1 = 15.12$, $\text{S.D.}_2 = 15.13$

  c. $s_{D_M} = \sqrt{\dfrac{\Sigma x_1^2 + \Sigma x_2^2}{N_1 + N_2 - 2} \left(\dfrac{N_1 + N_2}{N_1 \cdot N_2} \right)}$ (Formula 8.16)

$$\Sigma x_1^2 = (\text{S.D.}_1)^2 N_1 = (15.12)^2 \cdot 20 = 4572.29$$

$$\Sigma x_2^2 = (\text{S.D.}_2)^2 N_2 = (15.13)^2 \cdot 18 = 4120.5$$

$$s_{D_M} = \sqrt{\dfrac{4572.29 + 4120.5}{20 + 18 - 2} \left(\dfrac{20 + 18}{(20)(18)} \right)} = \sqrt{(241.47)(.106)}$$

$$= \sqrt{25.488} = 5.049$$

  d. $M_1 = 226/20 = 11.3$ $M_2 = 275/18 = 15.28$

  e. $t = M_1 - M_2 - 0/s_{D_M} = 11.3 - 15.28/5.05 = -0.788$

  f. $\dfrac{df}{\alpha} = 0.05 = N_1 + N_2 - 2 = 20 + 18 - 2 = 36$

$$t_{.05} = 2.03$$

  g. $.788 \not> 2.03$, so do not reject h_0.

8-10. $h_0 : \mu_E = \mu_C$, $h_1 : \mu_E \neq \mu_C$, $t_{.05} = 2.02$, two-tailed, so reject h_0. In light of the F-test, the distribution of variances and means are quite different from one another.

8-11. **a.** $h_0 : \mu_E = \mu_C$ $h_1 : \mu_E \neq \mu_C$

  b. $-3, -5, 0, -5, -2, -2, -4, -3$

  c. $M_d = -3$ $s_d = 1.69$

  d. $s_{d_M} = 1.69/\sqrt{8} = 0.598$ $df = N \text{ pairs} - 1 = 7$

  e. $t = -3/0.598 = -5.02$

  f. $t_{.05} = 2.36$

  g. $t = 5.02 > 2.36$, so reject h_0. There is a difference in the experimental and control groups.

8-12. $h_0 : \mu_E = \mu_C$, $h_1 : \mu_E \neq \mu_C$, $t_{.05} = 1.5/.572 = 2.62 > 2.20$, so reject h_0. Those receiving a memory course performed better than those given intellectual tasks.

$$t_{.05} = 2.20$$

8-13. **a.** $h_0 : \mu_{D_E} = \mu_{D_C}$ $h_1 : \mu_{D_E} \neq \mu_{D_C}$

  b. $D_E = 10, 10, 15, 0, 10, -5, 6, 6, 7, 9, 7, 7$

  $D_C = 12, -2, -5, -2, 5, 3, 1, 1, 6, -3, 2$

  c. $M_{D_E} = 82/12 = 6.83$ $M_{D_C} = 18/11 = 1.64$ $\Sigma X_E^2 = 289.67$ $\Sigma X_C^2 = 232.55$

  d. $s_{D_M} = \sqrt{\dfrac{289.67 + 232.55}{12 + 11 - 2} \left(\dfrac{12 + 11}{(12)(11)} \right)} = 2.08$

  e. $t = (6.83 - 1.64)/2.08 = 2.50$

  f. $df = 12 + 11 - 2 = 21$ $t_{.05} = 2.08$

  g. $t = 2.50 > 2.08$, so reject h_0

8-14. $h_0 : \mu_{D_E} = \mu_{D_C}$, $h_1 : \mu_{D_E} \neq \mu_{D_C}$, $t_{.05} = 3.74$, so reject h_0.

8-15. **a.** $h_0 : \mu_{D_E} = \mu_{D_C}$ $h_1 : \mu_{D_E} \neq \mu_{D_C}$

b. $D_E = 40, 36, 40, 47, 41, 36, 41, 43$
$D_C = 22, 32, 25, 29, 28, 19, 21, 31$

c. $d = 18, 4, 15, 18, 13, 17, 20, 12$

d. $M_d = \Sigma d / N \text{ pairs} = 117/8 = 14.625$

e. $\Sigma X^2 = \Sigma X^2 - (\Sigma X)^2 / N \text{ pairs} = \Sigma d^2 - (\Sigma d)^2 / N \text{ pairs}$
$= 1891 - 1711.125 = 179.875$
$s_d = 5.07$
$s_{d_M} = 5.07 / \sqrt{8} = 1.79$

f. $t = 14.625/1.79 = 8.17$ $df = 8 - 1 = 7$

g. $t_{.01} = 3.50$

h. $t = 8.17 > 3.50$, so reject h_0.

8-16. -14.23 to 6.27

8-17. **a.** $h_0 : \rho_1 = \rho_2$ $h_1 : \rho_1 \neq \rho_2$

b. $Z_1' = 0.377$ $Z_2' = 0.448$

c. $Z = \dfrac{0.377 - 0.448}{\sqrt{1/(20 - 3) + 1/(105 - 3)}} = \dfrac{-0.071}{0.262} = -0.27$

d. Look up in table $P(Z < -0.27) = 0.3936$

e. $0.3936 > 0.05$, so do not reject h_0. No significant difference.

8-18. -5.09 to -0.91

8-19. **a.** $h_0 : \rho = 0.60$ $h_1 : \rho \neq 0.60$

b. $\rho = 0.6$ $r = 0.75$ $N = 52$ $Z_0' = 0.693$ $Z' = 0.973$

c. $Z = (Z' - Z_0')/\sqrt{1/(N - 3)} = 0.973 - 0.693/\sqrt{1/49}$
$= 0.28/(1/7) = 7(0.28) = 1.96$

d. $2 \times p(Z > 1.96) = 2 \times 0.25 = 0.05$

e. $0.05 = 0.05$, so reject h_0. There is evidence that the true correlation is not 0.60

8-20. .244 to 2.76

8-21. **a.** $h_0 : \mu \geq 5 \text{ years}$ $h_1 : \mu < 5 \text{ years}$

b. $N = 6$ $\Sigma X = 25.9$ $\Sigma X^2 = 112.75$ $M = 4.32$ S.D. $= 0.398$ $s = 0.436$

c. $s_M = \text{S.D.}/\sqrt{N - 1} = 0.398/\sqrt{5} = 0.178$

d. $t = (M - \mu)/s_M = (4.32 - 5)/0.178 = -3.83$

e. $df = N - 1 = 5$ $t_{.01} = 3.75$ (one-tailed)

f. $t = |3.83| > 3.75$, so reject h_0. There is evidence that contradicts the manufacturer's claim.

8-22. $h_0 : \sigma \leq 0.89$, $h_1 : \sigma > 0.89$, $11.17 \not> 13.3$, so do not reject h_0. Not sufficient evidence to dispute the guarantee.

8-23. **a.** $h_0 : \rho = 0$ $h_1 : \rho \neq 0$
$h_0 : \rho = 0.23$ $h_1 : \rho \neq 0.23$

b. $\Sigma X = 426$ $\Sigma X^2 = 16224$ $\Sigma XY = 960$ $\Sigma Y = 27$ $\Sigma Y^2 = 113$ $N = 12$ $r = .0063$

c. $t = r\sqrt{N - 2}/\sqrt{1 - r^2} = .0063 \sqrt{12 - 2}/\sqrt{1 - (.0063)^2}$
$= 0.0198/0.999 = 0.02$
Convert 0.0063 and 0.23 to $Z' = 0.006$ and $Z_0' = 0.234$
$Z = Z' - Z_0'/\sqrt{1/N - 3} = 0.006 - 0.234/\sqrt{1/9} = -0.228/0.33$
$= -0.691$

d. $t_{.05} = 2.23$ $P(Z < -0.691) = 0.7549 - 0.500 = 0.2549$
$t = 0.02 \not> 2.23$, so do not reject h_0
$P(Z < -0.691) = 0.2549 > 0.05$, so do not reject h_0.

Chapter 9

9-1. **a.** $f_t = \quad 2.3 \quad + \quad 5.9 \quad + \quad 13.8 \quad = \quad 22$
$\qquad\qquad$ (110–119) (100–109) (90–99) (90–119)
$\qquad f_t = \quad 2.6 \quad + \quad 6.4 \quad + \quad 14.6 \quad = \quad 23.6$
$\qquad\qquad$ (20–29) (30–39) (40–49) (20–49)

$\quad$ **b.** $f_0 = 3 + 8 + 12 = 23 \quad f_0 = 2 + 7 + 16 = 25$

$\quad$ **c.**

X	f_0	f_t
90–119	23	22.0
80–89	20	24.5
70–79	36	32.6
60–69	33	33.0
50–59	24	25.3
20–49	25	23.6

$\quad$ **d.**

X	f_0	f_t	$(f_0 - f_t)$	$(f_0 - f_t)^2$	$(f_0 - f_t)^2/f_t$
90–119	23	22.0	1.0	1.00	0.05
80–89	20	24.5	−4.5	20.25	0.83
70–79	36	32.6	3.4	11.56	0.35
60–69	33	33.0	0.0	0.00	0.00
50–59	24	25.3	−1.3	1.69	0.07
20–49	25	23.6	1.4	1.96	0.08

$\quad$ **e.** $\chi^2 = 1.38$
$\quad$ **f.** $df = 6 - 1 = 5, \chi^2_{.05} = 11.1, \chi^2_{.01} = 15.1$
$\quad$ **g.** $\chi^2 = 1.38 \not> 11.1$. There is *not* a lack of fit.

9-2. $\chi^2 = 15.45$. There is evidence that the observed values do not fit the theoretical values.

9-3. **a.** h_0 : Organizational membership is independent of support functions.
$\qquad h_1$: There is a relationship between organizational and nonorganizational and support.

$\quad$ **b,c.**

	Org	Nonorg	Total
Go	15	11	26
No-go	11	39	50
Total	26	50	76

$\quad$ **d.** $f_{t_{11}} = (26 \times 26)/76 = 8.89 \quad f_{t_{12}} = (26 \times 50)/76 = 17.11$
$\qquad f_{t_{21}} = (26 \times 50)/76 = 17.11 \quad f_{t_{22}} = (50 \times 50)/76 = 32.89$

$\quad$ **e.** $f_{t_{11}} = 8.89 < 10$. Use Yates' correction.

$\quad$ **f.** $\chi^2 = \dfrac{(15-8.89-0.5)^2}{8.89} + \dfrac{(11-17.1-0.5)^2}{17.1} + \dfrac{(11-17.1-0.5)^2}{17.1} + \dfrac{(39-32.89-0.5)^2}{32.89} = 9.6$

$\quad$ **g.** $df = (\# \text{ rows} - 1)(\# \text{ of columns} - 1) = (2 - 1)(2 - 1) = 1df \quad \chi^2_{1df} = 3.84 \quad \alpha = 0.05$
$\quad$ **h.** $\chi^2 = 9.6 > 3.84$, so reject h_0. There is evidence that nonorgs are less supportive.

9-4. h_0: early marriages for males are stable
$\qquad h_1$: early marriages for males tend to be unstable
$\qquad \chi^2 = 14.28$, so reject h_0.

9-5. **a.** h_0: item should be included, i.e., males do not differ from females

      h_1: item should be excluded, i.e., males and females are different in responses

  b. $f_{t11} = (100)(70)/200 = 35$ $f_{t12} = (100)(30)/200 = 15$

      $f_{t13} = (100)(100)/200 = 50$ $f_{t21} = (100)(70)/200 = 35$

      $f_{t22} = 15$ $f_{t23} = 50$

  c. None < 10, so Yates' correction is not necessary

  d. $\chi^2 = \dfrac{(40-35)^2 + (30-35)^2}{35} + \dfrac{(20-15)^2 + (10-15)^2}{15} + \dfrac{(40-50)^2 + (60-50)^2}{50} = 8.76$

  e. $df = (2-1)(3-1) = 1 \times 2 = 2$ $\chi^2_{2df,.05} = 5.99$

  f. $\chi^2 = 8.76 > 5.99$, so reject h_0. The item should be excluded.

9-6. h_0: There is no relationship between age and opinion.

    h_1: There is a relationship between age and opinion. (Younger group likes movie more than older people)

    $\chi^2 = 33.8$, reject h_0. There is a relationship between age and opinion. The producer should appeal to the younger group.

9-7. **a.** h_0: the lanes are uniform

      h_1: the lanes are not uniform

  b. $N = 180$ 6 lanes $f_t = 180/6 = 30$

  c. $\chi^2 = \dfrac{(20-30)^2 + (22-30)^2 + (40-30)^2 + (45-30)^2 + (32-30)^2 + (21-30)^2}{30}$

        $= (100 + 64 + 100 + 225 + 4 + 81)/30 = 19.12$

  d. $df = 6 - 1 = 5$ $\chi^2_{5df} = 15.1$

  e. $\chi^2 = 19.12 > 15.1$, so reject h_0. There is a bias in the lanes. The center lanes are better than the outer lanes.

9-8. h_0: the die is fair

    h_1: the die is unfair ("loaded")

    $\chi^2 = 15.2$ so reject h_0. There is evidence that die is biased toward the "1" face.

9-9. **a.** h_0: no preference among soft drinks

      h_1: there is a preference for one soft drink over the others

  b. $N = 100$ # of softdrinks $= 5$ $f_t = 100/5 = 20$

  c. $\chi^2 = \dfrac{(27-20)^2 + (15-20)^2 + (22-20)^2 + (17-20)^2 + (19-20)^2}{20}$

        $= 49 + 25 + 4 + 9 + 1 / 20 = 88/20 = 4.40$

  d. $df = 5 - 1 = 4$ df $\chi^2_{4df,.05} = 9.49$

  e. $\chi^2 = 4.40 \not> 9.49$, so do not reject h_0. No preference indicated.

  f. Not quite. It would mean that soft drink X is the most preferred in comparison to some, but not all.

9-10. h_0: Attitudes are independent of status.

    h_1: Attitudes are related to status.

    $\chi^2 = 8.79$, reject h_0. Status is related to opinion. Student and faculty tend to be in favor while administration is not.

9-11. **a.** h_0: size of package is independent of sales.

      h_1: size of package and sales are dependent.

  b. Determine row and column totals

    High $= 237$ Small $= 195$

    Low $= 109$ Medium $= 89$

    N $= 346$ Large $= 62$

    $f_{t11} = (237)(195)/346 = 133.6$ $f_{t12} = (237)(89)/346 = 61$

    $f_{t13} = (237)(62)/346 = 42.5$ $f_{t21} = (109)(195)/346 = 61.4$

    $f_{t22} = (109)(89)/346 = 28.0$ $f_{t23} = (109)(62)/246 = 19.5$

  c. $\chi^2 = 1.87$

  d. $df = (\text{\# row} - 1) \times (\text{\# column} - 1) = (2-1) \times (3-1) = 2$ df

    $\chi^2_{2df,.05} = 5.99$ $\chi^2_{2df,.01} = 9.21$

  e. $\chi^2 = 1.87 \not> 9.21$, do not reject h_0. Not enough evidence to support that sales and package size are dependent.

9-12. h_0: ratio of classification is $9:3:3:1$.

    h_1: the ratio of classification is not $9:3:3:1$.

    $\chi^2 = 1.32$, so do not reject h_0.

9–13. **a.** h_0: the data size are consistent with the hypothesis.

h_1: the data are not consistent with the hypothesis.

b. $N = 200$

$f_{tA} = 200 \times 0.38 = 76$

$f_{tB} = 200 \times 0.27 = 54$

$f_{tC} = 200 \times 0.35 = 70$

c. $\chi^2 = (80 - 76)^2/76 + (50 - 54)^2/54 + (70 - 70)^2/70$

$= 16/76 + 16/54 + 0 = 0.21 + 0.30 = .51$

d. $df = 3 - 1 = 2 \; df \quad \chi^2_{2df,.01} = 9.21$

e. $\chi^2 = 0.51 \not> 9.21$, so do not reject h_0. The data appear consistent with the research hypothesis.

9–14. h_0: no relationship between treatment and improvement

h_1: relationship between treatment and improvement

$\chi^2 = 6.50$, so reject h_0. There is evidence that therapy and improvement are related.

9–15. **a.** Sex: M and F

Age: 10, 11, and 12

b. There are 2 sexes and 3 ages, so create a 2×3 table

c.

		Age			
		10	11	12	Total
	Male	5	3	2	10
Sex	Female	2	4	4	10
	Total	7	7	6	20

9–16. χ^2 cannot be calculated because f_t are less than 5.

9–17. **a.** Grade = 5 different types: A, B, C, D, F.

Sex = 2: M and F

b,c.

		Sex		
		M	F	Total
	A	2	1	3
	B	3	3	6
Grade	C	4	3	7
	D	0	2	2
	F	1	1	2
	Total	10	10	20

9–18. See page 274.

9–19. **a.** h_0: The observed data are consistent with the binomial distribution.

h_1: The observed data are not consistent with the binomial distribution.

b. $N = 1000$

$\text{freq}_t(X = 0) = .005 \times 1000 = 5$

$\text{freq}_t(X = 1) = .03 \times 1000 = 30$

$\text{freq}_t(X = 2) = .092 \times 1000 = 92$

$\text{freq}_t(X = 3) = .170 \times 1000 = 170$

$\text{freq}_t(X = 4) = .218 \times 1000 = 218$

$\text{freq}_t(X = 5) = .207 \times 1000 = 207$

$\text{freq}_t(X = 6) = .147 \times 1000 = 147$

$\text{freq}_t(X = 7) = .81 \times 1000 = 81$

$\text{freq}_t(X = 8) = .035 \times 1000 = 35$

$\text{freq}_t(X = 9) = .011 \times 1000 = 11$

c. There is one cell with a theoretical frequency of 5, but N is large.

$\chi^2 = 7.23$

d. $df = 10 - 1 = 9 \quad \chi^2_{9df,.05} = 16.9$

e. $\chi^2 = 7.23 \not> 16.9$, do not reject h_0. The observed data appears to be consistent with the binomial for $X = 0$ through $X = 9$.

9–20. h_0: no relationship between type of therapy and level of recovery.

h_1: type of therapy and level of recovery are related.

$\chi^2 = 17.84$, so reject h_0. There is a relationship between type of therapy and level of recovery.

Chapter 10·

10–1. **a.** h_0: no difference between the group means.

h_1: at least one pair of means is different.

b. $\Sigma X = 769 \quad \Sigma X^2 = 22469 \quad \Sigma x^2 = 2756.97$

c. $N = 30$

d. $\Sigma X_1 = 371 \quad \Sigma X_2 = 248 \quad \Sigma X_3 = 150$

$\Sigma X_1^2 = 13913 \quad \Sigma X_2^2 = 6244 \quad \Sigma X_3^2 = 2312$

e. $M_1 = 37.1 \quad M_2 = 24.8 \quad M_3 = 15 \quad M = 25.63$

f. $\Sigma x_1^2 = 148.9 \quad \Sigma x_2^2 = 93.6 \quad \Sigma x_3^2 = 62$

g. $SS_W = 148.9 + 93.6 + 62 = 304.5$

$N_T - k = 30 - 3 = 27$

h. $s_W^2 = 304.5/27 = 11.28$

i. $n_i(M_i - M)^2 = 1314.84 \quad n_2(M_2 - M)^2 = 6.94 \quad n_3(M_3 - M)^2 = 1130.68$

j. $\Sigma n_i(M_i - M)^2 = 1314.84 + 6.94 + 1130.68 = 2452.47$

k. $df = 2$

$s_B^2 = 1226.23$

l. $F = s_B^2/s_W^2 = 108.73$

m. $df = 2.27 \quad F_{.05} = 3.35 \quad F_{.01} = 5.49$

n. $F = 108.73 > F_{.01}$ and $> F_{.05}$, reject h_0.

o.

Source	Sum of Squares	df	Mean Square	F Ratio
Between-groups	2452.47	2	1226.23	108.73
Within-groups	304.50	27	11.28	
Total	2756.97	29		

10–2. h_0: no difference between the group means.

h_1: at least one pair of means is different.

$F = 2.41$, do not reject h_0.

10–3. **a.** $k = 3$

b. $S_1^2 = [\Sigma X_1^2 - (\Sigma X_1)^2/n_1]/(n_1 - 1) = 148.9/9 = 16.54$

$S_2^2 = [\Sigma X_2^2 - (\Sigma X_2)^2/n_2]/(n_2 - 1) = 93.6/9 = 10.4$

$S_3^2 = [\Sigma X_3^2 - (\Sigma X_3)^2/n_3]/(n_3 - 1) = 62.0/9 = 6.88$

c. $F_{12} = 16.54/10.4 = 1.59$

$F_{13} = 16.54/6.88 = 2.40$ largest

$F_{23} = 10.4/6.88 = 1.51$

d. $df = 9, 9 \quad F_{.05} = 3.18$

e. $F_{large} \not> F_{.05}$, the assumption is tenable.

10–4. $F_{.05} = 2.82$

$F_{large} \not> F_{.05}$, the assumption is tenable

10–5. **a.** F-test for question 10–1 was significant.

b. $t_{12} = \dfrac{37.1 - 24.8}{\sqrt{11.28[(10 + 10)/100]}} = 8.19$

$t_{13} = \dfrac{37.1 - 15.0}{\sqrt{11.28[(10 + 10)/100]}} = 14.71$

$t_{23} = \dfrac{24.8 - 15.0}{\sqrt{11.28(10 + 10)/100]}} = 6.52$

c. $t'_{.05} = \sqrt{2(3.35)} = 2.59 \quad t'_{.01} = \sqrt{2(5.49)} = 3.31$

d. $t_{12} > t'_{.01} \quad t_{13} > t'_{.01} \quad t_{23} > t'_{.01}$

All comparisons between two group means are significant at the 0.01 level.

10–6. Not a significant F-test for question 10–2. Computations are done as an exercise.
$t'_{.05} = 2.90$, $t'_{.01} = 3.58$
None are significant.

10–7.   **a.** h_0: no difference between the group means.
   h_1: at least one pair of means is different.
   b. $\Sigma X = 353$ $\Sigma X^2 = 4041$ $\Sigma x^2 = 4041 - (353)/33 = 264.97$
   c. $N = 12 + 11 + 10 = 33$
   d. $\Sigma X_1 = 152$ $\Sigma X_2 = 126$ $\Sigma X_3 = 75$ $\Sigma X_1^2 = 1958$ $\Sigma X_2^2 = 1472$ $\Sigma X_3^2 = 611$
   e. $M_1 = 12.67$ $M_2 = 11.45$ $M_3 = 7.5$ $M = 10.7$
   f. $\Sigma x_1^2 = 32.67$ $\Sigma x_2^2 = 28.73$ $\Sigma x_3^2 = 48.5$
   g. $SS_W = 32.67 + 28.73 + 48.5 = 109.89$ $df = N_T - k = 33 - 3 = 30$
   h. $s_W^2 = 109.89/30 = 3.66$
   i. $n_1(M_1 - M)^2 = 46.56$ $n_2(M_2 - M)^2 = 6.31$ $n_3(M_3 - M)^2 = 102.21$
   j. $\Sigma n_i(M_i - M)^2 = 46.56 + 6.31 + 102.21 = 155.08$
   k. $df = 3 - 1 = 2$ $s_B^2 = (155.08)/2 = 77.54$
   l. $F = s_B^2/s_W^2 = (77.54)/3.66 = 21.17$
   m. $df = 2, 30$ $F_{.05} = 3.32$ $F_{.01} = 5.39$
   n. $F = 21.17 > F_{.01}$, reject h_0.
   o.

Source	Sum of Squares	df	Mean Square	F Ratio
Between-groups	155.08	2	77.54	21.17
Within-groups	109.89	30	3.66	
Total	264.97	32		

10–8. $F = 19.50$, reject h_0.

10–9.   **a.** F-test for question 10–7 was significant.
   b. $t_{12} = \dfrac{12.67 - 11.45}{\sqrt{3.66(12 + 11)/(12.11)}} = 1.53$

   $t_{13} = \dfrac{12.67 - 7.5}{\sqrt{3.66(12 + 10)/(12.10)}} = 6.31$

   $t_{23} = \dfrac{11.45 - 7.5}{\sqrt{3.66(11 + 10)/(11.10)}} = 4.73$
   c. $t'_{.05} = \sqrt{2(3.32)} = 2.58$
   $t'_{.01} = \sqrt{2(5.39)} = 3.28$
   d. Only t_{13} and t_{23} are significant at the 0.01 level, t_{12} is not significant.

10–10. t_{13}, t_{14} and t_{24} are significant at the 0.01 level. All others are not significant.

10–11.   **a.** F-test was significant in question 10–1.
   b. $D_{12} = 12.3$
   $D_{13} = 22.1$
   $D_{23} = 9.8$
   c. $HSD_{.05} = 3.51 \sqrt{11.28/10} = 3.73$
   $HSD_{.01} = 4.50 \sqrt{11.28/10} = 4.78$
   d. D_{12}, D_{13}, and D_{23} are all significant at the 0.01 level.

10–12. None of the differences are significant.

10–13.   **a.** F-test in question 10–7 was significant.
   b. $D_{12} = 12.67 - 11.45 = 1.22$
   $D_{13} = 12.67 - 7.5 = 5.17$
   $D_{23} = 11.45 - 7.5 = 3.95$
   c. $HSD_{.05} = 3.49 \sqrt{3.66/10.94} = 2.01$
   $HSD_{.01} = 4.45 \sqrt{3.66/10.94} = 2.57$
   $n_h = 3 / [(1/12) + (1/11) + (1/10)] = 10.94$
   d. D_{13} and D_{23} were significant at the 0.01 level.

10–14. $HSD_{.05} = 3.74$
$HSD_{.01} = 4.66$
D_{13}, D_{14}, and D_{24} are significant at the 0.01 level.
D_{34} is significant at the 0.05 level.

10–15. **a.** h_0: no difference between the group means.
h_1: at least one pair of means is different.
b. $\Sigma X = 71$ $\Sigma X^2 = 425$ $\Sigma x^2 = 425 - (71)^2/12 = 4.92$
c. $N = 4 + 3 + 5 = 12$
d. $\Sigma X_1 = 24$ $\Sigma X_2 = 16$ $\Sigma X_3 = 31$
$\Sigma X_1^2 = 146$ $\Sigma X_2^2 = 86$ $\Sigma X_3^2 = 193$
e. $M_1 = 6$ $M_2 = 5.33$ $M_3 = 6.2$ $M = 5.92$
f. $\Sigma x_1^2 = 2$ $\Sigma x_2^2 = 0.67$ $\Sigma x_3^2 = 0.80$
g. $SS_W = 2 + 0.67 + 0.80 = 3.47$ $df = N_T - k = 12 - 3 = 9$
h. $s_W^2 = 3.47/9 = 0.39$
i. $n_1(M_1 - M) = 0.028$ $n_2(M_2 - M) = 1.02$ $n_3(M_3 - M) = 0.40$
j. $\Sigma n_i(M_i - M)^2 = 0.028 + 1.02 + 0.40 = 1.45$
k. $k - 1 = 3 - 1 = 2$ $s_B^2 = 1.45/2 = 0.725 = 0.73$
l. $F = s_B^2 / s_W^2 = 0.73/0.39 = 1.88$
m. $df = 2, 9$ $F_{.05} = 4.26$ $F_{.01} = 8.02$
n. $F = 1.88 \ngtr 4.26$, do not reject h_0.
o.

Source	Sum of Squares	df	Mean Square	F Ratio
Between-groups	1.45	2	0.72	1.88
Within-groups	3.47	9	0.39	
Total	4.92	11		

10–16. $HSD_{.05} = 1.26$
$HSD_{.01} = 1.73$
None of the differences are significant.

10–17. **a.** h_0: no difference between the group means.
h_1: at least one pair of means is different.
b. $\Sigma X = 706$ $\Sigma X^2 = 42446$ $\Sigma x^2 = 42446 - (706)^2/12 = 909.67$
c. $N = 4 + 3 + 5 = 12$
d. $\Sigma X_1 = 242$ $\Sigma X_2 = 144$ $\Sigma X_3 = 320$
$\Sigma X_1^2 = 14666$ $\Sigma X_2^2 = 7206$ $\Sigma X_3^2 = 20574$
e. $M_1 = 60.5$ $M_2 = 48$ $M_3 = 64$ $M = 58.83$
f. $\Sigma x_1^2 = 25$ $\Sigma x_2^2 = 294$ $\Sigma x_3^2 = 94$
g. $SS_W = 25 + 294 + 94 = 413$ $df = N_T - k = 12 - 3 = 9$
h. $s_W^2 = 413/9 = 45.89$
i. $n_1(M_1 - M)^2 = 11.11$ $n_2(M_2 - M)^2 = 352.08$ $n_3(M_3 - M)^2 = 133.47$
j. $\Sigma n_i(M_i - M)^2 = 11.11 + 352.08 + 133.47 = 496.67$
k. $df = k - 1 = 3 - 1 = 2$ $s_B^2 = 496.67/2 = 248.33$
l. $F = s_B^2 / s_W^2 = 248.33/45.89 = 5.41$
m. $df = 2, 9$ $F_{.05} = 4.26$ $F_{.01} = 8.02$
n. $F = 5.41 > F_{.05}$, reject h_0.
o.

Source	Sum of Squares	df	Mean Square	F Ratio
Between-groups	496.67	2	248.33	5.41
Within-groups	413.00	9	45.89	
Total	909.67	11		

10–18. $F = 0.06$, do not reject h_0.

10–19. **a.** h_0: no difference between the group means.

   h_1: at least one pair of means is different.

   b. $\Sigma X = 255 \quad \Sigma x^2 = 5507 \quad \Sigma x^2 = 5507 - (255)^2/12 = 88.25$

   c. $N = 4 + 4 + 4 = 12$

   d. $\Sigma X_1 = 96 \quad \Sigma X_2 = 77 \quad \Sigma X_3 = 82$
   $\Sigma X_1^2 = 2314 \quad \Sigma X_2^2 = 1491 \quad \Sigma X_3^2 = 1702$

   e. $M_1 = 24 \quad M_2 = 19.25 \quad M_3 = 20.5 \quad M = 21.25$

   f. $\Sigma x_1^2 = 10 \quad \Sigma x_2^2 = 8.75 \quad \Sigma x_3^2 = 21$

   g. $SS_W = 10 + 8.75 + 21 = 39.75 \quad df = N_T - k = 12 - 3 = 9$

   h. $s_W^2 = 39.75/9 = 4.42$

   i. $n_1(M_1 - M)^2 = 30.25 \quad n_2(M_2 - M)^2 = 16 \quad n_3(M_3 - M)^2 = 2.25$

   j. $\Sigma n_i(M_i - M)^2 = 30.25 + 16 + 2.25 = 48.50$

   k. $df = k - 1 = 3 - 1 = 2 \quad s_B^2 = 48.5/2 = 24.25$

   l. $F = s_B^2 / s_w^2 = 24.25/4.42 = 5.49$

   m. $df = 2, 9 \quad F_{.05} = 4.26 \quad F_{.01} = 8.02$

   n. $5.49 > F_{.05}$, reject h_0.

   o.

Source	Sum of Squares	df	Mean Square	F Ratio
Between-groups	48.50	2	24.25	5.49
Within-groups	39.75	9	4.42	
Total	88.25	11		

10–20. t_{12} is significant at the 0.05 level.

10–21. **a.** F-test for question 10–19 is significant.

   b. $D_{12} = 4.75 \quad D_{13} = 3.50 \quad D_{23} = 1.25$

   c. $HSD_{.05} = 3.95 \sqrt{4.42/4} = 4.15$
   $HSD_{.01} = 5.43 \sqrt{4.42/4} = 5.71$

   d. D_{12} is significant at the 0.05 level.

Chapter 11

11–1. **a.** h_0: no difference between males and females.

   h_1: males and females are different.

   h_0: no difference between reducing programs.

   h_1: at least one pair of programs is different.

   h_0: no interaction between sex and program.

   h_1: interaction exists.

   b. $\Sigma X = 731 \quad \Sigma X^2 = 10059 \quad \Sigma x^2 = 10059 - (731)^2/60 = 1152.98$

   c. $N = 10 + 10 + 10 + 10 + 10 + 10 = 60$

   d. $\Sigma X_{MD} = 126 \quad \Sigma X^2_{MD} = 1648 \quad \Sigma X_{ME} = 105 \quad \Sigma X^2_{ME} = 1161$
   $\Sigma X_{MDE} = 159 \quad \Sigma X^2_{MDE} = 2697 \quad \Sigma X_{FD} = 156 \quad \Sigma X^2_{FD} = 2558$
   $\Sigma X_{FE} = 67 \quad \Sigma X^2_{FE} = 515 \quad \Sigma X_{FDE} = 118 \quad \Sigma X^2_{FDE} = 1480$

   e. $\Sigma x^2_{MD} = 1648 - (126)^2/10 = 60.4; \ \Sigma x^2_{ME} = 1161 - (105)^2/10 = 58.5$
   $\Sigma x^2_{MDE} = 2697 - (159)^2/10 = 168.9; \ \Sigma x^2_{FD} = 2558 - (156)^2/10 = 124.4$
   $\Sigma x^2_{FE} = 515 - (67)^2/10 = 66.1; \ \Sigma x^2_{FDE} = 1480 - (118)^2/10 = 87.6$

   f. within-groups sum of squares = 565.9

   g. $df = 60 - 6 = 54 \quad s_w^2 = 565.9/54 = 10.48$

   h. $M_{MALE} = 13 \quad M_{FEMALE} = 11.37 \quad M_D = 14.1 \quad M_E = 8.6$
   $M_{DE} = 13.85 \quad M_{MD} = 12.6 \quad M_{ME} = 10.5 \quad M_{MDE} = 15.9$
   $M = 12.18 \quad M_{FD} = 15.6 \quad M_{FE} = 6.7 \quad M_{FDE} = 11.8$

   i. Row SS $= 298.28 - (24.37)^2/2 = 1.33$
   $= (1.33)(3)(10) = 39.85$

j. $df = 2 - 1 = 1$ $s_R^2 = 39.85/1 = 39.85$

k. Column SS $= (19.29)(10)(2) = 385.83$

l. $df = $ number of column $- 1 = 2$ $s_C^2 = 385.8/2 = 192.9$

m. $=$ Total SS $-$ Row SS $-$ Column SS $-$ Within SS
$= 1152.98 - 39.85 - 385.83 - 565.9 = 161.4$

n. $df = $ (number of columns $- 1$) $\times$ (number of rows $- 1$) $= 2 \times 1 = 2$
$s_{rxc}^2 = 161.4/2 = 80.7$

o. $F_{ROW} = 39.85/10.48 = 3.80$
$F_{COLUMN} = 192.9/10.48 = 18.41$
$F_{INTER} = 80.7/10.48 = 7.7$

p. $F_{.05,ROW} = 4.02$ $F_{.05,COLUMN} = 3.17$ $F_{.05,INTER} = 3.17$
$F_{.01,ROW} = 7.12$ $F_{.01,COLUMN} = 5.01$ $F_{.01,INTER} = 5.01$

q. $F_{COLUMNS} = 18.4 > 5.01$ significant at 0.01
$F_{INTERACTION} = 7.7 > 5.01$ significant at 0.01

r.

Source	Sum of Squares	df	Mean Square	F Value
Between-rows	39.9	1	39.90	3.81
Within-columns	385.8	2	192.90	18.41
Interaction	161.4	2	80.70	7.70
Within-groups	565.9	54	10.48	
Total	1153.0	59		

11-2. $F_{.05} = 4.06$
$F_{.01} = 7.24$
No effect is significant.

11-3. **a.** F-test on columns and interaction is significant.

b. $t_{D \text{ vs } E} = \dfrac{14.1 - 8.6}{\sqrt{10.48[(20+20)/(20)(20)]}} = 5.37$

$t_{D \text{ vs } DE} = \dfrac{14.1 - 13.85}{\sqrt{10.48[(20+20)/(20)(20)]}} = 0.24$

$t_{E \text{ vs } DE} = \dfrac{13.85 - 8.6}{\sqrt{10.48[(20+20)/(20)(20)]}} = 5.13$

c. $t'_{.05} = \sqrt{(6 - 1)(2.38)} = 3.45$ $t'_{.01} = \sqrt{(6 - 1)(3.37)} = 4.10$

d. $t_{D \text{ vs } E} = 5.37 > t'_{.01}$ M_D vs M_E
$t_{E \text{ vs } DE} = 5.13 > t'_{.01}$ M_{DE} vs M_E

11-4. $HSD_{.05} = 4.280$
$HSD_{.01} = 5.145$
Only $D_{M \text{ vs } E}$ is significant at 0.05 level.

11-5. **a.** Significant F-test for column and interaction.

b. $D_{D \text{ vs } E} = 5.5$ $D_{D \text{ vs } DE} = 0.25$ $D_{E \text{ vs } DE} = 5.25$

c. $HSD_{.05} = 4.181 \sqrt{10.48/10} = 4.280$
$HSD_{.01} = 5.026 \sqrt{10.48/10} = 5.145$

d. $D_{D \text{ vs } E} = 5.5 > HSD_{.01}$
$D_{E \text{ vs } DE} = 5.25 > 5.01 = HSD_{.01}$

11-6. No significant F-test.

11-7. **a.** h_0: no difference between classrooms.
h_1: one pair of classroom means is different.
h_0: no difference between methods.
h_1: at least one pair of means is different
h_0: no interaction.
h_1: interaction exists.

b. $\Sigma X = 2699$ $\Sigma X^2 = 204079$ $\Sigma X^2 = 204079 - (2699)^2/36 = 1728.97$

c. $N = 36$

d. $R = 3$ $C = 4$ $k = 3 \times 4 = 12$

e. $\Sigma X_{C_1T_1} = 192$ $\Sigma X^2_{C_1T_1} = 12296$ $\Sigma X_{C_1T_2} = 211$ $\Sigma X^2_{C_1T_2} = 14865$

  $\Sigma X_{C_1T_3} = 205$ $\Sigma X^2_{C_1T_3} = 14041$ $\Sigma X_{C_1T_4} = 231$ $\Sigma X^2_{C_1T_4} = 17813$

  $\Sigma X_{C_2T_1} = 247$ $\Sigma X^2_{C_2T_1} = 20361$ $\Sigma X_{C_2T_2} = 241$ $\Sigma X^2_{C_2T_2} = 19525$

  $\Sigma X_{C_2T_3} = 237$ $\Sigma X^2_{C_2T_3} = 18797$ $\Sigma X_{C_2T_4} = 235$ $\Sigma X^2_{C_2T_4} = 18507$

  $\Sigma X_{C_3T_1} = 247$ $\Sigma X^2_{C_3T_1} = 20349$ $\Sigma X_{C_3T_2} = 225$ $\Sigma X^2_{C_3T_2} = 16883$

  $\Sigma X_{C_3T_3} = 222$ $\Sigma X^2_{C_3T_3} = 16436$ $\Sigma X_{C_3T_4} = 206$ $\Sigma X^2_{C_3T_4} = 14206$

f. $\Sigma x^2_{C_1T_1} = 8$ $\Sigma x^2_{C_1T_2} = 24.67$ $\Sigma x^2_{C_1T_3} = 32.67$ $\Sigma x^2_{C_1T_4} = 26$

  $\Sigma x^2_{C_2T_1} = 24.67$ $\Sigma x^2_{C_2T_2} = 164.67$ $\Sigma x^2_{C_2T_3} = 74$ $\Sigma x^2_{C_2T_4} = 98.67$

  $\Sigma x^2_{C_3T_1} = 12.67$ $\Sigma x^2_{C_3T_2} = 8$ $\Sigma x^2_{C_3T_3} = 8$ $\Sigma x^2_{C_3T_4} = 60.67$

g. Within SS $= 542.67$

h. $df = 36 - 12 = 24$ $s_W^2 = 542.67/24 = 22.61$

i. $M = 74.97$

  $M_{C_1T_1} = 64$ $M_{C_1T_2} = 70.3$ $M_{C_1T_3} = 68.3$ $M_{C_1T_4} = 77$

  $M_{C_2T_1} = 82.3$ $M_{C_2T_2} = 80.3$ $M_{C_2T_3} = 79$ $M_{C_2T_4} = 78.3$

  $M_{C_3T_1} = 82.3$ $M_{C_3T_2} = 75$ $M_{C_3T_3} = 74$ $M_{C_3T_4} = 68.7$

  $M_{C_1} = 69.9$ $M_{C_2} = 80$ $M_{C_3} = 75$ $M_{T_1} = 76.2$

  $M_{T_2} = 75.2$ $M_{T_3} = 73.8$ $M_{T_4} = 74.7$

j. Row SS $= 12 \times [(69.92 - 74.97) + (80 - 74.97) + (75 - 74.97)] = 609.65$

k. $df = $ number of rows $- 1 = 2$ $s_R^2 = 609.65/2 = 304.83$

l. Column SS $= 9 \times [(76.2 - 74.97) + (75.2 - 74.97) + (73.8 - 74.97) + (74.67 - 74.97)]$
  $= 27.222$

m. $df = $ number of columns $- 1 = 4 - 1 = 3$
  $s_C^2 = 27.222/3 = 9.074$

n. $SS_{\text{Interaction}} = $ Total SS $-$ Row SS $-$ Column SS $-$ Within SS
  $= 1728.97 - 609.65 - 27.222 - 542.67 = 549.43$

o. $df = $ (number of columns $- 1) \times$ (number of rows $- 1) = 3 \times 2 = 6$
  $s_{R \times C}^2 = 549.43/6 = 91.57$

p. $F_{\text{Column}} = 9.074/22.61 = .40$
  $F_{\text{Row}} = 304.83/22.61 = 13.48$
  $F_{\text{Interaction}} = 91.57/22.61 = 4.05$

q. Column: $F_{.05} = 3.01$ $F_{.01} = 4.72$
  Row: $F_{.05} = 3.40$ $F_{.01} = 5.61$
  Interaction: $F_{.05} = 2.51$ $F_{.01} = 3.67$

r. Interaction significant at 0.01
  Row significant at 0.01

s.

Source	Sum of Squares	df	Mean Square	F Value
Between-rows	609.65	2	304.83	13.48
Between-columns	27.22	3	9.074	0.40
Interaction	549.43	6	91.57	4.05
Within-groups	542.67	24	22.61	
Total	1728.87	35		

Chapter 12

12-1.   **a.** h_0: no difference between the two appraisers.

        h_1: there is a difference between the two appraisers.

    b. 6 −7 7 0 −5 −3 5 10 −2 13 −1

    c. # of "−" = 6 # of "+" = 4 drop the one with "0"

    d. V = 4

    e. $n = 10$ critical value = 1

    f. V = 4 ≮ 1, do not reject h_0. There is not sufficient evidence to indicate any real differences between the two appraisers.

12-2. $U = 59.5$ ≮ 31 and ≯ 90, do not reject h_0. No evidence of a real difference between appraisers.

12-3.   **a.** h_0: the correlation is not different from zero.

        h_1: the correlation is different from zero.

    b.

	1	2	3	4	5	6	7	8	9	10	11
John	1	7	6	4	5	2	8	10	3	9	11
Bobbi	1.5	8.5	6	3	5	1.5	7	10	4	8.5	11

    c. $d = -.5$ −1.5 0 1 0 0.5 1 0 −1 0.5 0

       $\Sigma d = 0$

    d. $d^2 = 0.25$ 2.25 0 1 0 0.25 1 0 1 0.25 0

       $\Sigma d^2 = 6$

    e. Rho $= 1 - (6 \times 6)/11(121 - 1) = 1 - 36/1320 = 0.973$

    f. Critical value = 0.6091

    g. Rho $= 0.973 > 0.6091$ reject h_0. The rho value (correlation) is significantly different from zero. A real relationship exists between the two appraisers.

12-4. $R_t > C_{.025}$ $R_t < C_{.975}$, do not reject h_0. The sequence is random.

12-5.   **a.** h_0: no difference in the distribution of performance between the 3 colleges.

        h_1: there is a difference between the three colleges in academic performance.

    b.

College A	College B	College C
8	13	1
11.5	6	17
5	15.5	2
15.5	14	11.5
10	4	7
9		3
18		

$n_1 = 7$	$n_2 = 5$	$n_3 = 6$
$r_1 = 77$	$r_2 = 52.5$	$r_3 = 41.5$
$\bar{r}_1 = 11$	$\bar{r}_2 = 10.5$	$\bar{r}_3 = 6.92$

    c. $KW = \dfrac{12}{18 \times 19}(77 - 0.5 \times 7 \times 19) + (52.5 - 0.5 \times 5 \times 19) + (41.5 - 5 \times 6 \times 19)$

          $= (.053)(15.75 + 5 + 40.04) = 2.13$

    d. The n's are greater than or equal to 5, so table L of Appendix B cannot be used. Instead the χ^2 distribution is used with $k - 1$ degree of freedom.

       $k = 3$ $df = 2$ $\chi^2_{.05,2df} = 5.99$

    e. $KW = 2.13$ ≯ 5.99, do not reject h_0. Not sufficient evidence to claim a true difference between colleges.

12-6. $U = 30$ ≮ $U_{.025}$ and ≯ $U_{.975}$, do not reject h_0. Not enough evidence to suggest a difference between the two learning programs.

12–7. **a.** h_0 : the number of defective components produced by the two lines are equal or A produces more than B.

h_1 : Line B produces more defective components than line A.

b. -5 -6 -18 19 -33 16 -21 -20 -11 35

c. Number of minus $= 7$, number of pluses $= 3$

d. $V = 3$

e. $n = 10$, critical value 1

f. $V = 3 \not< 1$, do not reject h_0. Insufficient evidence to claim a difference between the two production lines.

12–8. $W = 20 \leq 9$ and ≥ 46, do not reject h_0. Insufficient evidence to indicate a real difference between line A and line B.

12–9. The dates are *paired* by days.

12–10. Rho $= 0.481 \geq 0.5179$, do not reject h_0. Not enough evidence to claim the correlation is different from zero.

h_0: the correlation is zero.

h_1: the correlation is not zero.

12–11. **a.** h_0: the ratings of the two commercials are the same.

h_1: the ratings of commercial 1 are different from the ratings of commercial 2.

b. $d = -3$ 4 1 0 1 -1 8 -2 0 -2 -1 1 1 -1 0 $N = 12$

c. $N = 12$, drop 3.

$d = $ 3 4 1 1 1 8 2 2 1 1 1 1

rank $=$ 10 11 4 4 4 12 8.5 8.5 4 4 4 4

d. $r = 10 + 4 + 8.5 + 8.5 + 4 + 4 = 39$

e. $r_t = 11 + 4 + 4 + 4 + 4 + 12 = 39$

f. $W = 39$

g. $N = 12$ $W_{.025} = 14$ $W_{.975} = (12 \times 13)/2 = 78 - 14 = 64$

h. $W = 39 \geq 14$ and ≤ 64, do not reject h_0. Not sufficient evidence of a real difference in ratings.

12–12. The ratings are independent, in that it is unlikely that the exact same students took courses from the same professor.

12–13. **a.** h_0: The ratings of the two professors are equal.

h_1: The ratings of the two professors are different.

b.

Professor T	8.5	13	4.5	6	8.5	2		
Professor S	4.5	12	11	7	3	1	14	10

c. $s = 8.5 + 13 + 4.5 + 6 + 8.5 + 2 = 42.5$

d. $U = 42.5 - (6 \times 7)/2 = 42.5 - 21 = 21.5$

e. Let $\alpha = .05$, $U_{.025} = 9$ $U_{.975} = 6 \times 8 - 9 = 39$

f. $U = 21.5 \leq 9$ and ≥ 39, do not reject h_0.

12–14. Sign test: $V = 4 \leq 0$, do not reject h_0. No difference between the judges.

Wilcoxon rank-sum test: $W = 15 \geq 4$ and ≤ 32, do not reject h_0. No real difference exists between the two judges' rating.

12–15. **a.** h_0 : the sequence is random.

h_1 : the sequence is not random.

b. Number of runs $= 10 = R$

c. $n -$ number of 'A' $= 15$ $n =$ number of 'B' $= 15$ $C_{.025} = 10$ $C_{.975} = 20$

d. $R_t = 10 \leq C_{.025}$, reject h_0, the sequence is not random. Failure rates are not random.

12–16. $U = 111 \geq U_{.025}$ and $\leq U_{.975}$, do not reject h_0. There is no evidence that the ranks are different between group A and B.

12–17. **a.** h_0 : weight reduction program not effective

h_1 : weight reduction program is effective.

b. Number pluses $= 8$, number of minuses $= 21$

c. $V = 8$

d. $n = 29$

critical value $= 8$

12–18. $R_t = 9 \leq C_{.025}$, reject h_0. The sequence of correct answers is not random.

12–19.   **a.** h_0 : the length of the small right fingers of Northern California men are longer or equal to that of Southern California men.

h_1 : Northern California men have a shorter small right finger than Southern California men.

b.

N. Cal	5	2.5	7	8	4	6	
S. Cal	10	1	13	2.5	12	11	9

c. $s = 5 + 2.5 + 7 + 8 + 4 + 6 = 32.5$

d. $U = 32.5 - (6 \times 7)/2 = 32.5 - 21 = 11.5$

e. $n_1 = 6$ $n_2 = 7$ $U_{.05} = 9$ $U_{.95} = (6 \times 7) - 9 = 42 - 9 = 33$

f. $U = 11.5 > U_{.05} < U_{.95}$, do not reject h_0. There is insufficient evidence to suggest that Northern California men have a shorter small right finger than Southern California men.

12–20. $W = 6.5 < W_{.025}$, reject h_0. There is evidence (who really cares) that the right small finger is larger than the left small finger in Caucasian men.

12–21.   **a.** h_0 : the % of women in the 3 areas are the same.

h_1 : the % of women in the 3 areas are not equal.

b.

Area A	6	3	15	16	7	10	$n_1 = 6$	$r_1 = 57$
Area B	1	14	18	2	8	12	$n_2 = 6$	$r_2 = 55$
Area C	9	4	17	13	11	5	$n_6 = 6$	$r_3 = 59$

c. $KW = \dfrac{12}{18 \times 19} \left[\dfrac{57^2}{6} + \dfrac{55^2}{6} + \dfrac{59^2}{6} \right] - 3 \times 19$

$= 57.05 - 57 = 0.05$

d. The n's are equal to 6. Table L of Appendix B does not have values higher than 5. Hence, the distribution is used with $k - 1$ *df*.

e. $df = 3 - 1 = 2$ $\chi^2_{.05} = 5.99$

e. $KW = 0.05 \geq 5.99$, do not reject h_0. There are no differences between the 3 areas.

Glossary

abscissa
The x-axis or horizontal axis in a two-dimensional plot. *132*

acceptance region
That portion of the area under the sampling distribution of a test statistic that corresponds to acceptance of the null hypothesis.

alpha (α)
The probability of making a Type I error (rejecting the null hypothesis when it is true). The significance level of a test. *177*

alternative hypothesis
The negative of the null hypothesis. It is symbolized as h_1. Also known as the research hypothesis. *177*

ANOVA
Analysis of variance. *296*

average
One of the measures of central tendency—mean, median, or mode. *60*

bar chart
See histogram

bell-shaped curve
See normal distribution

beta (β)
The probability of making a Type II error. *177*

between-subjects design
Also known as between-groups design. Design for testing the difference between two or more treatment groups where each group of different subjects receive a different treatment condition.

biased estimator
An estimator that does not approximate the parameter. *209*

bimodal
Having two modes. *47*

binomial distribution
The "histogram" of possible outcomes from a binomial process of some specified number of trials, with a fixed probability of "success" on each. *169*

binomial process
A sequence of trials performed, each of which has the same likelihood of success (1) or failure (0), and for which the outcome of one trial in no way affects the outcome of any other trial.

bivariate data
Data that occur in ordered pairs, such as height and weight; two-variable data.

CEEB scores
Linearly transformed scores with a mean of 500 and a standard deviation of 100. CEEB stands for College Entrance Examination Board. *103*

cells
The categories in a contingency table, or the categories in a two-way analysis of variance; the items in a matrix. *270*

central limit theorem
The mathematical theorem stating that for any population with finite variance, if the sample size is large enough, then the distribution of sample means will be approximately normal. *207*

change scores
See difference scores

chi square distribution
The distribution of the chi square statistic. *232*

chi square statistic
The statistic that measures the discrepancy between the observed values and the expected values in a contingency table. *266*

class
A collection of numbers, usually all close together, grouped for purposes of constructing a histogram of a population; for example, the class of between 37,000 and 38,000.

class boundaries
The highest and lowest actual values of observations that belong to a specific class.

class interval
The range of actual values that belong to a given class. *20*

class limits
The highest and lowest recorded values that can go into a specific class. *23*

coefficient of correlation
A statistical measure of the degree, or strength, of the association between two variables. *142*

coefficient of determination
A statistical measure of the proportion of the variation in the dependent variable that is explained by the regression equation. *143*

coefficient of rank correlation
A measure of association between two variables that is based on ranks.

coin toss
The result of an imaginary experiment of tossing a fair coin and allowing it to fall to rest so that the outcome of the coin facing heads or tails is due to chance and cannot be predicted in advance. *167*

confidence
The probability that a specified interval contains the population parameter of interest.

confidence interval
An interval of numbers that locates a parameter along the number line, together with a confidence level. *218*

confidence interval estimate
A range of values that includes the population parameter with a stated probability.

confidence level
The percentage of time that a statement about the location of some parameter will be correct. *174*

confidence limit
An endpoint of confidence interval.

constant
A quality that retains the same value throughout a series of calculations. *62*

contingency table
The rectangular array resulting from the partitioning of two variables, such as political preference and age, into classes. *271*

continuous variable
A variable that can assume any value within some range of values. Observations on a continuous variable are the result of measurements. *204*

correlated samples
See matched samples

correlation
The degree of linear relationship between bivariate data or a bivariate population.

correlation analysis
A statistical method for measuring the strength of association between any two random variables.

correlation coefficient
A statistic or parameter that measures the degree of correlation. *139*

critical region
That portion of the area under the sampling distribution of a test statistic that corresponds to rejection of the null hypothesis.

critical value
The value of a test statistic that divides the acceptance region from the critical region.

cumulative frequency distribution
A frequency distribution that indicates the proportion of observations that are less than or more than certain specified values.

degrees of freedom
A technical term measuring the number of freely, or independently, varying data. Applies to the chi square distribution, the F distribution, and the *t* distribution. *212, 267*

dependent variable
The variable whose value is determined when the value of the independent variable(s) is(are) known. Usually the variable of the vertical axis, or *y*-axis. *136*

descriptive statistics
Statistics concerned only with describing and summarizing sets of numerical data. *7*

d.f.
See degrees of freedom

difference scores
The resulting scores formed by subtracting *X* from *Y*, or *Y* from *X*, where *X* and *Y* are paired scores. Also known as change scores. *243*

discrete variable
A variable that is restricted to certain values within some range of values. Generally these values are integers.

distribution-free statistics
Statistical procedures that require no assumptions about the population distribution; this term is used interchangeably with nonparametric statistics.

distribution of sample means
The histogram that records where the sample means of a certain fixed sample size fall, and how often, assuming that the sampling procedure is repeated *n* times (sampling distribution).

error
Error in statistics is generally of three types: (1) in the formation of confidence intervals, the percentage of the time that we incorrectly announce the location of the parameter; (2) in hypotheses testing, false rejection or acceptance of a hypothesis (*see* Type I and Type II error); (3) in ANOVA, a source of variation in the data from within populations.

estimator
A statistic used for estimating the value of a population parameter.

event
Some specified collection of possible outcomes of an experiment, usually determined by some numerical values (for example, the event "40 or more successes out of 50 trials").

expectation
The mean of the distribution of statistic.

expected numbers
The numbers we observe, on the average, if the null hypothesis (of no relationship between the variables in the contingency table) is true.

experiment
Any well-defined action that leads to a single well-defined result.

factor
A variable of interest that is controlled in an analysis of variance. *334*

factorial
The product of the numbers 1, 2, 3, and so on, up to the given number. The given number must be a positive integer or 0. By definition, 0 factorial is 1. It is written as "*n*!" *167*

factor level
A value at which the factor is controlled; same as a treatment. *334*

fair coin
A coin for which the likelihood of heads and tails is the same. *176*

Fisher's z′ transformation
A logarithmic transformation of correlation coefficients, *r* to *z′*, where *z′* is approximately normally distributed with a standard error independent of rho, the population correlation. *226*

fourfold table
A square table consisting of two rows and two columns (four cells). *195, 272*

F ratio
A statistic used to test whether several populations all have the same mean. *See* analysis of variance. *229*

frequency
The number of observations that fall into a specified group.

frequency distribution
A tabular arrangement or grouping of numerical data into classes and a count of the number of observations that fall into each class. *18*

frequency polygon
A graphical representation of a frequency distribution in which frequencies are plotted by placing a dot above the class mark and connecting the dots with straight lines. *24*

F-test
Test of the hypothesis that there is no difference between two variances, using the *F*-ratio as the test statistic. *229*

function
A rule that relates the values of one variable, called the independent variable, to the values of another variable, called the dependent variable.

goodness-of-fit test
A test of the hypothesis that data observed from a sample belong to a stated population distribution. *269*

grand mean
The overall mean for all groups taken together. *298*

histogram
A diagram that indicates the relative frequency or percentage of different groupings of numbers, called classes, in the population. Often called a bar chart. *18*

hypothesis
A statement about some population parameter that is to be tested for its correctness.

hypothesis testing
The procedure of forming a hypothesis and testing it by the use of some statistic. *175*

independent groups (samples)
Two or more groups of subjects chosen from the population where no logical or meaningful relationship is assumed to exist between the members of each group. *243*

independent trials
Trials of an experiment in which the probability of a specific outcome on any trial is unaffected by the outcome of any other trial.

independent variable
The variable over which the experimenter has control, normally the variable of the horizontal axis, or *x*-axis. *136*

index of dispersion
A measure of variability used for qualitative data (categorical). *82*

inferential statistics
Statistics, which allow an inference to be made about the whole population from information about only a part of the population, or a sample. *7*

interval
See class

interval scale
A measurement scale with all the properties of an ordinal scale plus the property that distances between pairs of objects may be calculated and interpreted. Data measured on this scale lacks a meaningful absolute zero point. *9*

Kruskal-Wallis test
A direct extension of the Mann-Whitney *U* test for more than two independent samples (groups). A nonparametric equivalent to the one-way ANOVA. *377*

level of significance
See significance level

linear transformation
A transformation of data using a linear equation that changes only the scale of the scores (mean and/or S.D.) and not the shape of the distribution. The relative positions of the scores are retained. *100*

line-of-best-fit
A straight line fitted to the paired data either visually or mathematically to best represent the data linearly. *135*

location
The position along the number line, or *x*-axis.

Mann-Whitney *U* test
A nonparametric test to determine whether two independent groups exposed to different treatments differ significantly. *364*

matched samples
Data from two groups that are paired, where a correlation is assumed to exist between the pair. *See also* within-subjects design. *243*

maximum
The largest number in a population or sample.

mean
The average of a collection of numbers, obtained by adding all the numbers and then dividing by the total number of numbers. *60*

mean square
The sum of squares divided by the number of degrees of freedom; a sample variance.

median
The number such that half the population (or sample) is larger and half is smaller. *55*

method of least squares
The criterion that measures which straight line best fits the given data; refers to the minimum possible value of the sum of the squared residuals of predicted *y*-values from observed *y*-values. *135*

minimum
The smallest number is a population or sample.

mode
The most frequently occurring number in a population or sample. The mode of a population is a parameter. *46*

multiple regression
Statistical method for determining the degree of relationship between one dependent variable and two or more independent variables. *See also* regression analysis.

negative correlation
A relationship in which one variable decreases when the other variable increases. *140*

nominal scale
Objects are assigned numbers for the purpose of identification only. So-called measurements under this scale are used for naming or labeling only. No meaningful arithmetic operations can be performed on these measurements.   *46*

nonlinear transformation
Transformation of scores that changes the scale of the scores (mean and S.D.) and also the shape of the score distribution.   *97*

nonparametric statistics
Statistical procedures that involve no hypothesis about population parameters.   *360*

norm
A numerical standard used for the interpretation of test scores, usually based on scores derived from a nationwide sample.   *97*

normal approximation
Approximation of a given distribution, such as the DSM or binomial distribution, by a normal distribution.

normal curve
The normal distribution.   *9*

normal distribution
The pattern that was discovered from repeated measurements and that has the shape of the bell-curve.   *105*

normal table
The table of values of the bell-shaped curve.

null hypothesis
Represented by the symbol h_o, it is always a statement about a population parameter and states that some condition concerning the population parameter is true.   *176*

observed numbers
The values obtained from conducting the experiment.

one-tailed test
A hypothesis test in which the entire critical or rejection region is located in only one tail of the sampling distribution of the test statistic.   *179*

ordinal scale
Data measured using this scale represent order or rank. In comparing two values on this scale, one can be judged to be equal to, or smaller than the other. However, how much smaller or larger cannot be determined.   *7*

ordinate
The *y*-axis or vertical axis in a two-dimensional plot.   *132*

paired data
Two equal-sized sets of data that have a natural coupling.

parameter
A number derived from knowledge of the entire population, such as the minimum of the population, or the mean of the population.   *7, 166*

Pearson correlation
A correlation coefficient computed on two variables measured on a continuous scale.   *143*

percentile
One of the division points between 100 equal-sized pieces of the population when the population is arranged in numerical order. The 78th percentile is the number such that 78% of the population is smaller and 22% of the population is larger.   *92*

pooled estimate
A parameter estimate obtained from two or more samples by pooling or combining all of the sample data.

population
Any complete set of objects, usually numbers.   *7*

population or statistical population
The entire set of possible observations that may be made on the statistical universe. Although the term "universe" typically refers to the elementary units themselves and the term "population" refers to the observations made on the units, the two terms are frequently used interchangeably. In this book, the term population is used to refer to both the elementary units and the observation.

population parameter
See parameter

positive correlation
A relationship in which two variables increase or decrease together.   *139*

post-hoc tests
Tests used to compare group means following a significant *F*-test in ANOVA. Also known as "data snooping methods."

power
The probability that a test will reject the null hypothesis when it is false.   *179*

probability
The proportion or percentage of the time that an event occurs. The likelihood that something will happen.   *166*

quartile
One of the division points between four equal-sized pieces of the population when the population is arranged in numerical order. The second quartile is the number such that 2/4 of the population is smaller and 2/4 is larger.   *69*

randomizing
The process of placing in random order.

random numbers
Numbers that are "chance" outcomes of some process. Usually, these are uniformly distributed random digits, meaning that each digit (0, 1, 2, 3, 4, 5, 6, 7, 8, or 9) records the result of a process of rolling a 10-sided fair die. The likelihood of any single digit is the same as that of any other digit; similarly for pairs, triples, and so on.

random order
The result of ordering data in a random sequence, like shuffling cards.

random sample
A subset of the population where each element had an equal probability of being chosen.   *106, 229*

random variable
A variable whose values are determined by the outcome of an experiment.

range
A measure of variation determined by subtracting the lowest value in a set from the highest.   *19*

rank-sum
A statistic, such as the Mann-Whitney, which is based on ranking data.

ratio scale
The highest order of measurement scales. It has all the properties of the interval scale plus an absolute zero point.   *9*

raw data
Statistical data in its original form, before any statistical techniques are used to refine, process, or summarize. *92*

regression analysis
A statistical method for determining the type of relationship that exists among two or more variables and for using that relationship to make estimates or predictions.

regression coefficients
The constants in the regression equation.

regression equation
The equation of the regression line. *137*

regression line
The straight line that best fits the given data. *See also* least squares. *137*

rejection region
Same as the critical region.

relative frequency
The number of occurrences of a given class, divided by the total number of numbers in the population; a percentage, or proportion, of the whole.

residuals
Prediction error in regression analysis: $(Y - Y')$.

run
Succession of identical letters or labels that is followed by and preceded by a different letter or label or no letter or label at all. *375*

runs test
A nonparametric test used to determine if an order of individual items (e.g. runs) are randomly arranged. *375*

sample
A collection of numbers usually drawn at random from a population of numbers which represent a portion of the statistical population. Generally, a sample consists of fewer elementary units or observations than are contained in the population.

sample mean
The mean, or average, of a sample. *205*

sample proportion
The proportion of successes in a sample.

sample statistic
See statistic

scatter diagram
A chart in which pairs of values (x,y) are plotted to provide a visual display of the pattern of their relationship.

Scheffe' test
A very conservative post-hoc test used to compare pairs or combinations of group means following a significant *F*-test in ANOVA. *316*

semi-interquartile range
A measure of variability that is insensitive to extreme values. Generally calculated as $(Q_3 - Q_1)/2$, where Q_3 and Q_1 are 3d and 1st quartiles. Used along with the median for describing a distribution of scores. *70*

significance level
The proportion of the time that a hypothesis test will lead to a type 1 error, usually 0.10, 0.05, or 0.01. The probability at which an observed deviation can no longer be attributed to chance. *174*

sign test
A nonparametric test based on replacing all observed values by either + or −, since the values observed are larger or smaller than some specified value, and then testing the resulting binomial distribution. *360*

skewness
Departure from symmetry. The distribution with extreme scores at one end. *63*

slope
The coefficient of x in the regression equation; − representing the degree of change in the dependent variable as a function of the independent variable. *136*

small sample
Usually less than 30.

Spearman correlation
A correlation coefficient computed on two variables measured on an ordinal (rank order) scale. *371*

standard deviation
A measure of spread, applied either to a population or to a sample. Also the square root of variance. *79*

standard error of the difference between means
The standard deviation of a sampling distribution of the difference between two sample means. *236*

standard error of the estimate
A measure of the variation of the observed values about the regression line.

standard error of the mean
The standard deviation of the sampling distribution of sample means. *206, 300*

standard normal distribution
The normal distribution with mean 0 and variance 1.

Stanine
One digit normalized scores with a mean of 5 and a standard deviation of approximately 2. The distribution of stanines are approximately normal. *105*

statistic
A number derived from observing a sample from a population of numbers. A value computed from a sample. *166*

statistical inference
Statistics that are concerned with drawing conclusions about populations on the basis of incomplete data, that is, on the basis of data obtained from a sample.

statistical significance
The likelihood that an event has occurred by chance is less than some experimenter—established limit. Rejection of h_0 at α level. *197, 256*

statistics
The branch of applied mathematics that is concerned with the collection and analysis of numerical data. Also a body or set of numerical data.

sum of squares
The sum of the squared deviations of sample values about the sample mean. *300*

tally
Counting one for each number that falls in a given interval or class. *21*

tally sheet
A sheet of paper used to record (tally) into which class each number in the population falls. The tally sheet records how many numbers in the population fall into each class.

t **distribution**
A distribution based on forming confidence intervals when we must estimate the population variance from the variance of a small sample. *212*

test of independence
A test of the hypothesis that membership in a particular category of one classification variable has no effect on membership in a particular category of another classification variable.

test of significance
A test to determine whether or not the data supports the null or alternative hypothesis.

test statistic
The statistic used to test a hypothesis about a population parameter.

treatment
A value of a controlled factor. *See also* factor level.

treatment combination
A combination of one level of one factor with one level of another factor in a two-factor analysis of variance.

trial of an experiment
The execution of an experiment or observation process.

t-**score**
McCall scores rescaled to have a mean of 50 and a standard deviation of 10. *103*

t-**table**
A table of critical values of the *t* distribution.

t-**test**
A hypothesis test based on a small sample, leading to a *t*-score and examination of a critical value of the *t* distribution. *237*

Tukey HSD test
A post-hoc test used to compare pairs of group means following a significant *F*-test in ANOVA. *317*

two-tailed test
A test for which values can fall in either tail of the distribution and be cause for rejection of the null hypothesis. The critical or rejection region for the test is divided between the two tails of the sampling distribution of the test statistic. *180*

two-way ANOVA
The case when a model is formed that has two different treatments. *334*

Type I error
Rejecting the null hypothesis when it is correct. *177*

Type II error
Accepting the null hypothesis when it is false. *177*

unbiased estimator
An estimator whose expected value is equal to the value of the parameter to be estimated. *210*

universe, statistical universe
The entire body of elementary units that are of interest and are the subject of a statistical investigation.

U **statistic**
See rank-sum

variability
See variation

variable
An observable quantitative characteristic of an elementary unit that may vary from unit to unit. *11*

variance
The specific measure of variation that computes mean-squared deviations from the mean. The square of the standard deviation. *80*

variance ratio
See F ratio

variation
The degree to which elements in a population fluctuate when sampled. *See also* variance.

weighted arithmetic mean
A modified arithmetic mean that weights each value in a set according to its relative importance.

Wilcoxon rank-sum test
A nonparametric test used to determine whether data (paired) collected from a within-subjects design is statistically different. *367*

within-subjects design
An experimental design using the same subjects in both experimental and control groups. The subject serves as its own control and is measured at least twice. *244*

Yates' correction factor
A correction factor used to more accurately approximate the continuity of the chi square distribution. When N is relatively small and the expected value of any cell in a contingency table is less than 10. The factor reduces the absolute difference between observed and expected frequency before squaring. *274*

y-**intercept**
The point where the regression line (line-of-best-fit) intersects the *y*-axis. Usually considered as the additive constant in the regression equation. *136*

Z-**score**
The distance of the given value from the mean of the distribution or population, when measured in standard deviations. *101*

References

Elementary Statistics

Alder, H. L., & Roessler, E. B. (1977). *Introduction to probability and statistics* (6th ed.). San Francisco: W. H. Freeman & Co.

Bradley, J. I., & McClelland, J. (1978). *Basic statistical concepts: A self instructional text* (2d ed.). New York: Scott Foresman.

Chao, L. (1974). *Statistics: Methods and analysis* (2d ed.). New York: McGraw-Hill.

Chase, C. I. (1984). *Elementary statistical procedures* (3d ed.). New York: McGraw-Hill.

Cochran, W. G., & Snedecor, G. W. (1980). *Statistical methods* (7th ed.). Ames, Iowa: Iowa State University Press.

Dixon, W. J., & Massey, F. J. (1983). *Introduction to statistical analysis* (4th ed.). New York: McGraw-Hill.

Downie, N. M., & Heath, R. W. (1983). *Basic statistical methods* (5th ed.). New York: Harper and Row.

Ferguson, G. A. (1981). *Statistical analysis in psychology and education* (5th ed.). New York: McGraw-Hill.

Freund, J. E. (1984). *Modern elementary statistics* (6th ed.). Englewood Cliffs, N.J.: Prentice-Hall, Inc.

Guilford, J. P., & Fruchter, B. (1977). *Fundamental statistics in psychology and education* (6th ed.). New York: McGraw-Hill.

Hoel, P. G. (1983). *Elementary statistics* (4th ed.). New York: Wiley.

Hollander, M., & Wolfe, D. A. (1973). *Nonparametric statistical methods.* New York: Wiley.

Kirk, R. E. (1984). *Elementary statistics* (2d ed.). Belmont, CA: Brooks/Cole Publishing Co.

Kushner, H. W., & DeMaio, G. (1980). *Understanding basic statistics.* San Francisco: Holden Day.

Lovejoy, E. P. (1975). *Statistics for math haters.* New York: Harper and Row.

McCall, R. B. (1980). *Fundamental statistics for psychology* (3d ed.). New York: Harcourt, Brace and Jovanovich.

McCarthy, P. J. (1978). *Introduction to statistical reasoning.* Melbourne, FL: Krieger.

McNemar, Q. (1969). *Psychological statistics* (4th ed.). New York: Wiley.

Mendenhall, W., & McClave, J. T. (1977). *Statistics for psychology* (2d ed.). Boston: Duxbury Press.

Minium, E. W. (1986). *Statistical reasoning in psychology and education* (3d ed.). New York: Wiley.

Siegel, S. (1956). *Nonparametric statistics for the behavioral sciences.* New York: McGraw-Hill.

Weinberg, G. H., & Schumaker, J. A. (1980). *Statistics: An intuitive approach* (4th ed.). Belmont, CA: Brooks/Cole Publishing Co.

Wonnacott, T. H., & Wonnacott, R. J. (1985). *Introductory statistics* (3d ed.). New York: Wiley.

Advanced Statistics

Barber, T. X. (1976). *Pitfalls in human research.* New York: Pergamon Press.

Box, G. E. P., Hunter, W. G., & Hunter, J. S. (1978). *Statistics for experimenters.* New York: Wiley.

Brunk, H. D. (1965). *Mathematical statistics* (2d ed.). New York: Blaisdell.

Cohen, J. (1969). *Statistical power analysis for the behavioral sciences.* New York: Academic Press.

Comrey, A. L. (1973). *A first course in factor analysis.* New York: Academic Press, Inc. (Out of print)

Daniel, C. (1976). *Applications of statistics to industrial experimentation.* New York: Wiley.

Draper, N., & Smith, H. (1981). *Applied regression analysis* (2d ed.). New York: Wiley.

Dunnette, M. D. (1966). Fads, fashions and folderol in psychology. *American Psychologist, 21,* 343–352.

Edwards, A. L. (1984). *Experimental design in psychological research* (5th ed.). New York: Harper and Row.

Gallo, P. S., Jamieson, K., & Christian, K. *The strengths of experimental effects in the psychological literature.* Annual Convention, Western Psychological Association, Seattle, Washington, April 20–23, 1977.

Gorsuch, R. L. (1983). *Factor analysis* (2d ed.). New York: L. Erlbaum Associates.

Harman, H. H. (1976). *Modern factor analysis* (3d ed.). Chicago: University of Chicago Press.

Hays, W. L. (1981). *Statistics for the social sciences* (3d ed.). New York: Holt, Rinehart & Winston.

Hoel, P. G. (1971). *Introduction to mathematical statistics* (4th ed.). New York: Wiley.

Hoel, P. G. (1984). *Introduction to mathematical statistics* (5th ed.). New York: Wiley.

Hollander, M., & Wolfe, D. A. (1973). *Nonparametric statistical methods.* New York: Wiley.

Kenney, J. M. (1985). Hypothesis testing: An analogy with the criminal justice system. *Statistics Division Newsletter.* American Society for Quality Control, 6(2), 8–9.

Kerlinger, F. N., & Pedhauzer, E. J. (1973). *Multiple regression in behavioral research.* New York: Holt, Rinehart & Winston.

Kirk, R. E. (1982). *Experimental design: Procedures for the behavioral sciences* (2d ed.). Belmont, CA: Brooks/Cole Publishing Co.

Linton, M. S., & Gallo, P. S. (1975). *The practical statistician: Simplified handbook of statistics.* Monterey, CA: Brooks/Cole.

Lord, F. M., & Novick, M. R. (1968). *Statistical theories of mental test scores.* Reading, MA: Addison-Wesley.

Lykken, D. T. (1968). Statistical significance in psychological research. *Psychological Bulletin, 70,* 15–159.

Marascuilo, L. A., & Levin, J. R. (1983). *Multivariate statistics in the social sciences: A researcher's guide.* Belmont, CA: Brooks/Cole Publishing Co.

Mosteller, F., & Tukey, J. W. (1977). *Data analysis and regression.* Reading, MA: Addison-Wesley.

Neter, J., Wasserman, W., & Kutron, N. H. (1983). *Applied linear regression models.* Homewood, Illinois: Irwin.

Nunnally, J. (1978). *Psychometric theory* (2d ed.). New York: McGraw-Hill.

Peters, C. C., & VanVoorhis, W. R. (1940). *Statistical procedures and their mathematical bases.* New York: Greenwood.

Prokasy, W. F. (1972). Developments with the two phase model applied to human eyelid conditioning, in *Classical conditioning II.* Black, A. H., & Prokasy, W. F. (eds.). New York: Appleton-Century-Crofts, pp. 119–147.

Propst, A. L. (1988). The alpha-beta wars: Which risk are you willing to live with? *Statistics Division Newsletter.* American Society for Quality Control, 8(11), 7–9.

Siegel, S. (1956). *Nonparametric statistics for the behavioral sciences.* New York: McGraw-Hill.

Simon, C. W. *Analysis of human factors engineering experiments: Characteristics, results and applications.* Westlake Village, CA: Canyon Research Group, Inc., Technical Report No. CWS–02–76, August 1976, 104 pages.

Simon, C. W. Design, analysis and interpretation of screening studies for human factors engineering research. Westlake Village, CA: Canyon Research Group, Inc., Technical Report No. CWS–03–77. September 1977. 220 pages.

Simon. C. W. *Economical multifactor designs for human factors engineering experiments.* Hughes Aircraft Company, Technical Report No. P73–326. June 1973. 171 pages. (AD 767–739).

Skipper, J. K., Guenther, A. L., & Nass, G. (1967). The sacredness of .05: A note concerning the uses of statistical levels of significance in social science. *The American Sociologist, 2,* 16–18.

Tatsuoka, M. M. (1971). *Multivariate analysis: Techniques for education and psychological research.* New York: Wiley.

Ury, H. (1967). In response to Noether's letter, "Needed—a New Name." *American Statistician, 21*(4), 53.

Winer, B. J. (1971). *Statistical principles in experimental design* (2d ed.). New York: McGraw-Hill.

Yates, F. (1934). Contingency tables involving small numbers and the X test. *Supplemental Journal of the Royal Statistical Society, 1,* 217–235.

Index

A

Abscissa, 132
Alpha level, 177
Alternative hypothesis, 177
Analysis of variance, 296–353. *See also* Variance, analysis of
ANOVA, 296–353. *See also* Variance, analysis of
Area method, 118–20
Arithmetic mean, 60. *See also* Mean
Average, 46–64
    mean, 60–63
    median, 48–60
    mode, 46–48
Axes, 24

B

Barber, T.X. 175, 198
Bar graph, 18
Bell-shaped curves, 7, 105. *See also* Distribution, normal
Best-fitting regression line, 133–35
Beta, 177
Between-groups variance estimate, 298, 300, 307, 336–38
Biased estimator, 209–10
Bimodal distribution, 47. *See also* Distribution, bimodal
Binomial distribution, 167
    application of, 166
    application to social science research, 170–71
    and difference between correlated frequencies, 195–97
    formula for, 167, 169
        derivation of, 172
    in hypothesis testing, 173–84
    mean of, 188, 193–94
    and normal curve approximation in hypothesis testing, 184–92
    standard deviation of, 188, 194–95
    table for, 000
    use of, 166–97
    use to make statistical test, 173–84

Binomial expansion, 185
    formula for, 169, 173
Box, G. E. P., 334, 358

C

Centile, 92, 97–98
Centile rank, 96
    converted to raw scores, 96–97
    derived from raw scores, 92–96
    from frequency distribution, 93–95
Centile-score scale, 97
    characteristics of, 97–98
    distortion in, 97–99
    usefulness of, 98
Central limit theorem, 207
Central tendency, measure of, 46–64
Change scores, 254
Chi square, 232, 266
    categorization, in computing, 266
    degrees of freedom in, 267, 270, 274, 277, 285
    distribution of, 232, 266
    errors in use of, 278
    exact test, 274
    formula for, 232, 266
    in fourfold tables, 272–73
    independence of frequencies, 271–78
    interpretation of, 270, 277
    in larger contingency tables, 275–78
    with small theoretical frequencies, 274
    table of, 405, 406
    testing
        difference between obtained and theoretical frequencies, 266–78, 283–85
        null hypothesis, 268, 272
    theoretical frequencies, determination of, 272–73
    use to test normality of a frequency distribution, 283–85
    summary of steps in calculating, 270
    variance of population, 232–34
Christian, K., 344, 358

C

Class interval, 20
    bottom, location of, 21
    choice of, 20
    determination of, 20
    limits of, 23, 93
Coefficient of determination, 143
Cohen, J., 179, 198
Computers, use of, 11–15
Confidence interval defined, 218
Confidence interval for true mean, 218
    difference between two correlated means, 249–51
    difference between two independent means, 241–43
Confidence levels, 174
Constant, 62
    defined, 62
    effect on M and S.D. of adding or subtracting of, 86–87
    effect on M and S.D. of multiplying by, 87
Contingency tables, 271, 277
Continuity, correction for. *See* Yates' correction in chi square
Correlation, 139
    causation and, 142–43
    curvilinear relationships, 134, 143, 144
    and grouping errors, 143–45
    linear relationships, 143
    nonlinear, 143
Correlation coefficient, 139
    computation of by raw-score formula, 147
        summary of steps, 149
    defined, 139
    degrees of freedom in testing, 244–45
    formula for deviation-score method, 139
        derivation of, 149–50
    formula for raw-score method, 147
        derivation of, 149–50
    interpretation of, 142–45
    size of, 143–44
    t ratio for, 223–25